AS[illegible] [illegible]

DATA ANALYSIS IN QUALITATIVE RESEARCH

PRACTICAL AND THEORETICAL METHODOLOGIES WITH OPTIONAL USE OF A SOFTWARE TOOL

Data Analysis in Qualitative Research

Practical and theoretical Methodologies with optional use of a software tool

Asher Shkedi

Contact: asher.shkedi@mail.huji.ac.il

ISBN 9781790521739

DATA ANALYSIS IN QUALITATIVE RESEARCH

PRACTICAL AND THEORETICAL METHODOLOGIES WITH OPTIONAL USE OF A SOFTWARE TOOL

ASHER SHKEDI

Contents

PREFACE

Qualitative research and the methods of qualitative data analysis pose complex challenges to the researcher. As a means of dealing with the data analysis challenges, this book seeks to offer theoretical and practical instruction and supportive software to analyze data in qualitative research.

How do we actually deal with the complexity of qualitative research, with all its characteristic elements? How can we ensure that throughout the data analysis process, a balance will be maintained between the unique methodological elements of each of the research projects? Qualitative research is based on the principle of "the human as an instrument," which seeks to utilize the researcher's intuitive, analytical, and verbal discourse skills to conduct proper qualitative research. However, the concept of "the human as an instrument" contains weaknesses which researchers must overcome. These weaknesses stem from the character of intuitive skills, which respond to the research area quickly and instinctively, as well as from the nature of verbal discourse, which is broad in scope and characterized by ambiguity.

Maintaining the principle of "the human as an instrument" and at the same time overcoming its weaknesses is a difficult, if almost impossible task. It seems that in many cases, qualitative researchers find it difficult to control such a large amount of data, and become forced to relinquish control and the reflective principle of qualitative research methodology. These researchers find themselves relying mainly or solely on intuitive impressions. Relying solely on intuitive impressions, interesting and challenging as they may be, does not meet the standards of research, but completely blurs the line between research and literary, artistic or journalistic works. In other cases, the difficulty of dealing with the great expanse of data has led researchers to give up the principles of closeness, involvement and empathy, and thus lose the essence of qualitative research. Using supportive software for qualitative analysis will help the researchers cope with these difficulties.

It seems that maintaining a balance between all the methodological characteristics of qualitative research requires the help of external supportive tools, which allow researchers and research students to control the huge amount of data and analyze it without forfeiting the methodological characteristics of the study. Indeed, computers and computer software may offer us such a tool. Starting in the 1970's and 80's, the massive advance of the Computer Age also introduced a real change to support qualitative researchers (Denzin & Lincoln, 2005a). Software already accompanies quantitative researchers, and carries out the analysis work most efficiently. Thus, it is necessary to define the term "qualitative analysis software." Unlike quantitative research, there should not be software that analyzes qualitative data, but **supportive** analysis software. Although researchers are assisted by the software, they are the ones who carry out the analysis, according to the principle of "the human as an instrument."

> "Software can provide tools to help you analyze qualitative data, but it cannot do the analysis for you, not in the same sense in which a statistical package like SPSS or SAS can do, say, multiple regression. Thus it is particularly important to emphasize that using software cannot be a substitute for learning data analysis methods."
>
> ***(Weitzman, 2000, p. 805)***

Once I was invited to a conference devoted to the issue of evaluation, where I was asked to present the qualitative analysis software. The participants were quantitative researchers. It is doubtful if some of them had ever experienced qualitative research, and this group would certainly not have overwhelmingly championed its cause. However, most of them were aware of the increasing demand for integrating qualitative elements into evaluation research. As each knew, quantitative analysis software allows those who are not well enough versed in statistical analysis to deliver their data analysis to experts and receive the findings, thanks to the software process. Thus, their expectations were that qualitative software could enable them to master qualitative analysis, an unfamiliar domain. I was surprised to see that the hall was filled with curious quantitative researchers. But only moments after the session began, hearty expressions of disappointment were heard from all corners. The fact that the software does not execute the analysis but only helps researchers in the process of analysis did not tally with their world of research notions and expectations.

Therefore, it is important to reiterate that the character and uniqueness of qualitative research is not consistent with analysis software. Given that qualitative researchers need to "recruit" their intuitive ability to create situations of closeness, involvement and empathy with the subjects of the study, it is clear that no external tool can meet this requirement other than the human himself. Therefore, what is offered here is supportive software, which helps the researchers to assume control over the great amount of data required in the research, both raw data and data obtained at various stages of analysis, and help them to meet the required analytical criteria (Dey, 1993; Flick, 1998; Weitzman, 2000).

The supporting software allows researchers to keep the analysis process open and systematic, but also documented and transparent, so that at any time the researchers can examine their work, correct and clarify. The transparency also enables colleagues and other researchers to examine the analysis work, and help the researchers analyze through comments and questions, as part of the effort to assure an increased research quality (Fielding & Lee, 1998; Grbich, 2007). Therefore, the software allows the "birthplace" of qualitative research to be transparent, visible, and exposed to colleague-researchers, other interested parties, and even to the researchers themselves, to consider what stands behind the researcher's decisions at various stages of the qualitative analysis. The transformation of the qualitative research process to being visible, systematic, documented and transparent may further increase the acceptance of qualitative research in the academic community, part of which still does not relate to qualitative research as being equivalent to quantitative research (Fielding & Lee, 1998).

However, we must not hide the dangers of using the software. As the possibilities for using computer software grow, the temptation to offer sophisticated computational operations is growing in kind, threatening to break the balance between the methodological foundations of qualitative research and make the analysis process more and more mechanical. Thus, it should be clear that the most sophisticated software and advanced technology are not necessarily the best tools. There is no substitute for the researchers' attention and their constant commitment to go back and re-examine the entire analysis picture. Without such attention, researchers can easily lose the unique character of qualitative research (Charmaz, 2000).

I myself can testify that over the years that I have conducted qualitative research, I have always engaged the help of external tools to perform the work of analysis. In my early years as a qualitative researcher, I utilized scissors as a supportive tool for analyzing. I cut data segments according to the categories that I set, and I pieced together fragments of data according to their categorical connection.

When the wonderful era of word processing dawned, the scissors gave way to the word processor's "cut and paste" operations. The transformation of the "scissors generation" to the "word processor generation" is characterized mainly by improved efficiency. Nevertheless, the word processor as a tool for analysis was not efficient enough to cope with the growing amount of data. This prompted me to look for the next tool generation, the "analysis software generation." After trying several versions of analysis software, I concluded that my mission was to initiate and develop software that will meet the particular demands of qualitative research. For some years, I have been accompanying the development of supportive analysis software, the "**Narralizer**," whose characteristics and implementation will be presented in this book.

PART ONE

THE PRINCIPLES OF QUALITATIVE RESEARCH APPROACHES AND ANALYSIS

INTRODUCTION TO PART ONE

This book deals with the principles of qualitative research methodology, focusing primarily on methods of analysis. These methods are common types of qualitative research, and clearly differentiate it from quantitative research and other written work. Although various qualitative researchers offer a large range of terms to represent the approaches, genres, and methodologies and to distinguish between different types of qualitative research, this book seeks to present several common methodological properties of qualitative research. As will be described, qualitative research is not characterized by one global methodology, but by many methodologies. Yet one can point to several clear and distinguishing methodological elements that are common amongst all types of qualitative research.

Qualitative research is characterized by three elements:

- Research in the natural human language, in the context of natural human life
- Research based on the intuitive human research skills, focused on closeness, participation, and empathy with the investigated phenomena
- Using analytic human research skills, focused on distancing, reflection and control of the process

Qualitative research is characterized by using language as the medium of the research. The research does not focus solely upon literal data, but also upon the thinking process anchored in the language, and the words as literal descriptions of the research. In order to hone this property of qualitative research, it is often compared to quantitative research (Hammersely, 1996; Miles & Huberman, 1994;

Denzin & &, 2000, 2005). Verbal discourse reflects the culture and meaning that all participants assign to their world, and the differences between human beings and human societies.

> "Words are the way that most people come to understand their situations. We create our world with words. We explain ourselves with words. We defend and hide ourselves with words. The task of the qualitative researcher is to find patterns within those words (and actions) and to present those patterns for others [...]"
>
> ***(Makyut & Morehouse, 1994, p. 18)***

The first component of a qualitative research study carried out in the natural context of human beings is the use of the natural language of human beings, the language of words, a self-evident truth to anyone with a minimal familiarity with qualitative research. The two other elements needed for clarification, however, appear to be in tension and apparent conflict with each other.

In order to clarify, *Figure A1* illustrates the methodological elements of qualitative research:

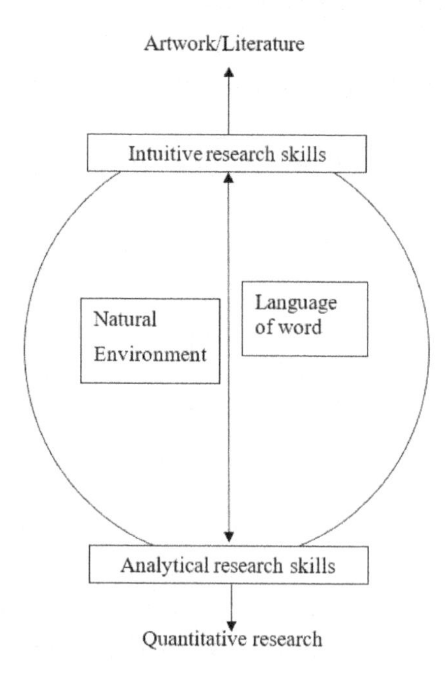

Figure A1: Methodological Elements of Qualitative Research

The area inside the circle chart represents the "zone" of qualitative research. As shown in Figure A1, the study takes place in the language of words and within the natural environment of the participants. This is a common characteristic of all different varieties of qualitative research. The chart also demonstrates that the two other components of qualitative research, intuitive and analytical research skills, are in opposition to each other and reflect the dimensions that pull in opposite directions. Those who restrict their research tools to intuitive research skills and do not employ the remainder of analytical research skills create an artistic or literary work, not a research work. The analytic components are essential to any research, and there is no a real research without them. On the other hand, those who restrict their research tools to analytic skills without employing intuitive research skills are actually facing towards quantitative research.

Any study or work, which contains all three components - the language of words and the natural environment, intuitive inquiry skills, and analytical research skills- can be seen as associated with qualitative research. If one of the three is missing, we cannot characterize the work as qualitative research.

As will be seen in the chapters that follow, the diverse types of qualitative research methodologies differ from one another with respect to the significance each methodology attaches to the three components of qualitative research. This is particularly apparent for the intuitive and the analytical components related to research skills. Qualitative studies are disparate in the significance assigned to each of the two components, but every research project which can be defined as qualitative research is an expression of the two components combined. One type of research may be more an expression of intuitive inquiry skills and less analytical research skills. Another study may contain contrasting characteristics. Others may maintain some kind of balance between the two research components. The nature of the language of words will largely depend on the importance attached to each of the two components of research skills. The language will be more narrative as the significance of the intuitive skill increases, and will be more focused and concise as the significance of analytical research skills grows.

The tension between these research skills, and the expectation that qualitative researchers will "recruit" their intuitive and analytic skills of inquiry, are compatible with the idea that humans have two systems of thought: one intuitive, fast, associative, automatic and relatively effortless, and the other analytical, slow, deliberate, strenuous and controlled by a set of rules (Kahneman, 2011; Kahneman

& Frederick, 2002; Sloman, 2002; Stanovich, 2004). The intuitive system encourages becoming close to the research objective phenomena, while the analytical system requires a certain amount of distancing. Both systems can operate simultaneously, as two experts and their responses can differ as well.

It seems that the language of words, with its inherent richness and wider use among all people, can better express the overall features of qualitative research methodologies and the tension in which they are steeped. On one side, they operate close to the participants' world and perspective, and on the other, a distance is required for a more profound examination.

CHAPTER 1

THE INTUITIVE RESEARCH SKILLS

The great Philosopher Martin Buber distinguished between two types of human relationships, "I –It" and "I –Thou." "I -It" expresses a situation in which a person is not really attentive to the existence of others, referring to others as objects rather than people. The human relationship of "I-Thou," however, is characterized by subjective attention to the existence of others - an attitude of empathy for others (Buber, 1964).

For our purposes as researchers, an "I-It" relationship reflects keeping a distance from others and attempting to examine properties from the outside, without trying to see the issues investigated through the participants' own eyes. The quantitative research tradition will argue that remoteness and alienation are necessary conditions for a rigorous, objective research. In contrast, qualitative research, as emphasized and clarified in this chapter, seeks not only to identify the characteristics of others, but to get involved with the subjects, see the phenomena under investigation through their eyes and develop empathy for them. Such a relationship would be defined by Buber as "I–Thou."

"...Quantitative studies emphasize the measurement and analysis of causal relationships between variables... from within a value-free framework." (Denzin & Lincoln, 2005a, p. 10) This reflects the tendency in quantitative research to maintain a distance from the objects under study to ensure a research process characterized by objectivity. To assure objectivity and the absence of bias, quantitative research equips the researchers with a set of external tools intended to guarantee the optimal management of research at a higher degree of objectivity. Qualitative research, which seeks to approach the investigated object as long as possible (Angrosino, 2005), does not only see the external set of research tools as the ultimate answer to the quality of research, but relies on human intuitive characteristics as the ultimate tool of the research.

The human brain's ability to identify others, to read their thoughts and feelings, and especially to develop empathy for them - characteristics that are clarified in this chapter - allows us to become more closely acquainted with the subjects and see the phenomena under study through their eyes. Qualitative research uses the ability of the human to get closer and closer to the world of the participants, so that Lincoln & Guba (1985) coined the concept "the human-as-an-instrument" to demonstrate the natural human inquiry skills. These skills are "recruited" for the existence of a proper process of qualitative research. As indicated by the title of this chapter and as will be explained throughout, these skills are intuitive human skills, but differ from human analytic skills, which are also "recruited" for qualitative research and which will be clarified in the next chapter.

Although qualitative research has become more prevalent in recent decades, it is still fighting for legitimacy, and not merely among researchers who prefer conventional quantitative research, but even among qualitative researchers. Much of the criticism against qualitative research is grounded from its dependence upon the human as a research instrument. Support from philosophy, psychology and researchers' personal experience is seemingly not enough. This section seeks to illuminate the human intuitive skills enabling a person to be a major tool of qualitative research—a tool that no objective instruments can replace. It seems that displaying these characteristics may help researchers and research students use their personal potential to carry out better qualitative research, without fear or apology. Over the past three decades, a great deal of knowledge about the brain, evolution, and human behavior has been amassed. This knowledge can be established and validate human characteristics, strengthening the arguments that claim the human is a major research tool.

The Human as a Research Instrument

People, that is to say, researchers (and also participants and readers of research), are perceived as being the only research tool flexible enough to accommodate the complexity, sensitivity and constant variability that characterize the human experience. The claim that human nature has the best possible research instruments with which no external tools can compete, is based on the assumption that human and social phenomena are dynamic and variable, and their components are not definite or fixed in advance. In its favor, the existence of research based on external methodological tools which are pre-selected or fixed is not possible, because "only the human instrument has the characteristics necessary to cope with an indeterminate situation "(Lincoln & Guba, 1985, p. 193). In light of this, Kemmis & McTaggart (2005, p. 562), coined the system based on the human as a research instrument as a "soft system" methodology, as opposed to the "hard system" methodology which characterizes quantitative research.

Humans have the ability to quickly respond to their surroundings and adapt to changes around them, and they are characterized by a sensitivity and ability to relate to cultural and personal environmental cues. People are able to simultaneously collect information from multiple factors (at some levels). They are able to understand multi-disciplinary concepts, and capable of putting the pieces into one holistic phenomenon. People have the ability to rapidly analyze information from the moment it becomes available, and to raise assumptions to test relevant hypotheses in those contexts. People are able to clarify and summarize, and to deal with surprising or unique reactions. They have the ability to function simultaneously and to change tacit knowledge into conscious knowledge (Greenwood & Levin, 2005; Polanyi, 1967). All these characteristics are important to qualitative research, and I doubt if there are any "objective-external" tools that can assemble all these characteristics simultaneously.

By definition, qualitative research refers to the social and cultural context that characterizes the subjects and their environment (Jorgensen,1989; Woods, 1996). The social system and its linguistic and cultural manifestations are characterized by the most rapid development and by great diversity. People belong simultaneously to different cultural systems that reflect their national and professional affiliations, their origins, status and more. The expression of each of these cultural systems in every person may be varied, and the relationship between these systems may constantly change. Thus, if we examine the linguistic expression of culture, we see that today's languages are not only different from the language that was acceptable in the past (for example, the language of the Bible or of Shakespeare), but even within one generation, language and culture reflect major changes that they have undergone.

Given that the social-cultural context is dynamic and subject to frequent changes, researchers need to approach their study with open responses in the context, and with a willingness to change the modes of study depending on the circumstances and in accordance with the insights arising during the study (Merriam, 1998). It seems that in light of this cultural dynamic which is reflected by subtle changes, often unexpected and unseen, there is an advantage to using the human as a research instrument. However, qualitative researchers "recruit" their intuitive personal characteristics and personal experience to realize the potential of the "human-as-a-research instrument" and to deal with the research task without barriers posed by external tools.

Eisner (1985) assumes that the human has the natural traits to be a "research tool," but he believes that they are potential properties which must be nurtured, enhanced and constantly improved. He said, "One can look without seeing, listen without hearing, eat without tasting, and touch without feeling" (Eisner, 1985, p. 151), but people have the ability to see (and not just to look), to pay attention (and not just listen), to taste and to feel. Sight, hearing, and other human senses contain abilities that enable the human brain to gather information and clues that are not necessarily picked up by using external tools and standard measuring techniques. It thus seems that the "human-as-a-research instrument" is the only tool for gathering information with the richness of dimensions that allows grasping the entire components of human social and cultural activities and life (Maykut & Morehouse, 1994).

Though Lincoln and Guba (1985) summarize, "There is no reason to believe that humans cannot approach a level of trustworthiness similar to that of ordinary standardized tests - and for certain purposes... even higher levels" (p. 195), we can assume that most people, certainly those who wish to engage in qualitative research, agree with this statement.

The Involvement of Researchers in the Research Process

As emphasized above, people and their actions are too complex to be fully understood by external devices. Qualitative researchers, who serve as research instruments, investigate the significant aspects of reality by their involvement in the reality under study. Experienced qualitative researchers recognize this situation and work with, rather than against it. The human being is different from all other living things by achieving the highest social research exploration and investigation of themselves, which no other living creature can equal. Qualitative research does not expect the researcher to develop a "research tool" out of nothing; qualitative researchers, whether consciously or unconsciously, "recruit" the social research capabilities that characterize people in their daily lives for the purpose of research.

It seems that in order to reach optimal research capabilities, it is impossible to separate researchers from their objects of investigation and to take an objective stance toward the phenomena under study. "...The turn is characterized as a movement away from a position of objectivity defined from the positivistic, realist perspective toward a research perspective focused on interpretation and the understanding of meaning" (Pinnegar & Daynes, 2007, p. 9). Involvement of researchers in research is essential to understanding the world and the phenomena we study (Guba & Lincoln, 1989, 1998). "The viewer, then, is part of what is viewed rather than separate from it. What the viewer sees shapes what he or she will define, measure, and analyze." (Charmaz, 2000, p. 524). In the truest sense, the person is seen as having no existence outside the world, and the world seems to be of no existence outside the person (Maykut & Morehouse, 1994). Experience is the way in which we engage the world around us and within us, and is the foundation upon which we construct meaning. This meaning, therefore, is closely dependent on our ability to communicate with the world we live in (Simons, 1996). Researchers "are always in an inquiry relationship with participants' lives" (Clandinin & Rosiek, 2007, p. 69). To understand the phenomena under study as they are perceived by those who take part in them, researchers must remain as close as possible to the unique construction of the world of the participants who experience the phenomena under study (Denzin, 1995; Maykut & Morehouse, 1994).

From the psychological and philosophical assumption that reality is created by way of construction, one can assume that researchers cannot understand human action by an outside observation which sees merely the physical manifestations of these activities. Instead, researchers should see not just what the "players" do, but also capture what they mean by their actions from the perspective of the "players" themselves (Moss, 1996). Researchers can never be excluded, but they assimilate themselves in the social context and in the minds of the participants (Sciarra, 1999). Qualitative researchers try to understand how the phenomena under study are taken up by those who experience and are involved in them (Jorgensen, 1989). If reality is constructed, and if the object of knowledge is not separated from it, as qualitative researchers claim, then researchers perceive not only the actions of the participants and their intentions and perceptions, but also the values of all participants in the study, and these values are relevant to understanding the phenomenon under investigation (Angrosino, 2005; Merriam, 1998). Moreover, because the researchers themselves are involved in the phenomenon being studied, even their values can be involved in the study (Jorgensen, 1989; Seidel & Kelle, 1995).

Qualitative researchers try to absorb not only the conscious knowledge, but also the tacit knowledge of the participants (Greenwood & Levin, 2005; Polanyi, 1967). Tacit knowledge is the foundation upon which researchers build many of their insights and assumptions, which should ultimately support the research. Researchers' natural inquiry skills and their involvement in the phenomenon under study may allow them to discover that the participants' knowledge is broader than the participants themselves are aware. Consequently, qualitative researchers encourage the participants to look into the depth of their knowledge, and help them to transform tacit knowledge to conscious knowledge, so that the participants can pronounce and think about it explicitly, and pass it on to the researchers and others (Lincoln & Guba, 1985). People themselves are not always able to explain or describe everything they know and feel, especially their tacit knowledge. Tacit knowledge is acquired through assimilation into the environment under study. When researchers live in a situation, they learn to pay attention to what lies beyond the visible. We understand the meaning of things not just by looking at them, but also by assimilating them (Arksey & Knight, 1999; Bishop, 2005; Maykut & Morehouse, 1994).

The sections to follow present the natural research skills of the human, the sources of their development and position in the human mind, and examine why we should "recruit" them for the purpose of conducting proper research.

The Development of the Human as a Social Being

There is no dispute that Homo-sapiens rose above the other creatures in our universe. The question of why the human gained this advantage often evokes controversy. It is clear that with regard to physical traits, there are other creatures, which overshadow the human. Yet the human being's advantage lies in his cognitive and emotional characteristics. Many researchers distinguish the human first and foremost as a social being, and suggest that our brain is designed from the onset to be social and diligently attracted to an intimate brain-to-brain connection at every meeting with another person. We are wired through the nervous system to connect with each other. This neural bridge lets us affect the brain of others - and thus also their bodies - when we come in contact with them, just as it allows them to influence us (Goleman, 1995, 2006).

The human mind has evolved over the ages through the process of trial and error of natural selection. The mental components of modern man have passed the test of survival and development under difficult conditions. Our ancestors were almost never alone. Alone would be too risky for such a physically weak creature as our species, who could not cope with many predators. Being in a group gave them security, as they wandered in search of food in constant fear of predators, rival groups and the like. Our ancestors managed to survive and grow and become our ancestors because they had all they needed to survive and reproduce. Millions of hominids lost the competition and became extinct. (Harris, 2006).

Even today, after we became the living creatures that rule the world, we still carry along this genetic heritage. When we watch a football game on TV, for example, we prefer to do so with friends and have the experience with them. In fact, we are interdependent and need to be in touch and to interact with one another to live a sane, happy life. In the absence of contact with other people, we fall into distress and despair, and our health is negatively affected. We hug when we meet, when we wish each other good luck, when our football team wins, and when students receive a diploma, but also in times of sorrow. The need for physical contact is inherent in the depth of our being (De Waal, 2005).

The human is not the only social creature. Many mammals are characterized by a social ability. But the human has brought these capabilities to the highest degrees. Even if we find animals, especially the great apes, with social skills similar to ours, none have reached the total accumulation of human capabilities (Milo, 2009). According the accepted assumptions of evolutionary researchers, only about five or six million years ago did the human being part from the common ancestor of chimpanzees and bonobos (dwarf chimpanzees) to begin the evolution toward Homo-sapiens. Creatures walking on two legs, which after many vicissitudes became the human, are now well-equipped with social features. It seems that their unique life continually improved these social abilities (Harris, 2006).

Over a relatively short evolutionary period of about six million years, the human brain expanded fourfold and more. The question of what sparked this phenomenal growth spurt intrigues scientists, some of whom maintain that it was the tool-producing skills that encouraged the transition of the creature earlier called "hominid" to become more "human-like" and to earn the moniker "homo" (man), though still not "Homo-sapiens." However, an examination of the tools that man created and used shows that they have not significantly changed in a million years, while the brain grew exponentially. We also can't point to environmental or climatic upheavals to have provided the impetus for adapting brain development for new physical conditions and/or new climates. While researchers may be divided about the "Archimedean point" that sparked this growth of the human mind, they all point to the growth of man's social skills. The tendency among many researchers is therefore to explain the dramatic growth of the human brain as a result of the need to develop social skills (De Waal, 2005).

The Social Investigation of the Human Brain

In order to understand the human social-investigating skills which qualitative researchers can "recruit" to investigate social and human phenomena, we must understand the properties of the "social brain" and the origins of its development. Which social needs caused the human brain to develop in the way it has? It seems that the most likely explanation lies in the need for survival. This need spurred the constant growth of the human mind, which enabled the brain to provide social tools to facilitate the human's survival in competition against other living beings. Hunting is an excellent example of this social imperative. In the absence of a powerful rifle, a successful hunt depends not only on physical fitness and personal skills, but also requires cooperation. It is difficult to snare a large animal alone, especially since the prey is probably faster and stronger than you are. But the need for cooperation did not stop with the end of the hunt. Distribution of meat and other foods acted as a kind of insurance policy in large groups. If you couldn't catch anything today, others will share their food with you, assuming that tomorrow they will share your food. Thus our ancestors, who lived their entire lives in a group, had to decipher the intentions or desires of others and convince them to choose certain directives. They had to understand the social power structure of the group and their relationships within it. Creating alliances were vital to success (Pinker, 1997, 2002).

Reproductive needs and concerns for offspring also required social organization, and thus accelerated the development of social skills. Ancient human groups were built for "cooperative breeding," to use the biologist's language, that is, many people worked together in a cooperative effort to serve one another. Most of the time, women watched the children while the men were involved in jointly cooperative tasks, such as hunting and the protection of the group. Everyone had a personal interest in the results of this joint effort, that is, the assurance that one could bring children into the world and their family could be secure. Young offspring needed protection and imposed on mankind a high level of solidarity to provide mutual help. (Milo, 2009). The reality of life and the need to procure food by hunting distanced men from their family for days and weeks. In the absence of social cooperation, the sexual rivalry could be difficult for everyone. Construction of the core family gave almost every man the possibility to reproduce, and thus an incentive to work for the entire group (De Waal, 2005). Along with the development of social ability, an investigative capability also developed which focused the need to understand others, their wishes and intentions, to create social connections and avoid damaging the union. Social skills are planted deep in our past, but determine the nature of human society to this day.

The premise that the human is a social being has been questioned in light of the alleged fact that people are "outstanding" in social violence within their social structure, and even more against other social structures. These facts led the etiologist Konrad Lorenz (1966) to infer from chimpanzee life, known for its social violence, generalizations about the lives of people. Indeed, chimpanzees do not maintain friendly relations between different groups, and the relations within their group are hostile at various levels. But the fact is that people, besides being violent from time to time, also have peaceful relations within and outside groups, and have engaged in trade, transportation and even mixed-marriages in the past as today. Relations between groups of people are always ambivalent: a combination of the desire for peaceful relations along with hostility brewing under the surface. An example of this kind of behavior can be found in bonobos. Neighborly relations of bonobos are far from ideal. They take advantage of every opportunity to emphasize the boundaries between habitats. But at the same time, they always leave time open for relaxation and friendships. One can say that the relations between groups of people may be characterized in part by those of chimpanzees and of bonobos groups. When hostile relations exist between human societies they are worse than conflicts among chimpanzees, but in the case of good relations, they are better than bonobos can sustain. In fact, we are also competitive and cooperative. Our social skills allow us to examine the reality, both cognitively and emotionally, and respond accordingly. There are "militant cultures" and "peaceful cultures," and every group reaches an equilibrium, which often varies from period to period and from place to place (De Waal, 2005).

Many neuroscientists assume that neurologically normal people have a special mental mechanism to read the minds of other people. This mental mechanism has evolved through natural selection to solve specific adaptive problems of social life, as described above. According to this theory, human brains increased during hominid evolution primarily because of the need to process complex social information. These brains would have to give their owners the tools to assess the others' intentions and abilities, and to test with whom to cooperate or compete. Under these assumptions, the ability to process complex social information characterized the transition from a creature known as the "hominid" into a creature, which could be called a "human." Those who can understand the behavior of others watch what they do in a given situation, deceive them, and perhaps even influence their behavior- are equipped with a better ability to function successfully in the social relations network group of human beings (Harris, 2006).

Some scientists even reached the conclusion that the negative social forces expressed by Machiavellian behavior inspired a further increase in the growth path of the human brain. Based on this assumption, most of our psychological characteristics - envy, guilt, etc. - are seen as a design of natural selection, leading to improve our ability to deceive others, but also to identify cheaters and avoid being thought of as cheating. Especially interesting are "hidden cheaters" who appear to repay a favor, but actually return less than they received. Hard work in researching primates and other animals convinced researchers that the enlargement of the brain, and especially the cortex, is essential for developing effective memory and cognitive mechanisms appropriate to build social comprehensive knowledge (Dawkins, 1989; Winston, 2002).

It seems that the growth of the social brain lies not only in the natural selection pressures, but also due to sexual selection pressures, i.e. the need for continuity, which is expressed in finding a good spouse. Men and women developed large, more complex brains to know people and to choose the appropriate spouse according to his/her characteristics. It seems that a complex combination of three factors mentioned above – the need to maintain social relationships with others, the need to wisely use Machiavellian behavior, and the sexual selection - are responsible for the fact that we find ourselves with great, complex brains possessing investigative capabilities (Winston, 2002).

From Social Investigative Skills to Social Research Skills

Psychologists and biologists believe that the human mind controls many tools - organs or cognitive mechanisms or instincts - designed to perform specific investigative tasks (Blackmore, 1999). Among the most significant human tools are the identifiable mental tools which enable us to distinguish between people - tools that gather and store social information. Indeed, British evolutionary psychologist Robin Dunbar (1997) believes that the origin of the evolutionary growth of the human brain came from the need to collect and store social information. As babies are born with the desire to crawl, stand, walk, and learn language and use it, they are also born with a great interest in distinguishing between people via tools that allow them to identify by creating a large pool of people distinct from one another (Harris, 2006).

Thus, the human developed the ability to investigate both the people and society surrounding him, and this ability has contributed to his development. These traits allow people to investigate others, to meticulously recognize their characteristics and decide accordingly whether to join them or fight them. These are exactly the characteristics that give a person an advantage not only over any other creature, but also over all external research tools. Each of us experiences the human "achievement" over the sophisticated electronic technology when boarding a plane. We and our baggage are tested by special electronic instruments, which are constantly honed and improved. However, it appears that the critical distinction is precisely made by people - professional selectors whose skills of observation and investigation could probably not be replaced by any machine. Certainly, we don't argue that research in general, and qualitative research in particular, has served as the catalyst for the development of social human skills. However, researching people and human societies is the task of the qualitative researcher, and as such the characteristics of the "human as a research tool" constitute an effective, irreplaceable research tool.

This chapter points out the developmental sources of the social research capacity until it became a key component of human characteristics. Qualitative research which highlights the human as a research tool is based on optimizing the use of these human features. It seems that "recruitment" of these natural human attributes in qualitative research provides a methodological advantage over studies that rely on external research tools. We can identify three characteristics of social research which provide the human the ability to be an effective, reliable research tool. These capabilities can be illustrated in a hierarchy in which each capacity is based on its previous capacity and developed from it: the ability to identify people, animals and objects; the ability to read minds and human emotions; and the ability to be empathetic toward others.

The Ability to Identify People, Animals and Objects

Between the Senses and the Brain

The human mind is not a blank slate. In order to live, eat and reproduce, it is crucial to see objects, follow them, catch and identify them, to distinguish between male and female, and so on. We can do none of these tasks without mechanisms to identify and distinguish (Pinker, 1997). Our senses are the means of connection between us and the world. The sensory system is not what distinguishes the human; these senses exist in other organisms as well. Living things may have heightened senses in one area and are lacking in another, and even have the ability to sense areas in which the human has no sensory perception whatsoever. Moreover, the absorption capacity of the entire human sensory system is not on an equal level - some are more effective, while others are less efficient. People have specifically-developed eyesight. This is especially apparent when referring to a baby's eyes. At the moment of birth, a baby's eyes are almost as big as an adult's eye, while its feet and hands, for instance, are tiny and should develop with age (Etcoff, 1999).

However, the senses are not the direct tool linking us and the world. We do not see, hear, or feel the world directly through our senses. We see, hear and feel the brain's processing of sensory information. Our eyes do not present our minds with an objective picture of what happened outside us, nor do they show us an accurate movie of what happens over time (Dawkins, 2006). Our brain has a mechanism that integrates many details running from the senses, and compares them by way of interpretation and reference to previous knowledge stored in the brain. The reality around us, in every detail, is not received as an objective fact until the brain interprets and determines the nature of the absorbed facts. Our brain stores information on objects, plants, animals, events and everything around us. Different minds (of different people) may interpret the reality around us differently. Some of the information is learned and accumulated over the years, and part is innate. This information accumulated in the brain allows a rapid identification of the mass information received by the senses.

As mentioned above, the detection capability, based on the most subtle clues, exists not only in people but also in animals, and has great survival value. If creatures around me are showing signs of distress and fear, I may have good reason to worry. If a bird in a flock perched on the ground starts to fly, the entire flock will spread its wings and fly off immediately before the birds understand what is going on around them. Those who remain behind may become prey. This is why panic spreads so fast among people. I recall such an experience from a youth theatrical performance full of students in the hall. About ten minutes before the end of the show, a small group of boys sitting in the front row decided to leave the hall. They began to run toward the exit. Observing this, the other students also began a mad dash out, without discovering the reason for the hasty exit. Within minutes, the hall was emptied, while the actors stood on the stage in disbelief. It seems that we identify and act quickly without consideration, which may have been necessary in this case, but not necessary in other cases (such as fire or attack). Brain researchers in recent years have shown that the tendency for "emotional imitation" lies in the ancient parts of the brain, common to us and many other animals (De Waal, 2005).

Apparently, then, evolution has chosen a strategy of probability, not certainty, perhaps because the greatest enemy of survival is complete certainty. Evolution prefers rapid identification even if it turns out that the rapid detection was incorrect (the rustle in the bushes was not caused by a snake, but by a harmless crawling creature). Thus, filtering the information already begins in the eye, which is not a passive mechanism such as the camera. The combination of a smart eye with a commentator brain makes the eye a mechanism of an infinitely higher quality than the camera, telescope or microscope, although these devices have better physical properties.

In one classroom, I made observations using a video camera alongside writing my notes as an observer. The camera filmed everything that happened and was said in class, and I wrote down what I saw and heard. There were noises outside the classroom, but they did not distract my attention from observing what was happening in the class and what was being said. In this lesson, the students were very attentive and seemingly the outside noises did not bother them either. When I watched the video and listened to the soundtrack, I found that it was hard to absorb what had been said in class. Outside noises, especially a teacher's voice from a nearby classroom overshadowed all that was said in class. Her voice sounded more powerful than the observed teacher's voice. Indeed, in real time our minds were able to filter background sounds, allowing us to absorb what was being said in class, even though not all was remembered after the lesson ended. The video camera "remembered" everything that was said in class better than we did, but it was unable to distinguish between (unimportant) noise and the (important) classroom discourse, and was overshadowed by all the surrounding noises.

The combination of sensory activity with brain function gives us the ability to discover the world around us. A rapid detection capability, the ability to cite reference cues, to distinguish between the important and unimportant, between the central and peripheral, and the rapid transition from a focus on one object to another object, are qualities that a human possesses. Research skills are developed in people (and animals) in order to function properly in the world and survive. The fact that people are equipped with these research skills gives us a significant advantage when we conduct research focused on issues of humans and society, as will be seen later.

The Ability to Carefully Identify People's Faces

Not only do humans possess an innate ability to observe and comprehend the world around us, but this ability could be especially refined when we look at other people. Our greatest attention is given to human faces. We can see others' expressions of joy or anger, interest or apathy, concentration or confusion, as well as the slight changes that occur within them. We distinguish between people especially by watching. The assumption is that our brain contains a system that allows us to identify other's glances (Harris, 2006). Studies have found that two-month-old babies already have a tendency to look at people's faces, primarily people's eyes. During their first year, most babies become experts at tracking people's faces. If, for instance, they see somebody significant to them watching the corner of the room, they will soon turn their heads and look in the same direction (Johnson, 2004).

We have the ability to simultaneously see and interpret different parts of the face, and quickly connect the information to get a bigger picture of the whole face. The unique talent of the brain is its ability to interpret different things simultaneously. Our brain is specialized in seeing things at once: seeing all parts of the geometry and understanding the shape, or seeing all the elements of a situation and understanding their meaning. This is especially useful in interpreting facial expressions and understanding their meaning (Pink, 2005). Throughout 35 years of research, Paul Ekman (2003) has developed a series of pictures as a kind of atlas that contains probably all facial expressions that people around the world use to express such feelings as anger, sadness, fear, surprise, disgust, contempt, happiness, and so on, as shown in Figure 1A. These facial expressions were found all over the world, from California to New Guinea. Our face can express the full range of human emotions. The human brain can identify the other's feelings by expressions of the human face, in any culture and place all over the globe (Ekman, 2003).

Figure 1A: Expressions of the Human Face

We read another person's face as if reading a book. Not only do we have the ability to observe and absorb the details of the human face and to identify its feelings, we can also remember the people we've watched. Evidence suggests that a human's mind contains a mental lexicon of people, with a separate entry for any person we know. This lexicon is a product of our mental tools to acquire information about people that enables a fine distinction to be made between human faces. It seems that there is no limitation to the number of faces we can recognize. The tools to acquire information about people continue to work all our lives, even if we do not know or have forgotten a person's name, unless part of the brain is injured. The ability to recognize faces has served a survival need. During the evolutionary history of our species, it was benefi cial to know people and to identify their feelings by looking at their faces, to know what motivates them and what to expect from them. This capability can help people predict someone's behavior and decide whether to share with them, mate with them, trust them, fear them, defend them or fight them (Harris, 2006).

The qualitative researcher uses this feature of distinguishing between people and their identification for the task of data collection, especially for observing people and social activities. Participant observation, where researchers are involved in the lives of the participants as they watch them closely, perhaps brings the human watching ability to its best application in the qualitative research process. During participant observation, the researcher does not need any external instruments, with the concept of the "human as a research tool" coming to the fore. This situation of involvement and participation allows researchers to meet the participants almost intimately. The assumption is that something you've seen inside, you can't see from outside (Angrosino, 2005; Fetterman, 1989; Gall, Borg & Gall, 1996; Tedlock, 2005; Woods, 1996).

The Ability to Read the Minds and Emotions of Other People

People cannot only identify others and their activities and movements– abilities which other animals also possess - and not only identify the emotions reflected from the faces of other people, but also read the thoughts and feelings behind the actions and movements of others. As part of this interpretive ability, people also monitor the intonation of speech to catch the speaker's emotional state and intentions. We place ourselves mentality in the shoes of the other, which means that our brains are running a simulation of events in the other people's heads, trying to decipher their feelings (Johnson, 2004; Pinker, 2002; Restak, 2006).

Our common sense about other people is a kind of "intuitive psychology." We try to interpret people's beliefs and intentions from watching what they actually do, and trying to predict what they will do. The intuitive psychology skill to read other's minds, the ability to be in one's "mental shoes" is called in cognitive science "folk psychology" or the "theory of mind." That's what makes intuitive psychology an entirely useful tool for the study of people's behavior (Goldberg, 2007; Goleman, 2006; Pinker, 1997, 2002). This capability provides a human advantage over computers. Computers, even if programmed to distinguish between different faces, are hardly able to recognize all of the small, subtle expressions, and certainly not to assume the beliefs and intentions of those expressions. The computer can perform a million calculations per second – more than the fastest human brain. But even the world's fastest computers are not able to identify expressions and read emotions and thoughts as accurately as a little boy can do (Pink, 2005).

"Intuitive psychology," as expressed in our ability to read another person's mind, begins to develop in early childhood when we first look upon our parents. Babies begin to make eye contact with their parents at age six weeks. From a very young age, a normal baby can distinguish when someone is watching him - so young that we must conclude that this is an innate ability. Babies watch their mother when they look at an object, and her reaction helps them decide whether to approach the object or move away. In the middle of the second year, the infant can look at his mother to see how she looks when she says a word to him: he supposes that the word refers to the item she is watching. Children of age three or four can use a person's gaze direction together with a facial expression and infer what is going on in their mind. It seems that children as young as four have a theory of mind. We accept this mind-reading of young children as an obvious fact, even though it is a very complex mental mechanism (Harris, 1998).

There is another reason for the existence of mechanisms designed to read the mind of others, which is no less important. We are equipped with mechanisms to read each other's minds so that we can copy what others are doing or intend to do. The mechanism that allows us to copy the actions of others stems from a desire to benefit from their knowledge and thought. Hunter-gatherers, as were our ancestors, gained knowledge about ways to build tools, control fire, outsmart prey and neutralize the toxicity of plants, and thus managed to survive by virtue of the cumulative knowledge amassed and passed from generation to generation. Without this capability, our ancestors would have been forced to "reinvent the wheel every time," and this situation would not have enabled us to survive and continue to evolve (Pinker, 2002). A system of neurological mechanisms evolved to help us watch others, learn from them and utilize their experience (Johnson, 2004). These features and these mechanisms, derived evolutionarily from our ancestors, are also relevant to our lives in the 21st century.

The ability to observe and identify others while watching their external characteristics - as discussed in previous sections - serves qualitative research and may be a tool that quantitative researchers use as well. The ability to read the thoughts and feelings of others is probably unique to the interests and expectations of qualitative researchers, and apparently does not serve the quantitative researcher's colleagues. The credible qualitative research "recruits" these capabilities to carry out a proper research process.

The Ability to be Empathetic With People

Empathy: The Ability to Feel Others' Emotions

Not only are we are able to identify other people's variety of external characteristics and to read their thoughts and feelings, we can also empathize with others. Empathy is a natural, spontaneous connection to the thoughts and feelings of another person. This is the ability to imagine ourselves in someone else's place, and feel intuitively what they see in their eyes and feel in their heart. Empathy is not a comment on a small number of emotions of others, such as pain or sadness, but an understanding of the entire emotional repertoire and the emotional atmosphere of others. It is the ability to effortlessly enter into another human, manage a relationship with another person with sensitivity and without insulting or hurting him, to be attentive and have caring feelings (Pink, 2005).

The willingness to be interested in other people and to empathize with them is deeply rooted within us from birth and develops throughout our lives. We have all seen a baby begin to cry when he hears another baby crying. Thus, empathy begins. Studies have found that one-year-olds demonstrate concern and compassion before their speaking skills are developed. This resulted in the assumption that empathy develops before language (De Waal, 2005). Basically, empathy is a crucial human ability, as it prevents us from doing anything that could hurt the feelings of others (Baron-Cohen, 2003).

Empathy is distinct from reading thoughts and feelings. It is not only the knowledge of others, but something deeper that seeks to respond to feelings, thoughts and actions of others via an appropriate emotion. In a study aimed to provide taxonomy of emotions, Baron-Cohen (2003) found that there are 412 different distinct human emotions. Empathy sharpens the reading of minds, allowing one to understand others more deeply, to predict their behavior and to contact them with emotional resonance. It helps us understand the other side of a debate and to comfort someone in distress. It is characterized by a natural desire to care, to connect and identify people with each other, and provides the scaffolding of our morality (Pink, 2005). Empathy encourages the understanding that your worldview is not necessarily the only one, nor the only correct view (Johnson, 2004).

Reading Emotions as an Essential Component of Empathy

Antonio Damasio (2003) suggests a distinction between emotion and feeling. In everyday use, the word “emotion” tends to be regarded as synonymous with the term “feeling.” But it seems that in trying to understand the chain of processes that are characteristic of the mind and human behavior, one can see a complex process which starts with emotion and ends by feeling. According to Damasio, emotions are actions or movements, many of which are public and visible to others and reflected in the face, voice or certain behaviors. The emotional process also has components that are not “visible,” but we can make them “visible” through scientific tests. Feelings, in contrast, are always hidden from view, like any cognitive images that are not visible to anyone but their owners. Emotions play on the stage of a body; feelings play on a stage of the mind. Although emotions are mostly visible, people are not necessarily aware of their existence or source. Feelings, even if hidden from view, are conscious processes with the awareness of the person who feels them, even if they are not necessarily accompanied by a clear verbal expression.

Empathy refers mostly to feelings, feeling what the other feels. Unlike rational thought and considered opinion which are expressed in words, feelings are not expressed spontaneously in a verbal way (Goleman, 1995). With our identified capabilities, we can look at others and study from their faces, voices and posture (expressions of emotions) the world of their feelings, even if they do not express them verbally. Because empathy is related to feelings and feelings are not transferred in many cases literally, the way to reach the hearts of others begins with identifying physical clues, and these capabilities are the precondition for empathy (Restak, 2006).

Empathy is a unique ability of human beings, more than any other attribute. It can indeed find similar expressions in the animal world, such as emotional imitation, when an animal responds to the fear or joy of others (De Waal, 2005). However, empathy in people is of a higher degree. Not only do we show compassion for individual suffering, as chimps and some other nonhuman species would probably do, we are also aware that we feel compassion and empathy. Perhaps because this is conscious, we frequently act according to the circumstances underlying the events that sparked this empathy (Damasio, 2003). It is almost impossible to identify empathy by external research tools, such as advanced computer software. Computers have tremendous mathematical abilities, but when it comes to identifying human relationships, they are autistic. Despite all the progress, no software can compete with a sensitive and empathetic human. Voice recognition software can now make out words and make changes in intonation, but even the most advanced software cannot understand our feelings. We can improve empathy, but we cannot fake empathy. An unnatural smile differs, even slightly, from a real empathetic smile, and most people equipped with a high degree of empathy can identify the difference easily (Pink, 2005).

What are the mechanisms that allow us to read the mind of others so naturally and immediately, spontaneously and instinctively, and to empathize with them even without being conscious of these internal processes?Neuroscience has found that our brains are wired to connect with other people. (Goleman, 2006; Winston, 2002)

Brain studies conducted in the1990 's indicated the existence of "mirror neurons" in the brain (Rizzolatti & Arbib, 1998). These neurons allow us to simulate the movements and intentions of other people, such as virtual reality. For example, when we look at someone holding an object, the muscles in our hands become stretched, as if we are preparing them for some action. These mirror neurons can read other person's intentions and not just be imitated by their actions (Iacoboni, 2008).

Different Dimensions of Empathy

The ability to feel empathy for others has added moral values. The essence of morality is the recognition that others have interests just like us (Pinker, 2002). An ongoing discussion entitled "What is the human superiority?" has suggested that the uniqueness of human beings lies in their empathy skills. This is not only the ability to identify the feelings and thoughts of others, which exists in some animals, but also the ability to understand the other's mental condition. The ability to observe the situation from the point of view of another person changes the way individuals relate to each other (De Waal, 2005).

Steven Johnson (2004) proposes to distinguish between reading another person's mind and empathy. The ability to not only read the minds of others, but also be sensitive to the feelings of others, is characteristic of empathy (Damasio, 2003). Reading another's mind, however, is much faster and much less conscious than reading another's feelings. Conscious, accurate empathy is developed on the basis of the ability to unconsciously read the mind of another. It seems that an intuitive empathic ability characterizes the best of the experts dealing with social issues, such as social workers, teachers, physicians, psychologists, and others (Goleman, 2006). One can believe that this capability will also characterize the best of qualitative researchers.

Reading the mind of others and empathy are human skills, and like any other human skill, except for a few, all of us are endowed with them to varying degrees (Baron-Cohen, 2003). Individuals who are autistic have difficulties with empathy and mind reading. They have no ability to read other's minds, to fathom the world from someone else's eyes and respond appropriately to his emotions. Autistic persons ignore other people and treat them as objects. They can cope with physical representations such as maps and charts, but they cannot cope with mental representations. Autism is therefore a state of "mental-blindness" (Baron-Cohen, 2003;

Pinker, 1997, 2002). Simon Baron-Cohen (2003) considers that there is a sequence in autism: while some people obviously are clearly autistic, millions of others suffer from mild symptoms of blindness to thoughts and feelings.

Some researchers contend that we can find statistically significant differences between the male and female brain. Women are more sensitive to facial expressions. They do better in decoding non-verbal communication, noticing the subtleties of tone and facial expression, and using them to judge other characters. On an average, women use more social smiles and eye contact than the average man. Some may argue that the extreme male brain has no idea about understanding the other's mind. The extreme "feminine brain," however, has great empathy and understanding of thoughts and feelings of others (Baron-Cohen, 2003; Pink, 2005).

Some researchers suggest evolutionary reasons for the average difference between the degree of empathy of the male and female brain. Our ancestors' millions of years as hunter-gatherers - the period which designed and developed the human brain - gave a survival advantage to those who could make social treaties and receive social assistance in difficult times. A woman who had a high level of empathy was better prepared to establish a community of friends to help her care for her children. The empathic ability to connect to others gave her a significant advantage in her attempt to get support. Moreover, some developmental psychologists have suggested that the female brain's ability to connect through face reading, voice reading and absorbing nuances of emotion gave her an advantage in her role as a nursing mother. Empathetic skills allow the woman's brain to catch non-verbal signs of infants who have not yet acquired language. Many researchers have hypothesized that the ability to feel another's pain and to quickly read emotional expressions gave Stone Age women the ability to sense potential danger or aggressive behavior, and thus save themselves and protect their children (De Waal, 2005).

Intuition as a Thinking System

Capabilities discussed in the previous sections - the ability to identify people, animals and objects, the ability to read minds and emotions of people, and the ability to empathize with others - are all expressed intuitively in everyday life. Intuition is characterized by speedy perception and understanding accompanied by a sense of certainty, and it is contrary to deep-thinking processes and deliberation (Kahneman, 2011). As defined by Shirley & Langan-Fox (1996) intuition is "a feeling of knowing with certitude on the basis of inadequate information and without a conscious awareness of rational thinking". (p. 564). Intuitive processes are fast, automatic, effortless, associative, hidden from view, unexposed to self-examination, and are often emotionally charged. They also occur in many cases by virtue of habit, and therefore it is more difficult to control or change them (Stanovich & West, 2000). The perceptual system produces intuitive impressions regarding the features of the objects of perception and thinking. These impressions are not intended and not expressed explicitly in words (Kahneman, 2005). Intuitions are often experienced as a kind of instinct - they work automatically and unconsciously. However, intuitions differ from instincts: instincts are innate behaviors; intuitions, however, are learned behaviors, even if they include innate components (Hogarth, 2001).

In principle, aside from the intuitive reasoning ability, we also have the skills to look at things and consider them analytically in depth and over time. (These skills and their place in qualitative research will be discussed in the next chapter.) However, in everyday life we have plenty of situations where we must examine things quickly and immediately. Our intuitive exploration ability gives us the opportunity to evaluate and judge things quickly but reliably, if not always accurately. Intuitive ability was the most effective means of survival (if our ancestors didn't immediately grasp that there was a tiger lurking in the distance, their lives were in danger). And as we will see, intuitive ability is valuable even when we are faced with tasks that are unrelated to survival, such as conducting research.

Intuitions play an important role in the perception of social reality and the people with whom we meet and share our world, allowing us to make speedy interpretations that neatly bypass verbal and analytical analysis. These shortcuts are very essential in making quick decisions - to know who stands before us, what he means and how to deal with it, if he needs assistance, whether we can expect him to help in the future, and so on. In short, intuition allows us to make a quick, immediate exploration. If one were to control all information and pay attention to everything, our lives would be hard to cope with, and we could not meet the challenges of life (Hogarth, 2001).

Intuitive processes are characterized by a great involvement of the emotions and feelings. Gut feeling can prevent us from a choice that previously led to negative consequences. We respond to this gut feeling intuitively, even if we are unaware of it or can't define it. A recoil reaction can occur before logic will tell us exactly the same (or sometimes tell us exactly the opposite). The emotional signal can also encourage rapid selection of a particular option because our attempt is associated with a positive outcome. Often the emotional signal makes the logical process unnecessary – for example, when we reject the possibility that will lead to immediate disaster, or vice versa, when we opt for a good chance based on the high probability of success (Damasio, 2003).

People have a tendency not only to distinguish, identify, classify and generalize intuitively, but also to come intuitively to conclusions arising from these distinctions. People tend to analyze, investigate and quickly build connected systems between things. People search for a causal relationship between the elements as part of their experience to understand the system. Systemic perception - the tendency to link things and find a causal relationship - is rooted in our evolutionary past. This intuitive approach has allowed our ancestors to tie a causal connection between adjacent events - for example, the sound of breaking twigs and the presence of an animal (such as a snake). This connection is necessary for evolutionary survival, and natural selection was set in the minds of animals and humans in the early stages of evolution. However, our natural tendency to find order and logic among events and things that surround us and to find causal relationships between events sometimes leads us to the wrong and even non-rational conclusions. For example, we may imagine a connection between non-adjacent events to be causally related, such as bad luck following the sight of a black cat. (Baron-Cohen, 2003). The problematic implications of the intuitive processes towards the qualitative research processes will be discussed in the next chapter.

The Role of Intuitive Research Skills in Qualitative Research

As qualitative researchers, we are supposed to "recruit" our intuitive research capabilities for constructing a deep, sensitive qualitative research process. A detection capability, reading the minds of others, and empathy give us the possibility to reach a deeper, more meaningful understanding with the research participants. Empathy in itself is not a goal of qualitative research, but it significantly increases the ability of researchers to identify characteristics of study participants and their social-cultural context, and strengthens the ability of researchers to read the minds and feelings of the study participants (the subjects). Researchers who achieve an empathic relationship with the participants can approach the highest levels of identification and read the minds and feelings of the participants. An empathic connection allows us to reach an understanding that we are in a relationship not with objects, but with people - people with feelings whose feelings affect our feelings.

To gain empathy with the participants, qualitative researchers do not place themselves above or outside the world of the participants. "...The researcher's self is inextricably bound up with the research." (Woods, 1996, p. 51). The ability of researchers to reach a high level of empathy with the participants creates a different relationship of those in quantitative research. Qualitative researchers do not create a subject-object attitude with the research participants, but a subject-subject relationship (Sciarra, 1999). To reach a high level of involvement and empathy, to understand the hearts of the participants and step into their shoes, qualitative researchers do not use an objective-distance observation that could leave them with some degree of detachment, but listen to the participants' discourse and stories and use research tools such participant observation, involvement observation, deep interview or focus group and/or non-formal conversation (Angrosino, 2005; Fontana & Frey, 2000).

Being involved is to live among and within. Involvement is essential to understand the view of others as they see it, to see how they see others, to identify their problems and concerns, and thus to interpret their discourse and behavior. To achieve this, the research participants have to be accessible to develop a rapport, trust and friendship, connection, empathy, the ability to assess their feelings and their cognitive inclination (Maykut & Morehouse, 1994; Woods, 1996). If we are to understand social life, what motivates people, what their interests are, what connects and differentiates them from others, what their values and beliefs are, what's important to them, why they act the way they do, and how they perceive themselves and others, we should put ourselves in their situation and look at the world with them (Woods, 1996). "The move away from an acceptance of the researcher-researched relationship as an objective one toward a more relational view involves a reconceptualization of the status of the researched in the relationship." (Pinnegar & Daynes, 2007, p.11). This means that qualitative researchers should unite with the research participants, "go a mile in their shoes," or understand their point of view, the perspective of others from an empathetic position, and not only from a sympathetic position. "In fact, it is the ability to be with others that distinguishes the qualitative researcher." (Maykut & Morehouse, 1994, p.28). Qualitative researchers seek to experience the world of the subjects as participants in their experiences.

A high level of empathy as a condition for in-depth-research is necessary not only to researchers, but also to the readers of the study. In order to reach a proper level of research understanding, the readers must be in epistemological harmony with the story of the research. This means not only to understand the research, not only to grasp the characteristics displayed, but to achieve a level of engagement and empathy with the research participants (Lincoln & Guba, 2002 Merriam, 1985). To enable readers to get involved at an empathetic level, researchers are required to create the conditions that will allow the reader to reach this level, i.e., to write in a style that enables the involvement and empathy of the readers. Clifford Geertz (1973) calls this kind of writing "thick description." (Creswell, 1998; Josselson, 2004) Researchers have therefore committed to support the process by providing rich and detailed descriptions of the research story, with all its properties and components. The research description should allow readers to participate in the phenomenon under study, step into the shoes of the participants, and to experience empathy to what the research report describes (Denzin & Lincoln, 1994; Eisner, 1991; Stake, 1995).

CHAPTER 2

THE ANALYTICAL RESEARCH SKILLS

The previous chapter described the skills that grant humans the ability to serve as a primary research tool for the study of human beings and human societies. Qualitative research places the human as a primary research tool, thanks to his intuitive abilities to identify others, to identify their cognitive and emotional state, and to be empathetic. These attributes make the human a primary research tool that no external, objective tool can rival in terms of accessibility and sensitivity. The question is whether in addition to the intuitive properties, human beings also have limitations that block their effectiveness as a research tool, and if so, how to overcome such limitations to preserve the utilization of human beings as effective research tools.

The ability to achieve intimacy, involvement and empathy with others, reading their thoughts and feelings by "recruiting" our intuitive inquiry skills, does not necessarily mean that we're always reading the real thoughts and feelings of others. Often we find ourselves ascribing thoughts and feelings to others that are not actually their intention. In fact, we have no way to accurately read the thoughts and feelings of others, but only to evaluate and make inferences. We can bring good and reasonable guesses based on what they say, what we read between the lines, what we see from their faces and their eyes, and what can be learned from their behavior and their way of life. This is the most impressive talent of the human being. But reading the thoughts and feelings of others, as well as the mood-esteem of others, is not more than reading the image of our own thoughts, beliefs and feelings and projecting them upon the other. What matters is not the other person's position, but what we think about the position of the other. When we, for example, see someone suffer and feel anxious, when we look at a picture that describes horror, we often react to our own mental images (Restak, 2006; Pinker, 1997).

These reservations have significant methodological implications. Despite the crucial importance of intuitive research human tools that have a central place in the research process, we must recruit more skills in our possession, even if not intuitive and immediate. Facing the weaknesses of intuitive research skills, we "recruit" analytical research skills as another important component of the qualitative research process.

Two Systems of Thought

Research in cognitive psychology carried out over recent years gradually led to the adoption of the approach that two parallel systems of thought serve people in different contexts (Kahneman, 2011; Evans, 2003). Sloman (2002) defines the two systems of thought "intuitive" and "analytical." He names the first an "associative system," and the other a "rule-based system." These systems have also been known as "System 1" and "System 2." From all these options, we have chosen the terms "intuitive system," and "analytical system." In the previous chapter, we presented the intuitive system. This chapter presents the limitations of the intuitive system, while suggesting the analytical system as a qualitative research process that helps overcome these limitations.

The distinction between intuitive thinking and analytical thinking aroused great interest among cognitive psychologists. There is considerable agreement as to the characteristics that distinguish between these two types of cognitive processes. In general, the actions of the intuitive system, as we have shown in the previous chapter, are fast, automatic,effortless, associative, hidden from view, often emotionally-charged, and unaffected by self-examination. They also are made within the context of daily routine, and are therefore more difficult to control or change. In contrast, the actions of the analytic reasoning system are slower, more linear and more likely to be conscious and deliberately controlled, while also being relatively flexible and usually controllable by rules (Stanovich & West, 2000).

The differences between the two modes of thought are evident in the attempts to resolve apparently contradictory results obtained from studies of judgment under uncertainty (Kahneman, 2011). For example, in Figure 2A, the perceptual illusion known as the Muller-Lyer Illusion is demonstrated, in which two lines with arrowheads in opposite directions lie side by side:

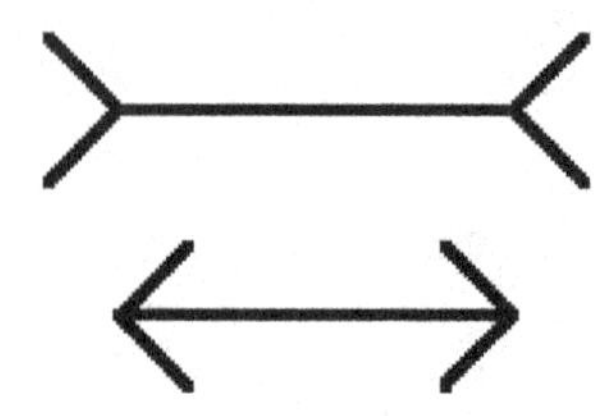

Figure 2A: Perceptual Illusion of Line Length

The perceptual-intuitive system causes us to judge one line as being longer than the other. However, if we measure the lines, the analytical system tells us that both lines are equal. Both these judgments coexist, and one does not cancel out the other. Yet in this case, it is clear that the intuitive system is misleading us, and we must use the analytical system to verify our judgment. Basically, these are not two alternative systems, but two systems existing side by side. The intuitive system is more rapid than the analytical system, and is generally operated first. The dependence allocated to each of the thinking systems varies from participant to participant, from task to task and from situation to situation, but we can activate two processes or two systems of thought and reach many decisions by combining the two systems or processes. Both systems can run simultaneously. They work as two experts, and their responses can be different (Kahneman & Frederick, 2002).

The only way to distinguish between the systems is by the degree of awareness of the process. When a response is generated by the intuitive system only, we are aware of the outcome rather than the process. When it is created by the analytical system, based on clear rules, we are aware of both outcome and process (Sloman, 2002). The analytic system "knows" part of the rules that the intuitive system tends to ignore, and sometimes the analytic system intervenes to correct or convert incorrect intuitive judgments. Therefore, intuition errors occur under two conditions: the intuitive system generates the error, and the analytical system does not correct it (Kahneman. & Fredrick, 2002).

Sloman (2002) argues that it is difficult to suppress the intuitive system, since it acts spontaneously and unconsciously. Moreover, yes, it is clear that it may disrupt the analytical system. Intuitive processes of the first system do consume cognitive resources, but compared to the analytical system, they are less affected by mental overload. With the development of training and experience, we can transform analytical thought processes based on rules into automatic intuitive processes.

As stated, qualitative research does not seek analytical thinking to overcome Intuitive thinking. On the contrary, qualitative research seeks to recruit the human intuitive inquiry skills in order to make a valid, worthy study. It is unlikely to expect that the deliberate, slow analytical thinking will replace the intuitive thinking. However, qualitative research is also cognizant of the weaknesses and limitations of intuitive thinking, suggesting that analytical research serves as a "monitor" of the intuitive system (but does not replace it) and ensures the quality of research. Proper integration of these two systems will assure processes valid and worthy of study.

Intuitive Thinking and Scientific-Analytical Thinking

Most qualitative researchers see qualitative research as scientific research (Lyons, 2007; Pinnegar & Daynes, 2007), although distinctly different in several types of science. There is a distinction between the terms "natural science" and the term "human sciences." "At the risk of oversimplification one might say that the difference between natural science and human science resides in what is studied: natural science studies "objects of nature" "things" "natural events" and "the way that objects behave."

Human science, in contrast, studies "persons" or beings that have "consciousness" and that "act purposefully." (Van Manen, 1990, p. 3). When Strauss and Corbin (1990) define the properties of the grounded theory, they perceive that "the method meets the criteria for doing 'good' science [...]" (p. 27). On this basis, there is a distinction between scientific methodology that characterizes qualitative research as "soft science," and scientific methodology that characterizes quantitative-positivistic research as "hard science" (Denzin & Lincoln, 2005).

Scientific theories are models of reality, not reality itself. Science does not come to determine what nature is, but what we think about nature. Unlike studies of the human and human societies, in natural sciences research, researchers lack the ability to maintain a dialogue with the objects under study, and there is no choice but to rely on earlier paradigmatic approaches to investigate the external objects. Historically these paradigms are changing, in accordance with the insights of the great scientists who have established new paradigms to look scientifically at the phenomena around us (Kuhn, 1962; Smith & Hodkinson, 2005).

Scientific thinking is fundamentally analytical thinking. Science is a method for collecting information about the world through observation, experiment and systematic deduction, via suitable modes of research designed to develop this knowledge. Systematic control is a cornerstone of science. By this definition, qualitative research can be seen as scientific research, and many researchers consider it as such. But it is clear that the scientific view that characterizes qualitative research in all its diversity will be different than that of the research in natural science. Qualitative research is not characterized by a process of experimentation, but there are definitely systematic, controlled methods of gathering and analyzing data. While qualitative research "recruits" the human intuitive skills as a main tool of research, the study of natural sciences (including physics, biology, astronomy, etc.) expects researchers and readers of research to position conceptual scientific hypotheses against their intuitive perceptions.

Although we don't suggest qualitative research as a methodology for the natural sciences, the debate over the tension between intuitive and analytical thinking in the context of natural science may illuminate the problem, even in contexts relevant to qualitative research.

The basis of the scientific approach is a systemic conception which is not content with identification and classification, but rather seeks to find relationships and causality between the elements under study. However, a gap and tension exist between the scientific concepts and intuitive perceptions. Good science is very costly, and requires a relatively long duration. Such a science had only a minor chance for helping groups such as those of our ancestors, who had to forage for food or go hunting frequently and immediately. We can expect that "scientific research" was not part of the basic nature of human intuition. Innate and/or intuitive natural abilities evolved in the minds of our ancestors, giving priority to rapid processes, (immediate, though often vague) over the more precise, accurate and long-term processes. Real science is famous for its ability to transcend intuitive feelings and set the rules found in the foundation of things (Pinker, 1997).

Intuition is often a major obstacle in the discovery of truth in the scientific method. The reality created by scientific evidence challenges the perceived intuitiveness of what humans deduce through their senses. Therefore, natural science cannot see intuitions and impressions of humanity as the ultimate research tool. Not so the research on human and human societies, which focuses on "creatures" with language and consciousness. Therefore, we need to set, a clear distinction between research on the human and human societies, as opposed to natural science research.

However, we must understand that in order to ensure the "scientific" aspect of the qualitative research (as a "soft science"), it is necessary to "recruit" the human analytical research skills as well as the intuitive research skills. At the same time, we need to understand that analytical research principles, which are necessary for research in natural sciences, are not necessarily the same in the qualitative research focused on humankind and society.

Intuition and Social Decision-Making

In 2002, Daniel Kahneman won the Nobel Prize for research he had executed in the 1970s, together with his late colleague Amos Tversky. The studies were focused on the issue of decision- making under uncertainty, and subjective assessments of uncertainty. Unlike in the natural sciences and exact sciences referred to in the previous section, Kahneman and Tversky's research dealt with the issue of decision-making in economics, psychology and other fields of the social sciences. Their studies dealt with the world of people and societies, areas that were addressed with qualitative research. The work of Kahneman and Tversky turned our attention to the fact that contrary to what was necessitated by intellect and logic, people who come to make decisions in situations of uncertainty tend to reach a quick verdict based on a limited amount of information. They found that people rely on a limited number of heuristic principles, that is, a few general simple rules of thought, offering quick and uncomplicated ways to get a simple answer, at the price of low accuracy.

Kahneman and Tversky's studies suggest that in decision-making processes, people reduce the complex task of assessing probabilities and predicting values into simpler judicial actions. These heuristics are usually quite useful and equal for all. The scientists showed that people solve problems in everyday life through the use of their implicit understanding of the world, an understanding often based on making intelligent guesses rather than on formal principles of logic. These judgments are all based on data whose validity is limited, and in many cases these thought processes led to severe and systematic errors (Kahneman, 2011; Tversky & Kahneman, 1974).

It seems that the intuitive system plays an important role in making such decisions, much more than rational models. Intuitions play a vital role in the perception of reality because they allow us to create swift interpretations that bypass tedious analysis and verbal processes. As shown in the previous chapter, these shortcuts are essential for quick decisions, to create connections between partial components of reality and fill in holes in the information received through the senses. However, they come at a price: the results are sometimes far from what is solved by the rational model. The language of the researchers coins it as "bias." These biases, like visual biases, are typical to the intuitive mechanisms of human being's judgments (Hogarth, 2001).

Bias means that most respondents do not come to an incorrect assessment because they do not know the norm and so they guess, but because of the shared, biased way of thinking for most people. Kahneman and Tversky suggest various explanations for biases, all of which rely on the intuitive thinking system, or as Kahneman prefers to call it, "System 1" (Kahneman, 2011; Kahneman & Frederick, 2002). For example, when people are asked to judge the relative frequency of objects or the likelihood of certain events, they are sometimes influenced by the relative availability of those objects or events in their actual experience. Such criteria of availability may be effective and accurate if it indeed reflects the available, real prevalence. This experience is highly influenced, for example, by the media's emphasis on certain topics which thus increases the sense of availability and also the estimate of their frequency (byte-Merom and Shahar, 2007).

Another example presented by Tversky and Kahneman (1974) is referred to as the "anchoring and adjustment heuristic." When people are required to estimate the amount or probability of something, they estimate to a great extent according to internal or external sources of information they already possess. This information is used as an anchor for estimating and reliance, and only then people may change it partially, if at all. Bias is caused by having relied on information that is not directly related to the objects being estimated. For example, a physician may interpret symptoms in patients on the basis of his experience with other patients, or information he has in connection with similar symptoms. However, this interpretation might be wrong. Psychological studies offer an explanation of this bias: people try to find information to verify the anchor, because their existing information gives them confidence in their evaluation, thus creating a biased outcome.

Emotions, too, affect the quality of assessments and decision-making, and can lead to bias. Decisions can be affected by emotional feeling towards the objects and bring a person to a biased assessment. For example, a person who shows negative feelings toward another person will reveal a biased tendency to interpret the behavior of the other without examining the specific case in depth. In other cases, decisions can be affected by the immediate emotions of the decider, such as mood, etc., and not necessarily be related to actual issues. Persons in an emotional state of joy may calmly relate to a certain phenomenon, yet in a different emotional state, they would treat it with great concern. Although emotions may be part of the analytical thinking process, emotions are often activated intuitively and automatically, unconsciously and without being put into words. (Damasio,2003; Loewenstein & Lerner, 2002).

Degree of Frequency of Intuitive Biases

Biased perceptions and assessments are not limited to laymen. Tversky and Kahneman (1974) examined thoroughly-regarded psychologists' judgments and found that they tended to be concluded from very small samples. The tendency of psychologists to generalize from the findings of very small samples is not completely commensurate with the general statistics that were known to each of these researchers. As we know, even though Daniel Kahneman is a psychologist, he accepted the Nobel Prize for his contribution to the economy. The research findings of Tversky and Kahneman have challenged the idea that people, including professional economists, make decisions based on analyzing economic rationality, on flow charts or various tables of predictions. Even the best professionals suffer from biased considerations and irrational decisions.

We can understand these biases in situations where people need to reach instant decisions without time to deliberate, for example in the heat of battle or (for a change) during classroom lessons. But what's interesting is that these studies show that even without time pressures, people employ method of intuitive thinking. Kahneman and Tversky's work indicates that the supposedly objective reasoning we use for our daily life decisions is much less common than we think and what is presumed to be apparent. Their research disproves the conventional wisdom that people make decisions and formulate judgments based on facts and rational considerations. (Damasio, 2003; Tversky & Kahneman, 1974).

The biases mentioned in previous sections are also typical to the processes of perception and the thinking of scholars in general and qualitative researchers in particular. Researchers may also be affected by prior information and from experience with similar situations in the past. Researchers, too, may be biased because of their momentary mood or their emotional attitude toward the phenomenon under investigation. Yet it's clear that these biases are likely to characterize researchers encountering time pressures of rapidly-flowing data. Qualitative researchers face part of the research processes relatively free of time pressures, allowing them to consider and examine a process in depth (in the stage of data analysis or of study design). In many other situations, qualitative researchers do face situations with time pressures exacerbated by the rapid flow of data, allowing limited time to accurately absorb (for example, during observations or interviews). It's clear that in all these situations, researchers, like other human beings, may rely upon biases and distortions.

In summary, intuition operates in the absence of conscious logic and plays an important role in the perception of reality. Moreover, we are often motivated by emotions rather than logic. When people judge or make decisions based on intuition, their brains do not work simply by way of a standard algorithm that gives significance to the facts and produces the best solution (a process known as an "analytic reasoning system"). Much of our intuition about what is likely and what is not reasonable is revealed to be misconceptions. Our advanced mental mechanism for identifying, reading thoughts and feelings, empathy, skepticism, and subjective probability often concocts huge mistakes (Dawkins, 1986; Tattersall, 1998; Rich Harris, 1998; Hogarth, 2001). The following sections will detail a number of intuitive processes that create a false image, indicating human limitations as a research tool and underscoring the need to accompany qualitative research with an analytical control system.

Classification, Generalization and Stereotypical Perceptions

We classify and generalize all the time. So our brain is built. We view certain plants as trees and other groups as bushes, even if within each group there are very dissimilar items. We treat many animals, such as dogs, as identical, even if their appearance is very different. We may refer to all snakes as one group, and immediately run away from any snake, even if it's not poisonous or harmful in any way. Since there is no effective way to learn how to handle every object, animal, or person individually, we categorize them and then apply what we've learned about one thing to others of the same type. The tendency to generalize and classify is instantaneous, and precedes the need to see the difference, if any. Classification and generalization are routine activities of human beings (Dey, 1993, 1999); classification and generalization are our mind's abilities by its very nature. They are also innate: three-month-old infants can classify the people around them by age and by gender (Harris, 1998; Hawkins & Blankeslee, 2004).

The intuitive ability to classify and generalize is undoubtedly a useful feature of life and research, but it carries a high cost, such as in our tendency to stereotype generalizations about people and situations. For most people, the word stereotype arouses negative connotations: it implies a negative bias. But, in fact, stereotypes do not differ in principle from all other generalizations; generalizations about groups of people are not necessarily always negative. Intuitively and quickly, we mentally sort things into groups based on what we perceive the differences between them to be, and that is the basis for stereotyping. Only afterwards do we examine (or not examine) more evidence of how things are differentiated, and the degree and significance of the variations. Our brain performs these tasks efficiently and automatically, usually without our awareness. The real danger of stereotypes is not their inaccuracy, but their lack of flexibility and their tendency to be preserved, even when we have enough time to stop and consider (Harris, 2006; Tversky & Kahneman, 1974).

Since we have the natural skills to identify people, read their thoughts and feelings, and to reach states of empathy with other people (see previous section), we tend to classify individuals relatively quickly and determine our relationship to them according to this classification. Not only do we classify people by gender, age, personality, character traits, and the like, but we are equipped with a mental mechanism that causes us to see a person's personality as being relatively fixed. Just as we determine a lump of coal as black and snowball as white, we attribute "friendly" to one person and "nasty" to another. With our intuition, we tend to judge someone's personality based on a too-small sample of that person's behaviors, even if the sample is random and may not reflect the total behavior (Harris, 2006; Tversky & Kahneman, 1974). Tversky and Kahneman (1974) call this unjustifiable bias an "illusion of validity." This illusion is not limited to nonprofessionals, and it remains unchanged even when the person becomes aware of the limiting factors of his assumptions. Many studies have found that when conducting interviews, even professional psychologists often show great confidence in their assumptions based on partial and ill-founded information, even when they are aware that the professional literature they are dealing with is inadequate. (Kahneman, 2005).

The very fact that our tendencies to categorize, generalize, and see people and situations stereotypically is a universal phenomenon and a part of our mental makeup raises the possibility that these biases actually helped our ancestors. These features allowed people to react immediately when they found themselves in dangerous situations, even if later it turned out that their perception was incorrect. In this way, they could judge the people around them and instantly determine who to trust, even with a margin of error. The purpose of this ability is to tell us how to behave if we meet that person again, with rapid generalization and stereotyping giving us a guess based on information that we possess. While these features often reflected a true picture, they frequently contained little errors within. These features were probably more beneficial than harmful to our ancestors, and serve all of us in everyday life then as now (Harris, 2006).

The tendency to immediately and intuitively generalize and classify may be helpful to qualitative researchers, who attempt to explore the human experience as it passes and moves quickly ahead. Yet the flood of many details may prevent researchers from seeing the whole picture. The ability to sort and distinguish draws our attention to the essential, and to gathering many details of the whole picture. The human as a research instrument can respond quickly in real time to observations or interviews. At the same time, the researcher can generalize, classify, and distinguish between individuals, between the trivial and the more or less important. However, this tendency bears a serious danger for the nature of the study. Yet the question is, how do these intuitive skills, which are often useful to researchers, not act as an obstacle to the qualitative research process? We can make interpretations which are not sufficiently grounded, and place more importance on impressions than on data. The challenge before us is how to deal with these risks without losing the quality inherent in being intuitive researchers.

Processes of Biased and Misleading Perception

Biases arise with the processes of the human's perception of the world around him. We do not see the world directly as it absorbed by the senses, but as the ultimate processing of sensory information in the brain. The facts surrounding us are not facts unless the brain decides that they indeed exist and determines how they exist. The world around us is real, humans are real, and animals are real, but the brain process determines the way we perceive and understand them. Some of this process involves deleting certain data that the brain decides is not essential (Rock, 2004).

The world inside our heads is shaped by the nature of information flowing from the outside world into the inner world, and by the ability of the inner world to process the information and the willingness of the brain to absorb and store the information. We can be aware of only what the brain has absorbed from the outside and the data it has deduced to be maintained. (Leakey, 1994; Hawkins & Blakeslee, 2004).

The brain and its mechanisms of absorption and memory have limited capacity and cannot absorb everything at once. The brain performs intentional selection processes of absorption and memory. We benefit by recording and remembering certain things that interact with what we already know or with the prevailing, accepted opinion. Comforting ideas have a better chance to be absorbed than scary ideas. Popular ideas and praise are better absorbed than the contrary. New theories that do not correspond with existing ideas or with intuitive concepts are usually rejected. False theories and erroneous ideas may be absorbed and stored easily, just because they sound pleasant, interact with known quantities, and are easy to grasp (many politicians and advertising professionals take full advantage of the feature). In contrast, threatening ideas are rejected out of hand (Blackmore, 1999).

Our brain is organized so that if we repeat claims frequently, it tends to eventually accept them as facts. The basis for this is familiarity: the more often something is repeated, the more familiar it becomes. Research has shown that if a lie is repeated enough times, listeners begin to believe it. It does not matter if the lie is presented as a fact or opinion, orally or in writing. People will believe it, even if it was stated from the start to be a lie. (Restak, 2006). We must be aware of our mind's great strength to create false memories. We are capable of being convinced about events that we only apparently experienced, if such external forces as the media or other people present them to us over and over again. In the same way, we can associate ourselves with the experiences of others if we hear them repeatedly, and/or if these experiences relate in one way or another with our personal experiences (Dawkins, 2006).

Brain research of recent years points to a fast, vigorous nervous arousal when a person looks at a face acknowledged as being beautiful. This raises the hypothesis that patterns of beauty are essentially innate patterns. Sometimes this awakening does not reach our consciousness, and even allows a subject to tell investigators that beauty is not important at all. While it's not nice for us to consider beauty as a virtue when wisdom is of higher worth, we still continue to respond quickly and strongly to beauty, according to a brain scanner. Even newborn babies look three times more often at a beautiful face than at others (Etcoff, 1999; Lampert, 2007). There are other elements that unconsciously appeal to our attention, such as the appearance or sound or letters of our name, big exceptional things, happy, radiant faces, and so on (Grandin & Johnson, 2005).

These biases are extremely relevant to the qualitative research process. Even the most open, responsive researchers might hear certain things, while other things just disappear from their ears. The memory will keep certain items while others will be discarded. I know from my experience as a researcher that in the process of observations and interviews, many things have been preserved in my mind while others, maybe important, are not retained at all. Or that certain things caught in my memory are completely out of proportion compared to other significant things. The fact that I was helped by external devices such as a tape recorder for interviews and a video camera for observations, allowed me to go back and examine the records and realize how perception and memory have led me astray.

Memory Processes and Deceptive Predictors

The issue of memory and recall is relevant, then, to qualitative research, particularly with respect to methods of observations and interviews, the key tools of qualitative research (Fontana & Frey, 2005; Clandinin, 2007). For a long time, scientists believed that memories are like books in the library. They assumed that when our brain wanted to remember something, it had to "find" the right "book" stored in the "library" of the brain and then "read out" the relevant section. The prevailing view today is that our memories do not work like an open book, a tape recorder or VCR. Rather, our memories work associatively, constantly being structured by the strong influence of the new meaning we give to events. This meaning does not precisely reflect the original conditions in which we remembered. Our memories rewrite themselves every time they are reawakened (Hogarth, 2001).

The brain reconstitutes the memory in a new associated context. In a sense, all recollection is recreating a memory. Every time we recall an object, a face or some event, we do not get an exact reproduction but an interpretation, a new version and reconstruction of the original. What we perceive and how we perceive, everything we see, feel and hear, depends on our memory. (LeDoux, 1998; Johnson, 2004; Damasio, 2003). These facts are utilized by researchers and clinicians in interviewing people who have suffered oppression, for example, to reconstruct their memory and thus make the renewed experience a part of the re-releases (Grabich, 2007). The human brain stores information on objects, plants and other animals, part innate and part learned. This information allows us to save valuable time in identifying the events, objects and organisms from a mass of information in sensory input. Memories stored in our brains – real memories or false memories - are used to constantly produce predictions about everything we see, feel and hear. When we look around us, our brain uses memories to create predictions about what it expects to experience before we experience it.

There is a cost to the superiority of mind over the senses: it can prevent us from absorbing significant events which are not already stored in our memory, even if they should be. As the brain is certain of its knowledge of reality, it tends to follow stereotyped and perceptual biases, and doesn't relate to facts that "confuse it." The issue is that we see not what the eye sees, but what the mind is set to see. In other words, we don't see the world directly, but via the last processing of the sensory information in the brain. It seems that this image challenges the qualitative researchers that "recruit" the intuitive thinking system as a significant factor in the study. Avoidance of placing control in analytical systems against the intuitive research system may weaken and even invalidate the quality of the research.

In regard to this issue, it seems important to address the issue of human ideologies. There are many definitions of the concept of ideology, but most all agree with the general definition of ideology as a belief system which focuses on action (Eagelton, 1991). This means a system of beliefs which are conceptions of knowledge stored in the brains of humans and constitute the perspective through which people look at the world phenomena, interpret them and act on the basis of this belief system. These beliefs deal with facts and with personal assessments. (Van Dijk, 1998).

Ideology, then, guides our perception of the world without our even being aware or having any control over this phenomenon. The status of ideologies in our mind is so powerful that ideological changes rarely take place, and usually come as an extreme response to crisis situations. In this context, Clandinin & Rosiek, (2007). claim that ideology refers to "a system of thought and practice that gives rise to false consciousness in individuals and communities." False consciousness is a condition in which a person acquires a habit of thinking and feeling that prevents him or her from noticing and analyzing the real causes of his or her oppression. [...]" (p. 47) It is therefore clear that an awareness of the power of beliefs and ideologies must guide qualitative researchers when they relate to the world of the research participants.

Expectations that Affect Perception

The issue of bias resulting from expectations has become a core of numerous studies. The most famous research on self-fulfilling expectations, "Pygmalion in the Classroom," was conducted by Rosenthal and Jacobson (1968). Teachers were fed misinformation about 20% of the children, purporting that there was a chance that they could improve their learning achievements. Those 20% then indeed blossomed as their IQ rose to higher averages compared to the rest of the class. Apparently the teachers formed expectations about this group of students according to the information received from the researchers. In another experiment that examined a similar issue, students received an experimental psychology research task with maze-learning rats. All the rats came from the same genetic strain and were distributed randomly to the students. However, students were told that some rats were selected for fast learning, while others were selected for slow learning. According to reports filed by students, the so-called fast-learning rats actually learned the maze significantly quicker than the other rat population. It thus seems that the so-called "faster rats" performed better because the students that ran them through the maze expected them to succeed. Compared to the "Pygmalion in the Classroom" study described above, there is no way the rats could know how much they're expected to succeed, so it can be assumed that the students unintentionally tipped the scales in favor of the results they expected to receive (Harris, 2006).

It seems that often we see what we expect to see no less than what is actually before our eyes. Our brain complements missing or corrupted parts with what it thinks should be there. Most people's perception is such that they see the things they're used to seeing. They suffer from an attention deficiency toward the unfamiliar. For example, when listening to others, we often think that we know what they are going to say before they speak. So we often do not listen at all and are even convinced that we hear what we expect to hear (Hawkins & Blakeslee, 2004).

Legend tells that when the Spanish conqueror Hernan Cortes landed on the beaches of Mexico, he was not identified by the natives as a real person, but rather characterized as a superhuman, consistent with the prevailing myths of their Indian society. Cortes was considered by the Indians to be a messenger of a legend-god, who according to Aztec lore will return once in 52 years. Cortes' appearance and the date of his appearance led the Indians to believe that he was not a hostile occupier, but a god or a messenger of the god (Megged, 2009). By the time the Indians understood their mistake, it was too late. No wonder that the Spanish conquered relatively easily, although they were a few hundred against millions of Indians. Thus we see what can happen when perception is influenced solely by information found in the memory.

In one famous experiment, as shown in Figure 2B, a group of subjects watched a film in which two groups of players passed a ball to each other. The subjects were asked to count the ball deliveries. After some minutes, while the subjects were focusing on their counting task, a woman dressed in a gorilla costume walked inside the court, beat her chest, and walked out several seconds later. Despite the unusual appearance, 50% of viewers did not notice her.

Figure 2B: Gorilla in the Ball Game

This phenomenon, although surprising, is very typical of how human beings perceive the world. People are made to see what they expect to see, and find it hard to expect to see something they've never seen or have not expected. New things, different and strange, are just not noticed. Most people do not consciously see something unless they look at it in a very focused manner. Ordinary people are blind to what they are watching without intently focusing. People are not attracted to innovations, but they are also not exposed to many innovations, not necessarily because innovations do not exist in their immediate environment or are not shown to them. In many cases, people simply do not notice innovation, at least not immediately. (Grandin & Johnson, 2005; Harris, 2006).

As we have seen, human beings cannot consciously experience the raw data that exists outside of them, but only the image created by their minds. Do we, as qualitative researchers, not notice data and events that are significant to the phenomenon under study? The answer is definitely yes. The appropriate qualitative research methodology has to suggest research analytical tools to ensure that these biases will be reduced to a minimum.

The Tendency to Find Causality and Intent

Attributing any action to linear causality is a habit that the human mind is specifically addicted to. Human beings deal with linear causality in a way that is seen as almost obsessive, even inventing ridiculous myths in an attempt to preserve a linear causality (Ridley, 2003). Many people, especially children (as well as members of primitive tribes) ascribe causality of human behavior to weather upheavals, movement of waves and water currents, falling rocks and other natural phenomena. Most of us, in fact, tend to take much the same attitude toward cars, usually when they disappoint us. Often, if our car gets stuck instead of checking out the engine for failure, we immediately and unconsciously attribute it to the car's intention. We talk to the car, and hope to cajole it to start to move. It seems that humans have a psychological bias to personify inanimate objects and view them as agents of action (Dawkins, 2006). The source of this fact reveals a tendency to assume the other's intentions, and sometimes fear them. We find it difficult to see something that is not driven by the causality caused by the person's behavior.

Just as every action is attributed to causality, so all is attributed to a purpose. It is believed that there are evolutionary reasons deep-rooted in our past for the conception of causality and intentions, and this tendency has survival value as a brain mechanism for decision-making in critical circumstances. When we see a lion in the distance, it is better not to stop and consider predictions about its expected behavior, but to relate to this lion's intention to eat us, and the appropriate reflex to flee. Similarly, when we encounter strange people, we often discover a tendency to attribute bad intentions to them, and immediately avoid any proximity or contact with them as a defense against possible problems. (Dawkins, 2006).

Intuitive inclination to treat intent and causality among others may constitute a significant hindrance to the quality of qualitative research. Researchers observing people see them in their routine operation. In many cases, researchers act as if they not only see the action, but also know its cause (since apparently, every action has an obvious linear clear cause). In one of the courses on qualitative research methods, I introduced a video tape of observation into one classroom lesson. In the video, we saw a teacher who introduces a series of questions to his pupils. The teacher asked a question and allowed several pupils to answer. He was very attentive and gave the respondents a nod, without any extra words. But there were a certain number of pupils who received a verbal response such as "very beautiful," "very good," and the like. Usually after their answer, the teacher moved on to the next question. The university students who watched the video were not bothered at all by the issue of why the teacher conducted the lesson in such a manner. The lesson process seemed obvious to the students: the general opinion amongst the students was that the teacher was waiting for a "correct" answer, and upon hearing it, he reinforced the pupil who replied and moved on to the next question. Although the teacher was present in the university class, none of the students asked him why he behaved as he did. But the teacher's explanation was quite different. He explained that the pupils who received encouraging responses were those who needed positive feedback, and he just waited until they would ask to speak, let them talk, and reinforced them. Only then could he continue to the next question.

This causal and intent reaction of attribution to others, usually derived from our experience and/or our beliefs, is a common intuitive response. As researchers, we likely fail in the same way as described in the university course above. Many times we feel that there is no need to listen to our research participants, since their actions speak for themselves and the reasons are inseparable from their actions. It is likely that the university students attributed the teacher's intentions according to their own experiences as pupils at the school. But this intuitive response does not necessarily reflect the world of the research participants. Automatic, quick and intuitive skills that exist in our day-to-day experiences are frequently justified, if not occasionally life-saving. But such responses in other cases may obstruct proper research. Qualitative researchers who recruit their intuitive research capabilities for the study should be aware of this tendency, and ask themselves whether it is necessary to use the analytical system to control the intuitive reaction.

The Brain's Ability to Complete Patterns

Not only are we often blind to what we see, showing a strong preference to see some things and not others, our brains also have the ability to complete the patterns absorbed from the partial version. We do not usually realize that we are continually complementing patterns, but that is a common property and basic in the way memories are stored in the brain. The previous chapter emphasized the importance of rapid absorption and completing missing patterns, a feature that gives us an advantage in gaining control of rapidly-flowing information. Now our attention will be directed to the problems associated with this feature. For example, during a conversation, we often can't hear all the words over the background noise. Our brain intuitively completes what is missing with what it expects to hear. In many cases, even though we can hear the words of others, we tend not to listen attentively to all that is being said. Sometimes we seal our ears, listen selectively, or even interrupt the other person, with the firm belief that we know what he/she would say because our brain completed what it would expect to hear. This does not guarantee that what the brain completed is indeed compatible with the intention of the speaker - or what he would actually say.

A similar phenomenon occurs when a stimulus causes us to recall past experiences. In this situation, if we remember only a fragment of a thought, the entire memory may arise and flood our consciousness. But it might be that the series of memory bites is not related precisely to the original piece of memory, and we have connected inappropriate or irrelevant segments. This need of the brain to predict in advance, to fill in the missing gaps between the particles in the images we see, is also the cause of our tendency to be misled by optical illusions. Figure 2C illustrates one example. We perceive this figure intuitively as a drawing of a Star of David, although analytical examination reveals otherwise.

Figure 2C: Perceptual Illusion of Filling the Gaps

We think we see something in a place we expect it to be found, despite it not being there, because our prediction patterns tell us it was there. Magicians, for example, use this characteristic of the brain quite adeptly to base many of their tricks on optical illusions, thereby misleading the audience (Hawkins & Blakeslee, 2004; Ratey, 2002).

The Brain as a Story Constructor

The tendency to supplement things and connect isolated segments reflects the brain's tendency to create stories with integrity, to weave our lives into a coherent tale: stories that reflect our self-perception. The mind creates explanations based on our self-image, memories of the past, expectations of the future, and the present social and physical environment in which we act. The brain takes all random and even contradictory information on what we did or remember at that moment, and gathers everything into a consistent, logical story. If there are details that do not fit, changes or edits may be created. In its quest to give meaning to the sequential information, the brain looks for clues of order and logic in this information, in order to categorize them in a narrative form. To this end, the brain generates theories and sometimes even fabricates explanations to interpret obscure events. All these combine together into a seemingly coherent story (Grandin & Johnson, 2005; LeDoux, 1998).

Brain researchers have raised the argument that there is a neural system (supposedly located commonly in the brain's left hemisphere), which continually seeks to provide explanations for our perception and experience. When the brain does not find the correct explanation, it simply invents something with the information at hand, just like the dreaming brain puts together varying components into strange combinations. Our brain constantly creates an ongoing narrative of our actions, our feelings, our thoughts and dreams. This is the glue that unites our story and creates our sense that we are rational, whole creatures. Apparently the system which joins all things into a whole story seems coherent, even if it is based on fictions, and gives people an advantage. Sometimes the story builds up the memory more than it is built up from the memory: we try to recover memories and adapt them to the actual reality. (Baiet-Marom & Shahar, 2007; Rock, 2004).

It is conceivable that our ability to interpret reality and assemble it into a continuous story that is logical, rather than based on factual information, is one of the most important mechanisms of evolution. Constructing a story from countless elements of reality involves making the distinction between what we see as essential and what we see as subordinate. On one hand, it allows us to ignore the tremendous amount of irrelevant information. On the other hand, it allows us to construct a reasonable picture of reality from partial information. It is difficult to function intellectually and emotionally in an environment where every little thing distracts us from the point. Thus, the brain allows us to absorb relevant information, ignore other, and create a meaningful narrative whole. Apparently, like the previous examples we have presented, evolution has chosen a strategy of probability rather than certainty (Grandin & Johnson, 2005).

As qualitative researchers who adopt the conception of the human as a main research tool, based on intuitive ability, we must be aware not only of the benefits of rapid, sharp and sensitive perception, but also of the major disadvantages that these traits carry. Our natural tendency to create a narrative whole, to distinguish between the important and unimportant, may prevent us in some cases from seeing the reality that reflects the world of the participants, acting instead on our mind's perceived reality. In qualitative research, which is characterized by a holistic conception of the phenomena under study, there is a constant tension between the reference to the entire picture and the place of each of the myriad of details. Sometimes, the fear of harming the holistic perspective makes researchers avoid scrutinizing the details but offering instead a number of themes that reflect their general impressions of the phenomenon under investigation. This was brought about by means of impression, without challenging the themes with systematic, analytical examination. As will be described below, qualitative methodologies offer methods utilizing analytical thinking processes that help researchers create a story based on research data obtained from the participants' world, without the addition of any unchecked data, while using all the advantages of researchers' intuitive skills.

Between Research and Non-Research: the Use of Analytical Research Methods

Qualitative research, as noted, seeks to recruit the properties of human intuition abilities for the research of humans and human societies. Qualitative research seeks to utilize these characteristics as a preferred research tool, as presented in detail in previous chapter. However, worthy qualitative research should not only raise the properties of human intuition, but also deal with the weaknesses of intuitive skills, as presented in this chapter. The appropriate qualitative research process is not based solely on the natural intuitive characteristics of the researchers. Literature, art, and even the media, in their various forms, tell the story of human beings based on their own impressions and talents. We should draw a clear line between art, literature, and research, and between journalism and research. Qualitative research is not another genre in the family of literature and arts, and certainly there is a clear boundary between journalism and research. Qualitative research, like its name, is first and foremost a research, and it must meet the criteria that justify the concept of research.

When I started teaching qualitative research methods at the Hebrew University of Jerusalem in the late 1980's, qualitative research was not yet being widely used in our department. I was often put in situations of defending the method, and I could accept the fact that I had to protect the quality of qualitative research with colleagues who mainly used quantitative methods. However, even the students would tell me-- before they'd learned anything-- that this was an unacceptable method. Although I've often heard unfounded arguments, I remember one student that challenged me by asking, "Is qualitative research indeed a real research?" When I replied affirmatively, she proceeded to divide the word "research" into the two components "re" and "search," which means the ability to verify the search again and gain. She continued, "The ability to search back is the essence of research, and this is what distinguishes between research and non-research." I nodded in agreement, as she added, "Quantitative research is based on a methodology that ensures repeated searches. Is this also guaranteed in qualitative research?"

It seems to me that this student put her finger on a critical issue in understanding qualitative research by alluding to the distinction between what might be called research and what is unworthy to be considered as research. Maintaining this distinction will assure that we not only take advantage of the characteristics of the human intuitive research tool, but refrain from falling into the lair of the weaknesses of human intuitive abilities. This is done by "recruiting" two systems of thought, the intuitive system and the analytic system.

Researchers must accompany their intuitive inquiry process by control mechanisms of analytical thinking to vouchsafe that the process achieves a suitable level of "research." We need to use a methodology enabling us to argue that it is characterized by a procedure that allows re-examination. This methodology must assure the process of transparency, allow and control reflection, and maintain a point of view of distance from the phenomena under study as an integral part of the research process.

Practical Implications of Qualitative Methodology

A. Criteria for Research Quality

Qualitative researchers are divided over the question whether to use the terms of validity and reliability, which are conventional in quantitative research, or to offer unique terms of qualitative research. Although I do not see why not to use the terms of validity and reliability for determining the quality of the study, while underscoring their own specific meaning in qualitative research, I see no reason to oppose the alternative unique expressions . Yvonna Lincoln & Egon Guba (1985) proposed the concept of "trustworthiness" as a standard for quality design and execution of qualitative research. This term focuses on the question of to what extent we can rely upon the results of the study. (Maykut & Morehouse, 1994). Trust is not a matter of personal assessment or giving personal credit to the researchers, but it's connected to the way in which the researchers collect, analyze and report data, namely, the nature of the methodology that they utilize . However, a number of researchers like Kirk & Miller (1986), Stake (1995), Greenwood & Levin (2005) and I myself in earlier books (Shkedi, 2003, 2005) prefer the use of conventional terms such as reliability, validity and generalization, with emphasis on the unique reference to the terms of qualitative research contexts.

No matter which terms are selected, it is essential to maintain a process to bring the researchers to base their research upon a methodology that includes an endless process of analytical thinking, examination and control. A main emphasis should be placed upon the research process, with all its parts exposed to the researchers and transparent to the readers. Researchers need to consider the methodological issues and clarify them in detail for themselves, to colleagues and to the readers of the research report (Merrick, 1999). To advance the nature of qualitative research which seeks the involvement of the researchers and their intuitive ability, it is important to ensure that the process will be conducted with sufficient distance. Intimacy and empathy, combined with distance and control must be integrated into all stages of the research. It seems that a research process which is based on transparency and documentation throughout all stages of the research may be required for research that allows examination and reflection.

As part of the analytical control, researchers should be able to identify the conceptions that they harbor in regard to the phenomenon under study. Qualitative researchers never come to a research field as a "blank slate" (Pinker, 2002). All researchers come to the field of the study with a theoretical perspective (conceptual perceptions, beliefs, ideology, etc.), whether or not they are aware of it, or even if they believe that they are supposedly free of any preconceptions. Therefore, researchers should find out and become aware of the research perspective they carry during the research process, and be able to explore its origins and validity from time to time (Strauss & Corbin, 1990). The manner in which researchers carry out their research work depends largely on their personal characteristics, experiences, interests, values, group reference, personal emotional orientation to the participants, and commitment to the principles involved in the study, etc. (Woods, 1996; Mason, 1996). Analytical control procedures that accompany the study are aimed to secure the quality of the research, helping researchers to identify the extent of their intuitive perceptions, and guarantee that the research will be carried out with proper research integrity and with minimal bias caused by the preconceptions of the researchers.

B. Data Collection

All components of qualitative research can be expressed as a tension between intuitive inquiry skills characterized by closeness and empathy, and analytical research skills characterized by detachment and control. At each stage of the study, in different ways, and in any methodology to be adopted by the researchers, the relationship between these two components will be disparate. (This issue will be discussed in great detail in all chapters of Part Two.) Obviously, in the process of data collection, there is prominent credence given to the intuitive skills of the researchers and the emphasis on the "human as a research tool," using a flexible approach, being sensitive and able to use these features simultaneously in a holistic manner. However, for that very reason, researchers are also aware of the dangers of relying on the intuitive research skills and know to control the intuitive human tools. Woods (1996) points to the dangers that might result from the intimate involvement of researchers in the study, accompanied by a tendency to give a romantic character to the study and its participants, and view them through rose-colored glasses. He proposes to develop an analytical distance as a security against these dangers.

To vouchsafe that the data will reflect the world of study participants, researchers use a number of research tools. These tools assure that the data will be absorbed as closely as possible to the version presented by the participants, rather than relying on the perception and illusory memory of the researchers. The researchers use notebooks to write comments, tape recorders, cameras, video cameras, computers, software, databases, and more (Fetterman, 1989; Creswell, 1998). These documentation instruments safeguard that researchers do not rely solely on intuition, personal integration and their personal selections, but that all events and/or what is said will be documented in a re-examining procedure. Based on this condition, the researcher can further interpret and make a deliberate selection when necessary.

The interviews are used as a key tool, perhaps the most central one, for gathering data in qualitative research. Maintaining proper tension between involvement and distance will be expressed during the interviews, as the researchers-interviewers will focus on listening and watching the interviewees, allowing them to tell their stories freely, while remaining focused on the research questions and/or focus points (Dey, 1993; Fontana & Frey, 2005). These two requirements seem contradictory and different from one another. It is important that the interview be recorded, and that all which is said be accurately maintained, thus researchers and their colleagues (and the readers of the study) will examine how the interview encounters these complex challenges. The assumption is that each spoken word reflects the opinions, views and feelings of the participants, and that converting it to another word often distorts the intentions of the participants to give expression to the researchers' perceptions and not to that of the participants. (Seidman, 1991).

There are certain qualitative research approaches and methodological patterns, and each requires research tools for data collection that correspond to the principles of the particular approach and methodology. Nevertheless, the tension between the intuitive and the analytical research process characterizes all the qualitative methodologies. Consequently, even if the qualitative methodology pattern is based to some extent on external criteria for research, there's still room to ensure the preservation of intuitive, unplanned, spontaneous responses of the interviewers and interviewees. At the same time, to avoid a situation where interviewers will rely solely on spontaneous intuitive perceptions, the interview should be accompanied by control tools, such as a list of topics the researchers wish to treat during the interview (Fetterman, 1989).

In addition to interviews, qualitative studies often make observations. We can indicate different types of observations, where the diversity between them focuses upon the place and role of the research observer. These range from the concept of the researcher as an external observer, distant from the phenomenon under investigation, to being an internal observer involved and participating in the research phenomenon. A characteristic of qualitative research is that the observer tries to participate (and not just to be involved) in the life of the phenomenon under investigation and to function as a participating observer, although different qualitative approach and patterns of methodologies offer different quantities for the continuum between involvement or participation and maintaining a distance. (Jorgenson, 1989). This issue will be expanded in Part Three.

One of the fundamental benefits of participation while watching is the opportunity to experience the everyday world from the inside, taking advantage of the intuitive characteristics of human beings. Integration into the environment being studied allows researchers to hear, see and begin to experience the reality of the participants' research experience. (Tedlock, 2005; Marshal & Roseman, 1989). The potential for misunderstanding and inaccurate observation increases as researchers remain distant, both physically and socially, from the subjects of the study. (Mason, 1996). However, observation that is based on engagement and participation without control and reflection will create a simplistic reading of the research picture. Again, using recording equipment, photographic and/or audio, may reduce the degree of selective, distorted perception. These tools will allow researchers to repeatedly reexamine the observed phenomena and to study the authentic world of the participants.

We can indeed preserve every word from the interview and the observation, but interpret it according to the concepts of our academic culture. Often, symbols that seem insignificant to us as researchers, from our professional and personal culture, are those with the most significance to the participants. It is important, therefore, that researchers understand and recognize the culture and the language of the participants. It is necessary "to acknowledge the importance of culture and cultural differences as key components in successful research practice and understandings." (Bishop, 2005, p. 110) Observation and interview without control could lead researchers to interpret the world of the participants according to their own cultural perception. Therefore, researchers must try to see these cultural symbols from the perspective of the participants, rather than to impose the frameworks and understandings of other cultures, which can attribute other meanings to these symbols. This requires a constant process of analytical thinking and distance control. (Woods, 1996).

C. Analysis of Data

If the data collection process deems particular significance to the intuitive inquiry process, the analysis phase will increase the clout of analytical thinking, although the role of intuitive inquiry process will be preserved. In recent years, I have had the occasion to read and examine a good many studies of various research projects conducted by qualitative research methodology. Much to my surprise, even though the studies were interesting and invited a thorough reading, it was hard to be convinced that they were based on a process of careful analysis. It seemed that some of the researchers relied mainly on their impressions, rather than employing a strict, disciplined analysis based on transparent analytic rules. I have no doubt about the intellectual honesty of these researchers. However, an analysis carried out on the basis of impression does not allow the researchers, their colleagues or readers to follow the analysis process, and thus to re-examine it and to meet the criteria of qualitative reliability and validity that assures that the research is not a literary, art or journalistic work, but indeed research. To guarantee the proper research criteria, it is necessary to lead a process of systematic, deliberate analysis based on transparency in all components and research stages. (Huberman & Miles, 1994).

The main process that safeguards the proper balance between involvement and control is the categorization process. It should be noted that some researchers would prefer to use the term "theme" to distinguish between qualitative research and quantitative research, which is closely identified with the term "category." Regardless of which term researchers prefer, categorization (or finding themes or the encoding process, as others prefer) is in the heart of the data analysis process (Charmaz, 2005, 2000). Categorization, namely sorting and organizing data in an analytical order process, is what connects several data units, which we perceive as similar in some ways to categories or themes (Shkedi, 2010).

As noted earlier in this chapter, the human mind deals with classifications and generalizations by its very nature. But, as stated, this categorization, with all its advantages, is an intuitive process, based on personal knowledge and memory. Categorization based only on the impression process could lead to quicker conclusions and be unsubstantiated. What is required is a categorization process based on the best practices and providing for review and monitoring at all times. In other words, we need an analytical-reflective process, not only an impressive-intuitive one. Such categorization "forces" researchers to examine the meaning of the data in depth. (Ryan & Bernard, 2000).

Categorization is not a random distribution of data, but it actually reflects the relationship between the theoretical perspective of the researchers and the data collected (Araujo, 1995). In order to conduct a suitable categorization, the researcher should develop a "conversation" between the theoretical perspective and the data. We should not impose the categories upon the data, even if we have an intuitive feeling that there is a relationship between the data and the proposed categories. Selected categories should reflect the data, and should grow as a result of the analytical depth discussion. In this light, it is also clear that the relationship between data and theoretical perspectives should be transparent and visible to researchers, colleagues and readers of research. It is important to keep a degree of congruence between the theoretical framework, the categories and the data. (Strauss & Corbin, 1990).

One of the characteristics of qualitative research is that even when the data collected (interviews, observation, etc.) is divided into unit analysis (categories), we pay careful reference to the whole research picture and its associated context. Indeed, qualitative researchers throughout the entire research process accompany the tension between the whole and the parts. (Lieblich et al., 2010). However, some researchers argue that they refer only to the whole and therefore consider themselves free of the categorization process. Yet, we repeat, even those who declare frankly that they avoid categorization, and believe that this should be the proper way, are essentially intuitive and unconsciously use categorization in their research. They intuitively divide the whole into parts (even if not distinct parts), and attribute categories (or themes, or whatever the term) to the complete data picture. In such a case, we should make a conscious effort to keep the analytical and transparent processes and to conduct worthy research.

CHAPTER 3

THE HUMAN LANGUAGE OF WORDS IN THE NATURAL ENVIRONMENT

As Mentioned in previous chapters, the methodology of qualitative research is based on at least three principles: the intuitive human inquiry skills of people, based on a closeness, involvement, and empathy towards the participants and the phenomena under study; the human analytic skills, based on distance, reflection and control of the research process; and the language of words as the language of study and the natural environment for the context of the research. This chapter will be devoted to the language of words, its characteristics, origins, development and its place and importance in qualitative research.

In referring to the language of words as the language of research, it is important to emphasize that not only the data of the study is presented in the language of words (which may also be true in quantitative research), but the language of words serves as a means for the entire process and components of qualitative research, characterizing the thinking process of the researchers. It is also the language in which the study is presented to colleagues and the community of readers. Using the language of words - the natural human language - also expresses the fact that qualitative research takes place in the natural human environment. The combination of natural language and the natural environment is reflected in the definition offered by Denzin and Lincoln:

> "[...] qualitative researchers study things in their natural settings, attempting to make sense of, or interpret, phenomena in terms of the meanings people bring to them." (Denzin & Lincoln, 2005, p. 3)

A number of researchers, and certainly some who are not in the qualitative research arena, consider using the language of words as a research language to be a particular disadvantage, compared to studies that use the language of mathematics. The purpose of this chapter is to highlight the quality of the language of words, and to argue that in many ways the language of words is best suited to serve as the particular language of study in research dealing with human beings and human society. This chapter points to two distinct advantages of the language of words as the research language in the context of the human and human societies:

1. The language of words is the ultimate natural language of humans, and as such allows for communication between researchers and research participants, researchers and their colleagues, and researchers and the readers of the study.
2. The language of words is characterized by the ability to encrypt and symbolize, create endless combinations of verbal terminology, represent culture, cultivate thinking and construct stories with both precision and ambiguity, and to offer the proper balance between the intuitive and analytical thinking that characterizes rich, qualitative research.

People have many languages: the language of words (regardless of whether it is English, Hebrew or Chinese), the language of mathematics, language of gestures, language of music, and much more. Our culture was built using these languages. Qualitative research is based on the language of words. This constitutes its advantage, and for that reason, there are those who criticize it.

Before the Existence of Language

How can we perceive the human being's world before the advent of language? We cannot answer this question with absolute certainty. Language (or its absence) has left no tangible archaeological evidence, neither in writing (which developed much later than the development of language), nor in prehistoric implements, nor by cut-clear fossil evidence. Researchers have had no choice but to raise certain hypotheses consistent with the sparse evidence that we have. Most, if not all, scholars do not question the fact that when the Homo-sapiens began his migration from East Africa to the rest of the world, he possessed language to a significant degree of sophistication. The brain of Homo-sapiens stabilized to its current size at least 100,000 years before the exodus from Africa, estimated to have taken place some 60,000 years ago. The fact that not even one tribe without a language has ever been detected seems to indicate that our ancestors spoke before leaving Africa (Milo, 2009). So far as is known, it may well be that all languages spoken today originate from a single primeval language spoken about 100,000 to 200,000 years somewhere in Africa. The oldest historical language findings available to us are around 5,000 years old, a very short duration in the history of language, and therefore cannot shed light on the formation of language. (Doitscher, 2007).

Speech requires a highly-developed brain. But it seems that the human brain reached its current size long before the human began to speak. Apparently the brain is not the only part of the body that has changed for speech. The first conditions for verbalizing speech are anatomic changes in the throat and at the base of the skull. We need a mechanism that will allow us to breathe automatically, but also a means to allow us to bypass this mechanism during speech, which requires the brain to control the muscles. In light of this, and based on anthropological research, most researchers believe it unlikely that speech and modern language appeared before the time of the Homo–sapiens, a little more than 100,000 years ago. (Diamond, 1999; Tattersall, 1998; Milo, 2009; Blackmore, 1999).

Before the development of the anatomical basis for articulating speech sounds, did human beings lack a language? By the term "language," we do not mean the ability to produce sounds and calls for communication, or even a limited number of sounds that refer to specific items, which animals can also produce. By language, we refer to a sophisticated language of words as complex as ours. Many researchers believe that anatomical change was indeed the catalyst for the emergence of language.

Apparently, language came later in human history. (Pinker, 1997). But language does not necessarily denote a speaking voice. In recent years, a very special idea has been formulated, purporting that human language was originally a language of gestures, and not a language of speech and voice. This argument is based on the fact that the part of the brain responsible for language (Broca's area) is located on the left side of the brain, which is also a brain region that humans (and monkeys and apes) use for hand gestures, to hold and touch, and for movements of the face and tongue. Strengthening the contention for the primacy of gesture language over the human voice is the ability of people to create meaningful communication using hand movements rather than the voice. In fact, even when people speak aloud, they accompany their speech with gestures. This is true for people who talk on the phone, and even for those who are blind from birth. One of the advantages of the theory of hand gestures is that it contains an interesting explanation for why humans developed human language, while other apes did not. One of the first things that our ancestors did over five million years ago, following the split from the common ancestor of humans and chimpanzees, was to stand upright on two legs. Our ancestors freed their hands to grab things and to make gestures long before they began to think and speak differently than other apes. Standing on two feet freed human hands not only to carry things and to create tools, but also to talk to each other in the language of movement (Ridley, 2003; Stokoe, 2001). According to this theory, language was rooted deeply in the history of humankind, with sounds being added much later. Bolstering this argument is the fact that sign language for the deaf is really a language with internal grammar, just like spoken language.

The Great Leap Forward

Even if we accept the theory that the language of gestures preceded spoken language, the limits of the language of gestures as compared to spoken language are very clear. In order to communicate with others and convey or receive information through the language of gestures, one must remain in continuous eye contact. Hence, it is difficult to communicate in the language of gestures and simultaneously deal with other things, and certainly it is difficult to communicate concurrently with a large number of people. There is no doubt that the development of spoken language as we know it was a crucial turning point in the history of the human being. From that point on, people could create new worlds: the world of inner consciousness and the world we create and share with others, what we call "culture" (Milo, 2009; Leakey, 1994; Diamond ,1992.)

Until 60,000 years ago, the chimpanzee population in Africa greatly outnumbered the human one. For 99% of its history, the prehistoric man walked a fine line between survival and extinction. Although the human brain grew from 400cc to 1,400cc following the split from the common human/chimpanzee ancestor five to six million years ago, the population of our ancestors largely remained stable and smaller (probably no more than 100,000 individuals). Suddenly, the big break occurred, manifested in the population growth of the Homo-sapien to reach the numerical proportions of today. How is it that the owner of a mind so grand, who constructed tools and fire, would remain nearly extinct for millions of years? What led to the big break in the struggle for survival that made the Homo-sapiens become the world's ruler? What led people to travel other continents relatively quickly and to reach the world over? Many link this to the formation of spoken language and its implications upon the human being's mental and physical functions. (Milo, 2009).

Until the appearance of spoken language, civilization evolved for millions of years at a "snail's pace" dictated by the slow rhythm of genetic development and change. After the appearance of spoken language, cultural development ceased to be dependent on genetic change. Over the last tens of thousands of years, our anatomy has changed by only a negligible degree, but cultural evolution has made manifold progress over its rate millions of years hence. From the formation of spoken language, it took only a few tens of thousands of years for humankind to domesticate animals, develop agriculture and technology, and invent writing (Diamond, 1992).

To date, there is no archaeological evidence from before the estimated period of the formation of spoken language to indicate any cultural handiwork production by humans with mental capacities like that of today. About 35,000 years ago, Homo-sapiens began to leave evidence of burial, art, trade, stone tool implements, etc. Many archaeologists and anthropologists argued that the dramatic change occurred simultaneously with the sudden appearance of a fully-developed language of words. For this reason, there is broad agreement that the development of the language of words is a crowning event within the human development process. (Blackmore, 1999; Leakey, 1994). Qualitative researchers often feel the need to justify the fact that they base their research in the language of words, and not in the language of mathematics, as is true in the study of natural sciences and quantitative research. They can draw reinforcement from the fact that the development of the language of words is arguably a revolutionary event in the history of humankind.

The Language of Words as an Innate Human Ability

In 1959, the linguist Noam Chomsky brought about a revolution in the study of language when he raised the claim that language acquisition is an innate biological process, and that language is not an external body of knowledge that children study. He alleged that babies are born with built-in basic language rules. Language, Chomsky argued, is so complex that it is not plausible that infants learn it from the discourse they hear from their parents and surroundings. Every sentence that children learn to speak is actually a new combination of words, and it can't be assumed that children were able to learn all the rules to understand and to create a sentence purely from observation and listening. In order for babies to learn language quickly and skillfully, they must have a special ability to learn language. Therefore, according to Chomsky, in the mind of every child there is an innate "universal grammar" that forms the basis to specialize in the particular language spoken in his environment. Evidence supporting the argument that language is rooted in our minds comes from the fact that there are striking similarities between all languages in the world. They all have a similar basic grammar. There is also a universal timeframe throughout the world when infants acquire language and learn words and grammar. Babies who grow up in human society will say their first words within a year, suggesting that babies already have an innate brain mechanism of language (Hogarth, 2001; Ratey, 2002; Harris, 2006; Winston, 2002).

From an examination of how people speak, Chomsky concluded that there is a similarity between all languages, indicating the existence of a universal human grammar. We all know how to use it, despite being nearly completely unaware of this ability. This means that inevitably, some of the human brain has, by virtue of its genes, a special ability to learn language. Of course, vocabularies cannot be innate; otherwise, we would all speak the same single language. However, children may learn the nomenclature of their native society and interject those words into the innate grammar system.

Chomsky's innateness hypothesis was repeatedly verified successfully for decades, with evidence from many scientific disciplines. It was repeatedly found that all people speak in languages with a similar level of grammatical complexity. Although languages differ considerably from one another in the scope of vocabulary, there is little variation between them in terms of grammatical complexity. Children across the globe speak with correct grammar at around age three or four. (Blackmore, 1999). No conqueror, world-explorer, or even a team of anthropologists has ever encountered a society which does not speak at some level of wealth and sophistication. Since some human societies have had no contact with other societies since they parted ways tens of thousands years ago, it is clear that our ancestors carried the genome of language skill before leaving Africa, even those who remain isolated in the mountains of New Guinea from the Stone Age (Milo, 2009).

The concept that language is an innate genetic ability should come as no surprise. People are born with other innate mechanisms, such as the ability and desire to walk, to recognize the faces of others, and more. As we have the same body parts, so we have similar "mental organs," and the ability to acquire language skills is one of the most prominent human characteristics. Steven Pinker (1997, 2002) refers to this human ability to acquire language as the "language instinct." According to Pinker, language is not an artificial culture that we learn, like telling time, but is part of a unique biological structure of our brain. Language is a complex and unique skill that developed spontaneously, without effort and without formal training.

Many would agree that the language of words, which is a universal human ability, characterizes us more than our other distinctive features, and distinguishes us from other animals. Since Aristotle defined humans as "talking animals," we are the only animal that can say something. As it says in the Bible, "And Adam gave names to all cattle, and to the fowl of the air and to every beast of the field,"(Genesis, 2: 20) the definition of man as a talking animal is the most common. Darwin also ranked language as first among the preliminary virtues of the human. Therefore, many would agree that the language of words is the greatest advantage of the human over animals (Milo, 2009).

Language as a Means for Social Unity

It has been emphasized in the first chapter that the human being is a social creature with social investigative skills, such as identification, reading the mind of others, and empathy. These skills have given the human being an advantage for survival. It seems that language played a significant role in the development and fortification of human social elements. In the absence of language, there could be no conveyance of information and messages, which constitute the social glue between people (Diamond, 1999). When we live in a group, as humans have lived since prehistoric time, the value of having good information is magnified, because information is the only commodity we can give others while at the same to keep it, since the information also remains in our possession (Pinker, 1997).

British psychologist Robin Dunbar (1997) argues that language function in people plays a role of in supporting and developing social relationships, like the role of delousing in monkeys. As we know, monkeys and apes engage many hours a day in delousing each other. The Gelada baboon monkeys, for example, engage in mutual cleaning about 20% of their waking hours. Dunbar argues that the role of delousing is to unite the social groups. Troops of monkeys and apes number a maximum of 55. In a group of this size, when spending a fifth of the day to delouse, each animal can reach the other, and remember those who did what, when, etc., thus ensuring a relationship with everyone. As the group grows, the demand for delousing becomes impossible, simply because there are not enough hours in the day.

According to Dunbar (1997), human societies that reached around 150 people in a group were unable to achieve cohesion through similar activities to those of monkeys. The reason that our ancestors lived in large groups, despite the potential for difficulties inherent in such groups, is that they faced a growing danger of predators lurking around them as they left the African forests for the grasslands of the savannah. Denver suggests that in this situation, language took the place of mutual care and cleaning. In many studies, Dunbar and his colleagues show that people use language not just to provide useful information, but also - and even primarily – for social exchanges. The findings of Dunbar and his colleagues indicate that much of people's speech is devoted to gossip. The researchers argue that gossip is a substitute or equivalent to delousing in monkeys. Using the language of words, we can reach more people, talk simultaneously with more than one person, and convey information about cheaters and swindlers, or tell stories about upstanding, trustworthy people-- and all this in an efficient, concise manner that the language of words can do so well.

Any human society has developed its own language, different from other languages of near or distant societies. Its unique language serves the social unity of each society. In Papua New Guinea, for example, there are 800 different languages even today, with each tongue unique to the tribe in which it developed, and different from the others. The detached state between the various groups caused them to maintain the uniqueness of each and every language. Presently, the world contains about 6,000 different languages, down from previous highs. Today we are witnessing the formation of a small number of dominant languages used by residents of many countries, such as English and Spanish, and this tendency is increasing as the connections between continents and countries grow stronger, and as the world—at least the Western world—has become one "global village." Language continues to be a means of social cohesion, which now encompasses entire countries and even continents.

The Invention of Writing

When we talk about research in general and qualitative research in particular, we connote not only the ability to use oral language, but also the ability of written expression. However, spoken language and written language did not grow side by side. In the annals of humankind, the invention of writing was a very late development. While spoken language has existed for thousands of years, the ability to represent the sounds with written symbols--making it easier for us to retain information, disseminate it between groups, and pass it on to future generations – is only about 5,000 years old. Moreover, only from the 20th century have a substantial proportion of different populations known how to read and write. While the ability to use spoken language is naturally acquired by people and relatively easy at quite a young age, reading and writing are acquired through a complex learning process. Even reading and writing are not automatically affiliated, although there is a connection between the two abilities. (Ratey, 2002).

Researchers estimate that with the development of human language, the lifespan of the human species has also lengthened. Humans belong to those relatively few species whose average lifetimes extend far beyond the age of fertility. What is the evolutionary advantage to extend the lives of people who have nothing to contribute to their biological gender reproduction? One possibility is that older people have an essential contribution to make towards the survival of the species by other means, in particular by accumulating knowledge and passing it on to future generations by such cultural tools as language. With the development of language, the scope of information available for transfer has greatly increased. Until the invention of writing, knowledge and experience were stored and passed on via the memories of old people-- a role that they play in tribal societies until today. In the life circumstances of hunter-gatherers, even information stored in the memory of one individual over the age of 70 can save an entire tribe or doom them to death by starvation, if that old man can provide information on similar past situations and guide the tribe to protect themselves. Long life was, therefore, important to our emergence from the status of the animal to the status of the human being. (Diamond, 1992; Goldberg, 2007).

It is supposed that writing developed relatively late because the human did not need it beyond the use of oral language. Early literary works were passed on orally from person to person and from generation to generation. The *Iliad* and *Odyssey* were composed and read by poets to listeners, and not written down until the development of the Greek alphabet centuries later. Writing did not develop by itself and was never adopted by societies of hunter-gatherers, because they had no need for an institutional use of writing. Writing developed independently in several places: the Fertile Crescent, Mexico, and probably in Egypt and China, because these were the first areas where agriculture and food production developed. It is believed that writing served the needs of political institutions, and that officials who used it were mainly engaged in the business of making and supplying food. (Diamond,

1999; Ryan & Pitman, 2000).

Knowledge gives power. Certainly, writing empowers modern societies, facilitating the transfer of knowledge from distant places and distant times, more accurately and in more detail than speech. Writing provided a great advantage for the "New World" over the "Old World," which was almost bereft of the ability to write. The Spaniards had a written language, while the Incan Empire did not. Thus, when written information returned to Spain from the New World, it inspired Spaniards to flock across the ocean. In the New World, however, the ability to write was limited to a small elite group among certain peoples of Mexico and the surrounding areas. Therefore, although in 1510 the Spanish conquered Panama, which lies 950 km north of the Incan border, it appears that the lack of written communication blocked the Incans' knowledge of the existence of the Spaniards until Pizarro landed in 1527. (Diamond, 1999).

For most of history, writing skills were limited to a special scribe who was specifically trained for this purpose. This sounds logical, because writing gave power to the rulers. The invention of the printing press in the 15th century was a prominent milestone in the spread of writing. Once books became available at an affordable price, information could multiply and change. Today we have the fastest, most efficient means to distribute written information than at any time. Written information sources used today, such as the Internet, computer software, text messages (SMS), faxes, television, newspapers and more, preserve and disseminate information at a faster speed than ever possible in our evolutionary history.

The Advantage of the Language of Words as a Research Language

Surprisingly, some researchers suggest that the use of language as a system of communication between people appeared only as a secondary function (Jacob, 2004). The language of words is much more than a means for communication. Language is a means of conceptualizing, thinking, data compression (allowing us to represent complex information in a compact code), collecting stories, and constructing and conveying culture. In order to examine the suitability of the language of words for research purposes, we must determine whether the structure of language and its components contain the features we need to express the complexity of the human and social phenomena under study. The sections that follow are devoted to this issue.

1. Means to Create Endless Combinations, Encryption, and Symbols

People possess not only vocabularies of thousands of words with different meanings, but also grammatical rules that allow us to construct a finite number of words into an infinite number of combinations (Diamond, 1992). A person uses a limited number of phonemes (e.g. "da" or "ba") to create thousands of words, which are built upon different combinations of the phonemes. By the end of high school, we know between 45,000 to 60,000 words. These words can create countless combinations of phrases, each of which has a different meaning. People are able to create words from smaller units of consonants and vowels, and thus establish a modular linguistic organization. Several dozen units reconstructed in various forms are sufficient to create a very large number of words. Two-year-old children in all human societies, who advance spontaneously from one word to two words, then to the multiword level, demonstrate this principle. (Diamond, 1992; Tattersall, 1998).

The unique nature of language is largely derived from the fact that it allows the coding and uploading of images from recognition. People shape their "reality" with words and phrases no less than they use their eyes and hearing sense. We have gained the ability to symbolize and encode cognitive representations in new ways. According to this assumption, the primary function of language is to allow symbolic representation, which gives a detailed, sharp, bright and richer picture of the world. Deacon (1997) calls the human being the "symbolic species." He argues that the use of arbitrary symbols to represent something else was a key factor in the development of the human brain.

The ability to identify objects and events from memory, months or even years after they actually occurred, requires a sophisticated coding system, which translates the visual and auditory world into representative symbols with sufficient accuracy and detail. Language serves this role, improving our ability to retain things in the memory. Grammar, which is uniquely human, helps to increase the volume of information stored in the memory. When we use a group of words (symbols) to create a large number of combinations of different meanings, the amount of symbolized information saved in the brain is significantly increased. In this way, we can easily keep, remember and repeat a dozen-word phrase, stories, and entire conversations. Many cultures have bequeathed their history to future generations via the oral transfer of long stories and myths. The symbolic nature and flexibility of human language also makes it a peerless instrument for developing the imagination.

Thanks to the endless combinations of symbols, it's always possible to come up with new and different worlds (Jacob, 2004; Blackmore, 1999).

Once information is preserved through language, it is also transferable and receivable from others. The mechanism that allows rapid delivery and absorption while maintaining high reliability is called "imitation." This does not imply a physical imitation, which is also part of some animals' capabilities, but a symbolic imitation. Symbolic imitation is a natural, easy action which even babies do. We imitate each other all the time and we are able to absorb and retain words and phrases we've heard from others.

Imitation, like vision, is done without effort, so we are completely unaware of it. Symbolic imitation means the dissemination of information from person to person by the use of language, reading and teaching, and behaviors. When we imitate someone else, a part of him comes to us (Blackmore, 1999). We can call what travels from person to person an "idea," "order," "behavior," "piece of information," or, as Dawkins (1989) called it, a "meme" - a term derived from the Greek "memory." The term "meme" rhymes with the name of the genetic replicator "gene," to emphasize the ability of cultural transmission which intensified with the emergence of the language of words. "Meme" means a unit of information that can be transferred from one brain to another. A meme is not necessarily a verbal unit, but can be any non-biological unit of information that can be transferred from person to person (e.g., pleasant music or graphic symbols). But the memes reach their peak effectiveness in verbal form.

2. Means of Constructing and Transferring Cultures

As human beings, we do not necessarily hold an advantage over animals, and certainly not by virtue of our physical abilities. Fish swim better, birds can fly, and the leopard and other predators are swifter. Nonetheless, the human is the most successful species in the universe. Despite our limited physical ability, we have conquered all the realms of the animals. No doubt the advent of the language of words gave the impetus for positioning the high status of the human being. Most of what distinguishes humans from the rest of the animal kingdom can be summarized in one single word: "culture." (Ratey, 2002; Weiner, 1994).

What distinguishes us as human beings is the powerful ability to convey a variety of formats and details of information from person to person and from generation to generation. The language of words can overcome the limitations of the creative power of a single individual's mind. Language is a cultural tool of unimaginable complexity and degree of diversification. Unlike other species, human beings are spared the hardships of having to discover our world from the beginning, and not required in every generation to "reinvent the wheel." We enjoy the impact of knowledge that has been gradually stored in society over thousands of years. This knowledge is stored in all kinds of cultural resources, primarily through the language of words, and delivered with its help from generation to generation. Access to this knowledge automatically enhances the cognition of every single individual in human society, for it makes him or her a partner to the society's collective wisdom. In this way, each of us can acquire wisdom which far exceeds the capacity of any single mind. This is a unique property of the language of words, of human society, and a powerful tool which has played a vital role in the success of our species. (Goldberg, 2007).

Humans learned to organize the accumulated knowledge systems—the accumulated cultural creativity – into disciplinary frameworks that were broadened and split apart. These entities have become a formal organization of our unique collective memory, which allows each generation to build on what it learned from previous generations (Kuhn, 1962). One could argue, as many do, that genetics are negligible compared to the advance of human culture. A child born in the world inherits not just a collection of genes, but also the work, thoughts and tools developed by other people, far distant in space and time. These are transferred to us primarily through the language of words. (Ridley, 2003).

3. Language, Thought and Emotion

Some argue that the language of words is identical to our mental functioning, and that words are actually base-units of thoughts. Even if language and thinking are not exactly the same thing, they are clearly intertwined. It seems that language shapes the way we understand reality and ourselves in the world. Language allows us to rearrange our ideas, and to take a break in order to shape our thoughts, rather than function haphazardly. Abstract thinking is impossible without language. Mapping thoughts and turning them into symbols allows us to define ourselves, function in the social world, evaluate our feelings, and change our attitudes and behavior (Ratey, 2002).

The language of words plays a significant role in their organization and their identification of our feelings and experiences. Feelings are generated when we are aware of the activities of our brain's emotion system. Many times, we classify and label our experiences by linguistic terms, and store them to give access through language. However, feelings in a brain that could sort the world in a linguistic way and classify different experiences into words are different than a brain that is unable to do so. The fine linguistic distinction between fear, anxiety, dread, apprehension, and the like is impossible without the vehicle of language. However, no word at all has any meaning without the existence of a basic emotional system that produces the brain conditions and bodily expressions that these words signify. In this way, the presence of the language of words in the human being changes the brain significantly. (LeDoux, 1998). Words are connected to the most basic fears, and thus diminish the fears and calm people. A picture of something is really scarier than its a verbal description, perhaps a telling commentary on the phrase "a picture is worth a thousand words." (Grandin & Johnson, 2005)

The ability to use language for guidance and planning future operations lies at the heart of humanity. Only by using human language can we plan and become set for tomorrow. Apart from humans, there is not even one wild creature who can calculate more than a few hours ahead. With language, we can overcome taking immediate, impulsive action. Language improves and refines our thoughts; it allows us to transcend the present, retain symbolic objects in our minds, and consider a variety of possibilities before we take action. Language can shape not only what is now, but also what will be, what could be and what we want or do not want to be. Language allows us to create symbolic models- not of the world as it is, but of the world as we want it. (Goldberg, 2007) To plan the day after tomorrow and sail away on a long journey in a time machine is not possible without language. (Milo, 2009; Tattersall, 1998)

4. Language, Thought and Narrative

Jerome Bruner (1985, 1996) identifies two basic ways of thinking. Each provides tools for organizing experiences, constructing reality, filtering the world as perceived by the senses, and organizing and preserving memory. One Bruner calls a "paradigmatic mode of thinking" or "logico-scientific," and the other a "narrative mode of thinking." The first mode of thinking seeks to find the universal truth, while the other seeks to find possible truth in specific contexts. The first deals with knowledge, free from the restrictions of the context, while the other relates to things in a specific context. The paradigmatic way of thinking centers around the question of "how to know the truth," while the narrative mode seeks to uphold the meaning and interpretation of the human experience. The first way of thinking expresses itself at best by formal-mathematical descriptions and explanations, while the other expresses itself in the best manner through narratives. (Lyons, 2007) Qualitative research represents the turning point for the provision of legitimate expression of human narrative thinking, what Brunner calls the "interpretive turn." (Bruner, 1996)

Human language is not just a collection of words that reflect one sign or another. We can say that behind the separate words and phrases, there is actually a story that reflects the human world and the minds of its inhabitants. One of the main functions of the mind is to weave our lives into a coherent story. It does so by producing explanations of behavior on the basis of our self-image, our past memories, expectations for the future, and/or the present social and physical environment in which our behavior is created. (LeDoux, 1998) People create narratives about their lives – their past, present and future, and their encounters with the people in the world around them and those involved in the process of constructing personal and social narratives. The narrative is the landscape in which we function and find meaning in our functioning. (Clandinin & Connelly, 2000)

Narratives are interpretive tools containing a practical, but also a very selective viewpoint by which we look at the world around us and give it meaning. People use narratives as a heuristic device to organize relevant facts in some logical order. In everyday life, we interpret the world around us, with the help of narratives. (Gudmundsdottir, 1995) Our narratives are much more than cumulative summaries of our lives. They are actually an instrument by which we interpret our experiences. (Jorgensen, 1989; Marble, 1997) It is easier to remember narratives, because in many ways narratives are the way we remember. The narrative is our primary means to look into the future, predict, plan and explain. Most of our knowledge and our experience is organized through narratives. (Turner, 1996)

We live our lives in such a way that allows us to tell stories about our experiences and our actions. The meaning of life does not exist independently of our narratives about life. Narratives that tell the stories of life actually change it, and give it a special form. (Grimmett & Mackinnon, 1992; Widdershoven, 1993) Narratives are very common in our culture, so we can say that they create the reality in which we live. Generally, people express their experiences in narrative, and use stories to explain and justify their thoughts and actions. (Clandinin & Connelly, 2000; Connelly & Clandinin, 1988, 1990; Gudmundsdottir, 1995) We live our lives constantly learning about ourselves, and thus forming a hermeneutic circle – the circle of understanding between the ego and our self. (Fontana & Frey, 2005; Widdershoven, 1993)

Narratives do not only reflect a contemporary mode, but an escape from the past into the present and future. We tell ourselves stories that are historical, but in a way learn something about the future. The 'truth' of our narratives is not a historical or scientific truth, but what might be called a "narrative truth." (Bruner, 1990, p. 111; Freeman, 2007, p. 136) Experiences from the past are not buried in the ground as archaeological treasures waiting to be found and investigated, nor do they dwell in a kind of memory-library awaiting someone to open and read them. The past is created each time through the story (Gudmundsdottir, 1995). Guy Widdershoven 1993)) argues that the relationship between life and narrative is expressed in one of two ways: life either seems like something that can be described by narratives, or narratives are seen as ideals through whose light we try to paint life. Thus, one can say that life's experiences and the narrative interact with each other, and that life has meaning because we live it from a screenplay narrative. (Beattie, 1995)

Human beings are storytellers by nature. Spoken or written words are the major vehicle through which people tell their stories. But the human trait to tell stories apparently preceded their use of oral language, and certainly the advent of a written language. To tell stories without words is a natural thing. Language only augments the ability to tell stories and brings it to higher levels. The visual representation of a series of events as images in our brain is the raw material with which our stories are made. Storytelling, in the sense of listing of events in the form of brain maps, is probably a brain obsession that began relatively early in the human being's evolution. Creating visual stories probably came before the development of language, because it is, in fact, a prior condition for language development. This natural pre-word imaging may be the reason that we ultimately create drama and books, and that a large part of humanity is now connected to books, plays, movies and TV shows. (Damasio, 2003)

People become narrative autobiographers in the way they talk about their lives. (Riessman, 1993). Storytelling is a characteristic of human nature, rather than an expression of artistic skills. (Beattie, 1995; Lieblich, Tuval-Mashiach & Zilber, 1998) As historians tell about the past through stories, people tell their life and lives through stories. Telling stories about the experiences of the past is a universal human activity, a form of the first discourse we learn as children. (Riessman, 1993) People dream at night through narratives, dream during the day through narratives, remember, expect, hope, despair, believe, doubt, plan, edit, build, learn, hate and love by narrative. (Widdershoven, 1993). The story is a landscape in which we live as humans and function as people who search for meaning. (Elbaz, 1991) Michael Connelly and Jean Clandinin (1988, 1990) argue that people are basically life storytellers, who live in personal and social-storied life. This is an epistemological argument: our lives are organized by stories, and can be better understand through stories. (Carter, 1993)

Qualitative Research: the Language of Words versus Other Languages

While the hallmark of quantitative research is its formal and mathematical descriptions and numbers and formulas, the hallmark of qualitative research is words and stories. Qualitative description is not just a combination of words, but a combination of words to display a story, a meaningful description. Qualitative research seeks to understand in-depth the world of its participants and the meaning behind their words and stories. Qualitative research assumes that the phenomena in human life and experiences are best displayed by stories and in a narrative way.

However, there are some other natural human languages that characterize the human being, such as the language of gestures (voices and facial expressions) to which we referred in previous sections. There are other languages. Some are even innate languages, such as the language of art, music, dance and the like. However, the development of these languages requires some process of learning, and people vary in their natural talents for these languages. Some are more gifted than others in painting, music, and dance. There are also other human languages which are not innate and not acquired naturally, such as the language of mathematics, computers, logic, and more. Acquisition of these languages requires a strenuous process of learning, but this learning process is not uniform, as different people vary greatly in the acquisition difficulties they experience.

Some researchers, such as Pinker (1997), argue that the emergence of the language of words was a precondition to the emergence of other languages (except the language of gestures, which proceeded to the language of words). These researchers base this claim, among other things, on the fact that archaeological findings show no indication of the presence of expressions of other languages, such as art, before the period when the language of words was believed to have been formulated among humans.

The language of mathematics is the ultimate language of quantitative research. Why, then, would it not be the language that all valid research should strive to use? The constant appeal of the numerical discourse can be explained by considering its important strengths, such as information, generalization, rigor and objectivity. (Sfard, 2008) The numeric descriptive is an effective tool to convey information. The numerical label is seen as having wide application, and apparently reflects a general feature. Numeric labels have divided the world into clear categories differentiated from one another. The resulting image looks clean, meticulous, accurate and free of uncertainty. To all these attributes must be added the fact that a mathematic language is perceived as objective and supposedly free from any human bias. One may disagree with each of these assumptions about the attributes of the mathematical language in describing human life and social reality, but there is no dispute about the important role of mathematics in the design of the natural sciences. Descriptive mathematical language is a central component of science. Galileo, whom Einstein called "the Father of Modern Science," believed that it was possible to describe the universe in the language of mathematics, or in Galileo's words, "The Book of Nature is written in the mathematical language." This approach represented a milestone for the appearance of modern science, and characterized it as state of the art. (Gutfreund, 2010; Damasio, 2003)

Qualitative research, throughout a range of its methodological variations, is based on the assumption that the language of mathematics, although possessing the power to represent accuracy better than words, cannot be the best-suited language to represent the complexity, richness and ambiguity that characterize human life and human society, and that these would be better represented by the language of words. (Bruner, 1996; Lieblich, et al., 1998; Pinnegar & Daynes, 2007) Using the language of mathematics for the research of humanity and society, which are based on the discourse of people, would actually require a translation from the discourse carried out naturally in the language of words to the language of mathematics. Any translation, obviously, involves detachment and reduction, if not distortion. Thus, the meticulousness that characterizes the mathematical presentation may distort rather than clarify reality.

Moreover, the language of mathematics, as opposed to the language of words, is a relatively new language in human history, and began to develop only when the human became a permanent agricultural inhabitant. As hunter-gatherers, they could manage with the most basic concept of numbers. As described in Chapter 3, we have a basic, innate numerical conception. But using the language of mathematics as a language of study is based on formal mathematics, whose ability people acquire the hard way through study and whose penchant is not equally shared by all. Some have more talents than others to acquire mathematical skill. (Pinker, 2002) Researchers, participants or readers do not "speak" fluently, freely and naturally in the language of mathematics, and even those who know mathematics quite well use the language of words to describe and explain their personal and social life. Readers of the study, or most of them, connect freely with verbal discourse, and have difficulty using a sophisticated mathematical (and statistical) discourse.

Based on the assumption that the language of words better describes the human world and society than any other language, and that this is the natural language of all research participants and enables communication with the readers and general population, qualitative research prefers the language of words as the language of research, with occasional expressions from other languages, and/or numeric expressions. Qualitative research is characterized by the fact that the process of collecting the raw data and the product of the research-- the final report – is delivered primarily through descriptions using the language of words. (Bruner, 1996)

PART TWO

ANALYZING DATA WITH NARRALIZER SOFTWARE

INTRODUCTION TO PART TWO

The following two chapters offer detailed initial acquaintance guidance to the Narralizer software and qualitative research data analysis principles. Attached to this book is the **Narralizer**, the supportive software for qualitative data analysis. Throughout the upcoming chapters, readers can find detailed explanations of how to utilize the **Narralizer** at various stages of analysis. Since the **Narralizer**'s development several years ago, it has been used by thousands of researchers and graduate students all over the world, as well as by several academic institutions and research centers. Training workshops have been held in dozens of institutes and in several academic courses taught in universities and colleges. The direct contact with software users has led to changes and improvements in response to user demands and to remove "bugs" which emerged during the initial version of the software's use. The **Narralizer** software was designed according to two guiding principles:

A. First, to ensure that the software does not perform the work of analysis **in place of** the researchers themselves. In many cases, the users requested the development of automatic, mechanical solutions that facilitate and accelerate the work of analysis, which the developers could easily provide. However, in order to preserve the unique character of qualitative research, these requests were denied. The software developers were aware that certain impressive software exists in the world market, but felt that this software sometimes crosses the fine line between supportive software and automatic analysis software, thus undermining the unique character of qualitative research.

B. Secondly, **the software was designed to be completely user-friendly**. This is reflected, for example, in that much of the software activities and symbols are similar to commands given to in common word-processing applications. Thus, we can assume that those who are accustomed to using word processor software can work with the **Narralizer** almost naturally. The software developers were cognizant that adding many more software options may not only harm the principles of qualitative research, but also make the software non-friendly to the users. To maintain the software's user-friendliness, the developers focused on the needs of most users, and responded less to the special needs of narrow researcher populations. Thus, the software does not offer possibilities that may interest only one or two percent of the users. However, the **Narralizer** is very rich in tool options, and the study of its use requires full attention and experience.

CHAPTER 4

ACQUAINTANCE WITH THE NARRALIZER SOFTWARE

Narralizer software is of the "code-and-retrieve" variety (Grbich, 2007; Weitzman, 2000). This software allows dividing the data into categories, specifying relationships between data and the categories, writing memos and linking them into categories and text. The software allows the analyzed material to be viewed – as a whole or in part - by selecting categories, according to cases being investigated or under specific subjects. It enables scanning and finding words and phrases, finding the frequency of the occurrence of a word, phrase or a category in the analysis documents or in the entire text documents. It allows the user to split or to merge categories, to split analysis documents or merge them according to various criteria, to change the category order, and to recreate an analysis array around selected categories. The software can distinguish between the categories according to levels of conceptual and theoretical expressions, in a way that helps to build grounded theory. It can preserve the data in its raw state, even after it has been analyzed and divided into units, to preserve each stage of the analysis for later reflection and review, and much more. The software supports all languages that can be read by Microsoft® word processor systems - both those that are written from right to left and left to right (English, Spanish, Arabic, Russian, Chinese and many more languages).

Since the process of analysis, like other qualitative research processes, includes elements of intuition and impressions as well as processes of perception and unconscious thought, there is always the possibility that in the study and practice of analysis, some significant steps will be hidden from the researcher's awareness. The **Narralizer**, which breaks down the analysis processes into separate steps and is based on the records of each stage, will ensure the transparency of the process and the cognizance of the researchers to the process, thus avoiding a lack of awareness of important elements of analysis or the possibility that they will be "skipped." So it seems that using the **Narralizer** as part of learning the qualitative analysis methods may clarify the processes for the learner. **The following chapters were written to**

be used as a training guide for the study of the analysis process, with or without the software. Indeed, the **Narralizer** is already used as tool for teaching qualitative research analysis in academic courses.

Introducing the Narralizer Software

This book comes with a "learning software" version of the **Narralizer**, a supportive analysis software. This learning software includes all the features of the **Narralizer**, and is intended to be used at no cost for an unlimited time. The only limitation is the number of case studies, sub-cases, categories, segments of data, and documents. With the learning software, you may analyze one case study and up to three sub-cases. It is possible to define 50 categories, using 250 segments of data and 20 documents of information sources. This learning software does not allow research to be carried out with a broad scope of data, but it can be useful in teaching the full potential of the software. To get the free Narralizer "learning software" please visit www.narralizer.com and *download* the learning software (Trial version of Narralizer).

Installing the Software

Go to **www.narralizer.com** and *download* the **free learning software (Trial Version)**. If you prefer to buy the **full version of the narralizer in a reduced price,** go to **www.narralizer.com/Software/BuyNow.aspx**

When downloading the Narralizer software, a link to the software will then be sent to your email along with installation instructions.

Please note that some email services experience problems receiving "heavy" software. If you fail to receive your learning software, please contact **narralizer@gmail.com**.

In the event of problems with the installation, it is recommended to make several additional attempts. If the problem persists, please contact **Narralizer** support at www.narralizer.com or by email to **narralizer@gmail.com.** Our experience shows that the installation is simple and trouble-free. If you experience problems, it is possible that they are not **Narralizer**-related, but rather caused by a computer disorder.

Introducing the Work Screen

Double-click the **Narralizer** icon, and the work screen is opened. The **Work Screen** has three panes. When the software is launched, three panes appear which function as the user's work environment. As in Windows, the user can change the size of the panes, or scroll up or down to reveal information.

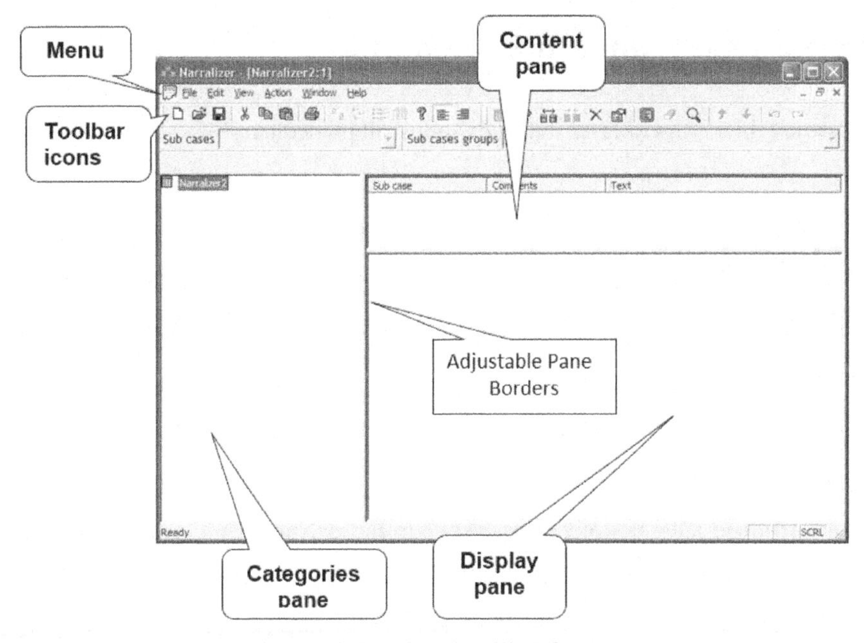

Figure 4A: The Narralizer **Work Screen**

The pane on the left will be referred to as the **Categories pane**. This is the long pane on the left of the **Work Screen**. It is used for listing and changing the list of categories.

The lower right pane will be called the **Display pane**. It is the largest pane in the **Work Screen** and displays the different phases of the analysis results.

The upper right pane is the **Content pane,** which allows us to modify the analysis array.

The upper section of the screen contains different **menus**: File, Edit, View, Action, Window, and Help.

The row beneath the menus consists of the **Toolbar icons**. Pointing to an icon reveals the name of its function. The menu and toolbar icons are activated by using the mouse.

You may enlarge or decrease the size of each pane by moving the inner border of the pane using the mouse, as shown in Figure 4A.

Performing Standard Procedures

There are three ways to carry out all or almost all procedures:

A. Through the menu

B. Through the icons on the toolbar

C. By right-clicking the mouse

Some actions can also be carried out using the keyboard.

Save

The **Narralizer** automatically saves every 5 minutes. Note that it is also possible to regularly save manually. Save by clicking the toolbar icon, clicking **Save** on the File menu, or pressing **Ctrl** + **S**.

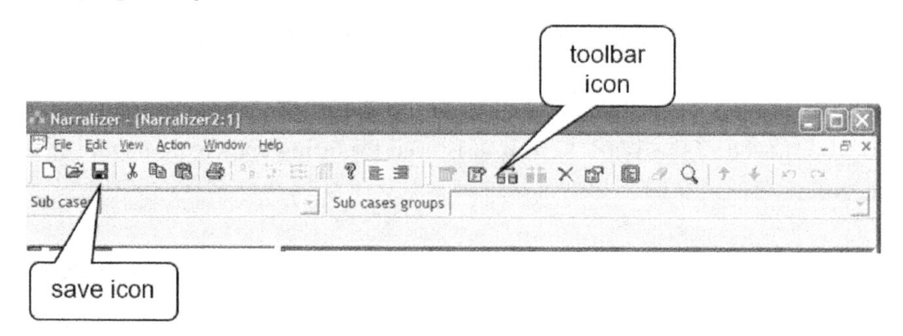

Figure 4B: The Save Operation

When saving, the program automatically saves a backup copy. The backup copy has the same name as the original document, with the additional wording "Backup of..."

Bear in mind! When reopening the analysis document, be sure to open the original document and not the file which says "Backup of..." If you open the backup copy, it becomes the original document and the corresponding backup copy will say "Backup of Backup of...." This can create confusion and a loss of control of the system for saving analysis documents.

Cut/Copy/Paste

Narralizer allows you to **Cut/Copy/Paste** different parts of the analysis document in the same way as with a word processor. This is achieved by clicking the Edit menu, right clicking the Toolbar icon, or pressing Ctrl + V = Paste, Ctrl + C = Copy, or Ctrl + X = Cut.

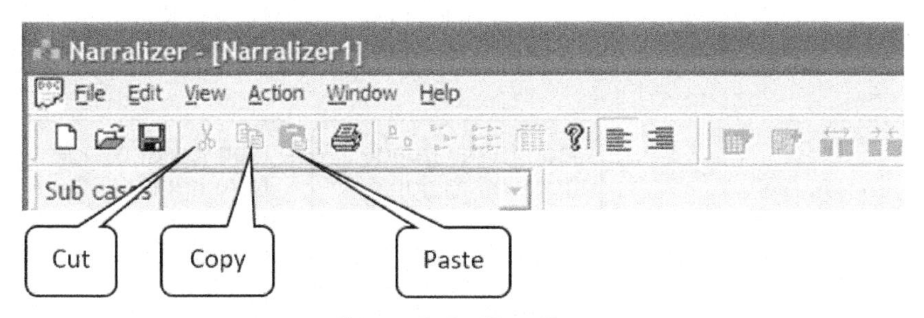

Figure 4C: Cut/Copy/Paste

Undo/Redo

To undo an action, click **Undo** on the Edit menu (or click the **Undo** icon).

The **Undo** command is accompanied by a description of the action to be undone (in the illustration below, the action to be undone is the "Add Category" command). You may also activate the **Undo** command by clicking the icon on the right side of the Toolbar. Pointing the cursor at the Undo icon on the toolbar reveals which action will be undone.

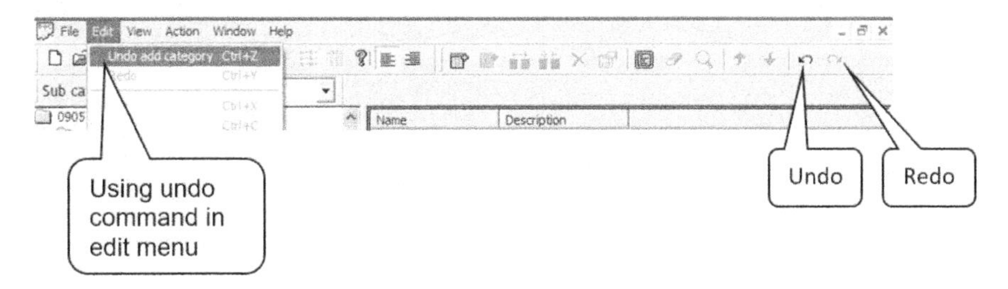

Figure 4D: Undo/Redo

Similarly, **Redo** will replace the action undone by the Undo command.

Undo and **Redo** can only apply to the last five actions taken. Also: Undo by pressing Ctrl + Z; Redo by pressing Ctrl + Y

Different Languages and Changing Languages

With **Narralizer**, you may work in languages written from left to right (e.g., English, Spanish, etc.) and right to left (e.g., Hebrew, Arabic, etc.). The software uses all fonts available in the Windows environment and any language used by the computer.

There are two icons on the Toolbar, one showing left alignment (for languages written from left to right), the other right alignment (for languages written from right to left). **Narralizer** allows you to simultaneously work with two or more languages in one document and to change languages any number of times.

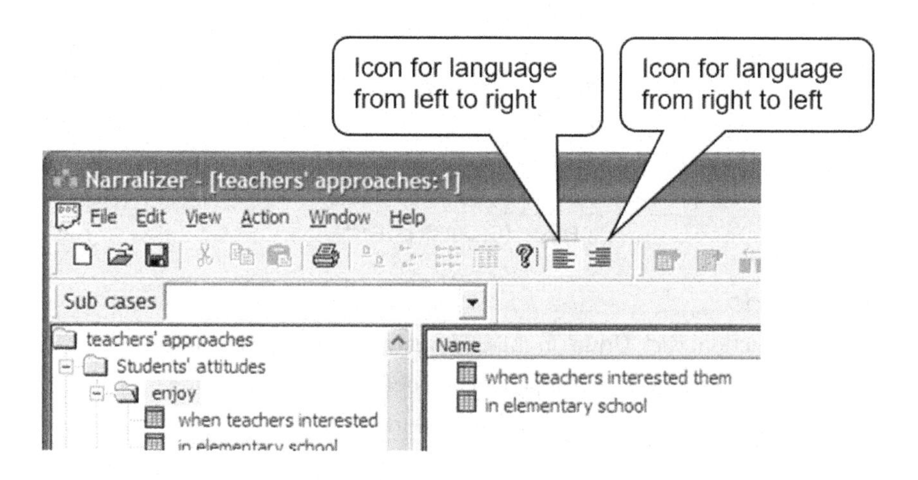

Figure 4E: Languages and Changing Languages

CHAPTER 5

READING AND ARRANGING THE DATA

While the previous chapter was an introduction and preparation for the process of analysis, this chapter begins the analysis process itself. This first stage of the analysis is relevant to all qualitative research approaches and methodologies.

Researchers arrive at the formal stage of data analysis after they have amassed a fair amount or even all of the data collected during the study. Although in the data collection stage researchers consciously or unconsciously analyzed the material gathered, this was still a non-formal analysis process, which did not allow the researcher to necessarily pay proper attention (Charmaz, 2000; Gall, Borg & Gall, 1996). The analysis process during data collection is essentially an intuitive process focused on impression, and in most cases, the researchers are not aware that they are analyzing the data. However, this non-formal analysis process does not replace the formal analytical process, which is described in this and subsequent chapters.

Before starting the formal process of analyzing, we must prepare the data. In this context, we can distinguish between two types of data. One is the intervention-based data for the purpose of the research, such as interviews and observations gathered in the data collection phase. This data type can also include field notes written by the researchers during the data collection or immediately thereafter. Ordinarily, this data is not available for immediate use, and must undergo a process of transcription, correction and editing. The other kind of data is that which is independent of the study and is available "as is" to the process of analyzing. Such data can include documents, diaries, etc., most being electronic texts, which are available for use in the analysis, process (such as word-processed documents, Internet-based texts, e-mail, or other such documents).

The analysis process in all qualitative approaches and methodologies begins by reading and organizing the data, for at least these two reasons:

- Qualitative researchers are generally faced with a large amount of data. Even researchers who reduce the study to a single case study find themselves confronted with a vast amount of data. Without proper organization, the researcher may lose control over the entire data.

- The analysis is essentially a process where the raw data is divided into segments and moved from its original context (its order in the observations, interviews or documents) to a new context that gives meaning to the data. There is a danger that each of the data segments will lose its authentic context and original meaning. Qualitative research is characterized, as mentioned, with significant reference to the context of the data (Araujo, 1995). The process of re-reading and organizing the data can ensure the fidelity of the data to its original context.

The initial reading of the data is actually the first stage of the analysis: on one hand, the goal is to prepare the ground for detailed analysis and segmentation for the analytical reorganization; on the other hand, at this stage the researchers identify the main features of the data (Maykut & Morehouse, 1994). In order to assure that the process will be meaningful, the researchers need to repeatedly read the transcript of the interviews and observations, the full range of reports and other data, and involve themselves in the details to get an overall picture and to feel the integrity of the data before separating it into units in the process of the formal analysis (Agar, 1980).

One efficient method for understanding the general picture of the data is to develop a comprehensive series of questions asking who, what, when, where, why and how. These questions, of course, are useful in any process of analysis and for anyone involved in analyzing data. Another question worth asking is "So what?" which leads us to think about the meaning of the data. These questions can lead to different and unexpected directions, and to opening interesting channels for investigating the data. Dey (1993) compares the first reading of the data to hoeing a garden. By digging the soil, we release the land and allow our analysis seeds to sprout and grow roots.

Distribution into Numbered Passages

In order to secure the ability to save the context of data even if it is divided into sections during the analysis process, we can number the text passages, **using the word processor software**, respective to their "authentic" sequence before analysis. In this context, a passage is defined as any part of the text which is formed in the word processor document after an "Enter" command and which appears in a new row.

The text with numbered passages will be used throughout the analysis process. In this way, it will always be possible to go back and locate the original position of the text segment in the entire data unit (interview, observation, etc.). In cases where the passages are too large, they can be divided into new short passages by using the "Enter" key before the word that will start a new passage.

Bear in mind! The division into numbering passages is only an important option. It is possible to skip it without any harm to the quality of the data analysis

Organization of the Data Document Directory

Researchers using a word processor to store all documental data would do well to create a special computer data document directory in which to store all the research data documents. The directory should be given a name that defines its identity in an easily-recognizable way, preferably the name of "case" under investigation. Experience with student-researchers has indicated that many of them are not careful to arrange the data in an accessible, clear manner, and consequently lose a large part of it within the range of documents in the computer's memory--thus creating unnecessary extra work. Among the purposes of using the **Narralizer** is to create order in the totality of data, as will be explained below.

Following the definition of the data document directory (the case under investigation) is the stage of placing all the data documents in the same directory. Each data document is identified by a name that will allow its characteristics to be recognized and to be distinguished from other documents, emphasizing the common and the varying traits. Each such document is identified as a "**sub-case**." Thus, there may be many subcases in one study project (in one case). Each sub-case will be identified by a short name to facilitate its quick identity and retrieval if necessary. The naming system will be uniform. For example, if we are dealing with a group of participants, the first sign that identifies them is likely to be their name, the second identifying sign may be the type of data (interview, observation and so on), and the third would be the date of receipt of the information. For example "Isaac, interview, 12.8," "Sylvia, observation No. 1," "Daniel, diary, 1st year," etc.

When the data analysis organization is carried out through the **Narralizer**, we can utilize word-processed documents, web documents, e-mail documents or any other texts (but not pictures of text). For the effectiveness of the analysis process, there is an option to bind together the raw data documents (which are named as "sub-cases") into an array of software directory databases.

However, researchers can easily and efficiently utilize the word processor documents stored in the Data Document Directory explained above. The researchers can import the data into the **Narralizer** research documents directly from the Word documents and from Internet-based texts, e-mail, or other such documents. Those who wish to use the **Narralizer** for creating the database, please skip to the appendix 1 for full instructions.

PART THREE

THE STRUCTURAL AND PARTIAL STRUCTURAL PATTERNS OF QUALITATIVE RESEARCH APPROACHES AND METHODOLOGY

INTRODUCTION TO PART THREE

Regarding the three characteristics of qualitative research, presented in chapter two of this book, we can point to at least two factors, which constitute the basis of the complexity of data analysis in qualitative research:

A. The delineation of qualitative research in the language of words, within the natural context of human life. The language of words, unlike the language of mathematics, is largely characterized by expressions of richness and ambiguity, and is probably far from the precision that characterizes the mathematical language and condensed formulas. These characteristics pose a challenge to those who wish to present overall verbal descriptions and explanations to illustrate the phenomena under study.

B. The establishment of the qualitative methodology for both the intuitive and analytical skills of the researchers. These two methodological components seem significantly contradictory to each other: the first is characterized by reliance on "the human as an instrument" (Lincoln & Guba, 1985) with a closeness, empathy and involvement with the phenomenon under investigation, while the second is characterized by distance, control and reflection. Qualitative research methodologies must give appropriate significance to each of the two characteristics. Actually, this is not an easy challenge to qualitative researchers, and seems more complex when it comes to the issue of data analysis.

Those who search their way in data analysis often find it difficult to position themselves and choose the appropriate type of analysis methods for their study. From conversations with student-researchers, I've realized that many times they adopt a method of analysis proposed in the literature or academic courses which is not exactly appropriate to their specific study. Researchers and student-researchers who wish to find their way in qualitative research and specifically the issue of data analysis, have difficulty being exposed to the full possibilities of data analysis. Even if they are exposed to a wide range, it is doubtful whether they are equipped to distinguish between methodological variances and to find the one best suited to their needs. This book seeks to provide researchers and student-researchers a "map" to allow them to find their way in the field of qualitative data analysis and to choose a unique analysis solution for their specific research.

We suggest to divide the qualitative approaches and methodologies into four patterns proposed and clarified in this part of the book (part 3) and in the following (part 4), which aims to guide the reader to design the most appropriate analysis methods. Before beginning the data analysis process, the researcher should define the approach and methodology (or methodologies in the case of mix-methodologies) that characterizes his or her specific research. This decision should probably be taken before embarking on the field research, and guide all phases prior to the formal stage of data analysis. However, since the dynamic of the research may lead to approach and methodological changes over the course of the study, it is important to go back and determine the actual research methodology being used. It seems that if researchers define their research approach and methodology as one of the four options proposed in this book (or at least close to one of these methodological patterns), this will contribute greatly to the effectiveness of the analysis process.

CHAPTER 6

THE CRITERIA-FOCUSED METHODOLOGY (ALMOST POSITIVIST RESEARCH APPROACH)

We open the presentation of qualitative approaches and the methodological patterns with the pattern of criteria-focused methodology, although it is less identified, compared to any other methodology represented here, with the spirit of the "interpretive turn" or "narrative turn" (Lyons, 2007, p. 611-613) which has characterized the study of the human and society in recent decades. Qualitative research based on the pattern of criteria-focused methodology is characterized more by analytical research skills than intuitive research skills (see part two of this book). This means that researchers operate with a set of external criteria (theory, principles, categories), which accompany them from the beginning of the study until its conclusion. This pattern is the closest to the assumptions of quantitative-positivist research, but does not equal in all its components. Figure 6A graphically illustrates the location of this approach and methodological pattern within the range of qualitative research

Figure 6A: the Criteria-focused Methodology Within the Range of Qualitative Research

Between the Pattern of Criteria-focused Methodology and the Positivist Approach

Some fully identify this qualitative methodological pattern with the positivist research approach, and refer to it as "positivist qualitative research." (Denzin & Lincoln, 2005a; Guba & Lincoln, 1998, 2005) Indeed, many of the methodological principles which guide researchers adhering to this methodology are characterized by the positivist approach: the researcher's perspective is focused upon a particular theory or on any external array of assumptions, which determines what and how the data will be collected, analyzed, and represented in the final report (Polkinghorne,1995). To use an expression common to the tradition of ethnography, the work gives expression to the etic rather than the emic (Gall, Borg & Gall, 1996; Stake, 2005). However, it is important to emphasize that there are significant methodological differences between this methodology pattern and the positivist research paradigm. Even if many assumptions of the positivist approach are also reflected in this methodological pattern, they are not absolute and exclusively identical. The choice of the language of words, with its complexity and vagueness, over the language of mathematics for the leading language of the research points to a deviation from the total adoption of positivist research assumptions. As with any other qualitative research methodology, there are also some expressions of the interior point of view-- closeness, involvement and empathy--that characterize the intuitive system. In the absence of these properties, we could not talk about qualitative research.

The positivist research approach is well-known as being based upon several philosophical assumptions (or as it is called, a "philosophical paradigm") representing the basic positions of the research. One is the ontological assumption about reality, about the nature of the phenomenon we wish to research. The positivist approach sees reality as being objective and as a certain fact that has an independent existence. Second is the epistemological assumption, which is engaged in the relationship between the knower and the known. The positivist approach sees the knower and the known – or in other words, the researchers and their research subjects, as being separated. The positivist approach claims to distinguish clearly between the objective and the subjective, and opts for the objectivist position (Arksey & Knight, 1999; Denzin & Lincoln, 2000, 2005a; Guba & Lincoln, 2005a; Heikkien, 2002; Lincoln & Guba, 2000; Shkedi, 2005).

Positivist assumptions lead to two sets of implications for research: the methodological issue and the language issue. The methodological issue deals with the question of what ways exist to find information. Positivist assumptions lead to selecting the proper research tools to assure objective research, such as closed questionnaires, structured observation and interviews. The researcher with a positivist approach would choose to control and influence the environment of the study and to isolate its components (or "variables" in the language of the study) which are relevant to the subject of the research and which may be approached objectively (Guba & Lincoln, 2005; Kirk & Miller, 1986; Maykut & Morehouse, 1994). The second implication places concern upon the language to be used to display the data in the research, to analyze, and to display the outcome of the study and its conclusions. The positivist approach, in its optimal expression, fulfills the ideal of a formal system, and is expressed in concise, precise mathematical language with clear and sharp formulas (Bruner, 1985, 1996). Even if the positivist research report is accompanied by a verbal explanation, it is not accompanied by urging the language of mathematics from the proscenium. Norman Denzin and Yvonna Lincoln (2005a) argue that in the first half of the twentieth century, researchers wrote "objective" verbal descriptions in their qualitative research projects that reflect the positivist approach.

With regard to the basic assumptions of the positivist approach, it is questionable whether those who chose the criteria-focused methodology indeed adopt in practice the inclusive assumptions of the positivist approach, with all its research implications. It is doubtful whether those who choose to use the rich but sometimes vague language of words as the research language indeed adopt the positivistic approach, even if many positivist assumptions are acceptable to them. The very choice of the language of words over the language of mathematics as the language of the research maintains the qualitative nature of the research, which is not identified with the pure positivist approach.

Using Criteria-focused Methodology without accepting its Assumptions

Even if we argue that criteria-focused methodology is not the same as the positivist research approach, it is likely that those who choose this methodology are related in their assumptions, consciously or unconsciously, to the positivist approach. Based on this argument, we should consider the following two tendencies:

In recent years, with the increasing acceptance of qualitative research in social sciences, many quantitative researchers have combined their studies with certain elements that they define as qualitative research. These quantitative researchers may include what they call an "open questionnaire" that invites the participants to answer written questions. Some even conduct a structured interview. In light of this, we should make clear the nature of the boundary between quantitative research and qualitative research (for mixed methodologies, see Chapter 19). The method of gathering verbal data in itself is not accepted as qualitative research, or even part of mixed methodology research. If researchers are seeking to quantify the verbal and convert it into quantitative data, if they strive to reach a concise formula and see the quantitative expression summary as the main expression of the study, or even if they manipulate their findings with sophisticated statistical ploys, we certainly cannot consider this research to be associated with the family of qualitative research. (Note that the option exists to use numeral signs for descriptive verbal data in qualitative research.) But if those quantitative researchers attempt to add a qualitative aspect to their study and use a verbal description which speaks for itself (even if a quantitative picture is displayed in parallel), we can definitely view this research as part of the family of qualitative research using criteria-focused methodology. It seems that this is a typical example of researchers who are far from the tradition of qualitative research, but connect to it by using the criteria-focused methodology pattern.

The second tendency refers to a situation in which qualitative researchers who do not see themselves as adherents of the criteria-focused methodology, and perhaps even distance themselves from it, have adopted this methodology. The following are several such examples:

A. Researchers engaged in qualitative evaluation may often adopt a-priori the criteria set that was chosen by the initiators of the project and its operators, and/or by those who commissioned the evaluation with the set goals. These researchers may use the criteria-focused methodology, but without necessarily adopting its research assumptions.

B. There are some qualitative researchers who do not adopt the assumptions of the criteria-focused methodology, but want to look at the phenomenon under investigation from the perspective that is accepted by the academic community or by the academic community to which they aspire to join.

C. In many cases, qualitative researchers adopt research assumptions that are completely different from those of the criteria-focused methodology (for example they adopt the participant-focused methodology, which focuses on the discourse of the participants). However, as part of their research, they may be interested in comparing their findings with the common theoretical perspective accepted by the academic community. In such cases, two sets of studies will be conducted side-by-side, one based on participant-focused methodology (or another research methodology pattern) and another based on criteria-focused methodology, without accepting their philosophic-methodology approaches.

D. There is also a case in which researchers adhere to a constructivist and/or critical approach (see upcoming chapters), and thus believe that the reality under study is constructed by the participants. Nevertheless, from a philosophical point they have concluded, "Because nothing is as it seems, the only things worth noticing are the terms, the formal structures by which things are perceived." (Clandinin & Connelly, 2000, p. 38) From the constructivist and/or critical perspective, they have adopted a formal system such as the focused –criteria methodology which is adopted by the academic community, for looking at the phenomenon under investigation. (Belsky and Alpert, 2007; Polkinghorne, 1995)

The Study's Theoretical Perspective

The approaches and methodological patterns of qualitative research differ from one another in their use of theories for guiding the research (Creswell, 1998). There are methodologies that have formulated theories before all other phases of the research, and others that formulate a theory in an ongoing process of the research. At this point, the focused-criteria methodological pattern is very clear by its definition: this methodological perspective advances the stage of the theoretical formulation of the research above all other research processes, formulating questions, setting the population, data collection, analysis, and any other research activity (Clandinin & Connelly, 2000). The study, then, begins at the definition of the criteria of the study. Qualitative researchers who choose this methodology pattern are committed from the start to a certain theoretical perspective, which guides them throughout the study. The theory is thus a source guide for all the research activities. (Carr & Kemmis, 1986)

Theories provide a point of view through which we can observe and interpret a means to comprehend the phenomenon under investigation. A good theory may provide research data categories that can be used to observe, collect, explain, predict and interpret the data and the phenomenon under study (Dey, 1999; Glaser, 1978). Theories are like conceptual networks that provide us "lenses" for perception, interpretation and action in the world. (Fetterman, 1989; Seidel & Kelle, 1995).

The main source to formulate a theoretical perspective is theoretical and research literature. Many times this literature is even the trigger for conducting the research. Therefore, a study based on criteria-focused methodology, like many other qualitative studies, will start by reviewing the literature. Through reviewing the literature, researchers will clarify for themselves, their colleagues and future readers of the research, which elements formulate the research theoretical perspective. This theoretical perspective will be preserved throughout the study and will guide it.

However, the researchers may—and perhaps even be requested to– increase and enrich (and perhaps change, to a limited extent) the literature review during the research process.

Formulation of Research Questions, Assumptions or Hypotheses

In criteria-focused methodology, the theoretical platform is the basis for reference to all the research phases. From the theoretical perspective, all other research processes are derived. A research work begins with an interest in a specific area or problem which has arisen, and in fact the theoretical perspective determines the concepts by which the research will be identified. What might seem vague in the stage prior to determining the theoretical perspective of the research becomes increasingly evident with the review of the theoretical literature (Clandinin & Connelly, 2000).

As stated, the criteria-focused methodological pattern is closer than other qualitative methodological patterns to positivist-quantitative research, close, but not identical. Criteria-focused methodology, like other qualitative research methodologies, and in contrast to quantitative methodology, describes the phenomenon under investigation mainly in words and not in a mathematical language. Even if the theoretical perspective is formulated at the beginning of the study, the researchers in all qualitative research processes open their eyes to a wider context of the phenomenon under study, not merely focusing on the predetermined research components. The outcome will be a bit broader than the predetermined research variables. (Pinnegar & Daynes, 2007; Saukko, 2005) The research questions and their assumptions or hypotheses will be formulated according the predetermined research theoretical perspective.

Research Questions

Research questions in criteria-focused methodology determine the exact direction of the research, and the elements and track of its development. The quality of the research will be reflected by the quality of the wording of the questions. The theoretical perspective determines how to articulate the research questions. Research questions will flow directly from the review of the relevant literature. In the case of evaluation studies, it is likely that the researcher-evaluator often receives the research questions from those who initiate the project and the evaluation. "Traditionally, qualitative inquiry has concerned itself with *what* and *how* questions. *Why* questions have been the hallmark of quantitative sociology" (Holstein & Gubrium, 2005, p. 498). However, as noted, the criteria-focused methodology is guided by theoretical and external criteria which provide the basis for asking questions like "why," or as we shall call them, "clarifying questions" as opposed to "descriptive questions" ("what" and "how"). (Marshall & Rossman, 1989; Yin, 1994)

Descriptive questions are research questions that yield a good description of the phenomenon under study, as reflected by the data collected in the field of the study. These questions will match with the study's theoretical perspective. Descriptive questions are aimed to derive the information directly from the research population, out of their discourse and from observations in their natural environment. Therefore, as the descriptive questions become clearer and the study variables become more specific, researchers will receive a better-focused picture. All such information will respond to the specific descriptive questions. This information may be a simple description of the subjects that are observed, a description of the stories, attitudes and opinions of the participants, but also relates to more complex information and even explanations of relationships of cause and effect.

Examples of Possible Description Research Questions

"How do citizens feel about the police department's effectiveness?" "How do residents view the changing economic situation?" "How do family physicians perceive their role?" or "What is the social worker's timetable?"

These and other such questions, limited or extensive, are phrased according to the research goals, which are expressed early in terms "what" "how" and other interrogative words.

There may also be descriptive questions that examine phenomena over time or compare two situations or parallel phenomena. They may be formulated in such questions as:

"Has there been a change in the attitude of citizens from the beginning of the year to the end, and if so, what kind of change?" Or a more focused question like, "Is there a difference between the perception of mathematic teachers and literature teachers regarding the issue of homework, and if so, how this is reflected in their work?"

The second type of research questions is the **clarifying question**, or the "why" questions. These questions deal with the theoretical conceptualization of the phenomenon under investigation, with theoretical explanations and correlations between the data regarding the theoretical perspective. In this kind of question, the theoretical perspective is not just a lens to describe the phenomena, such as in descriptive questions, but is also the basis by which the researchers explain the phenomena as well as the connections between their different components. The articulated research questions will use theoretical concepts. For example,

"What is the internal motivation and external motivation of social workers dealing with school students?" There are also more complex clarifying questions, such as: "Why has there been a change in the attitude of citizens from the beginning of the year to its end?"

The answer to such questions needs to link various descriptive elements obtained from the study, which can be clarified by the predetermined research perspective. Another kind of question might be, "What is the relationship between the social background of the citizens and their attitude toward the death penalty issue?" This question, as the previous one, examines a link between various components, or in other words, seeks correlations between them. In the absence of a properly articulated theoretical framework, it will be difficult to answer such research questions.

Research Hypotheses

Research questions in the criteria-focused methodology lead the researchers to formulate hypotheses. These hypotheses are distinct from the research questions by two characteristics: the hypotheses are more detailed than the questions, and they are formulated as a statement and not as a question. For example, following the research question, "Has the habit of watching TV from an early age caused students to prefer visual information over text information?" the appropriate hypothesis may be, "If children are exposed from an early age to many hours of TV viewing, their preference for visual information over text information will be greater," or even another version, "There is a positive correlation between the amount of hours that a child is exposed to TV and the degree of preference for visual information." Certainly, hypotheses are derived from the theoretical perspective as emphasized in previous sections.

According to the logic of the criteria-focused methodology, a research hypothesis which is based on a clear theory will lead the researchers to focus data collection, rigorous analysis and empirical valid proof (Bruner, 1996). By stating the hypothesis, the researchers direct their perspective so that they can collect relevant data (Kirk & Miler, 1986). When it comes to descriptive research questions, research hypotheses directly focus the researcher upon the relevant details of the information. The importance of hypotheses as a directed research tool will be greater in the case of clarifying research questions. There are different types of clarifying research hypotheses. One is a general hypothesis that claims a connection between variables, but without specifying the type of relationship and who influences whom. For example, "There is a connection between the education of citizens and their political opinions." This hypothesis relates to a research question like, "Is there a connection ..." but does not specify the type of connection. More specific hypotheses are, for example, "As the level of citizens' education increases, their tendency to express opposing views increases in kind," or "Citizens' exposure to international political viewpoints caused their tendency to express the opposition's views." There is no doubt that well-formulated research questions and their attendant hypotheses direct the research process and provide a significant expression to the researcher's analysis skills, more than their intuitive research skills -- an issue discussed at length in the first part of the book.

Making Changes during the Study

Although the criteria-focused methodology is characterized by dominant analytical thinking, with the theoretical perspective determined at the beginning of the study and directing the research until the end, this does not mean that adherents of this methodology will not change the theoretical perspective throughout the study. It seems that this stands in contrast to positivist-quantitative research, where research hypotheses aim to direct verification of theories in which the "truth" is defined as "the accurate representation of an independently existing reality." (Smith & Hodkinson, 2005, p. 916) In criteria-focused methodology, the aim of the hypotheses is often to direct the eyes of the researchers more than to verify theory. As noted above, the criteria-focused methodological pattern does not reflect loyalty--and is certainly not fully committed--to the positivist approach assumptions. Thus, as the research develops, the research perspective will possibly change, and may result in an alteration of the research questions and hypotheses. These are not revolutionary changes, but a sharpening and adapting of the components. If during the stage of data collection and analysis the researchers conclude that in light of the data collected they must significantly change the theoretical perspectives, the research questions and hypotheses, the consequences will be an abandonment of the criteria-focused methodology and a transition to another methodological pattern, as shown in upcoming chapters. Figure 6B (viewed from left to right) summarizes the connections between the components of the research methodology presented thus far.

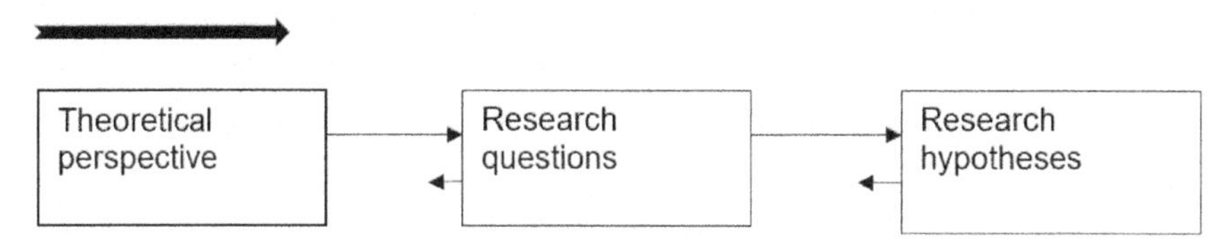

Figure 6B: The Connections between the Components of the Research in Criteria-focused Methodology

As illustrated in Figure 6B, the three components of research presented so far - the theoretical perspective, the research questions and hypotheses - are in a linear order, one by one. Arrows connecting the components are one-way and clearly indicate the order of the development of the research and the interdependence between the elements. Note also the short arrows (for example, from the research questions to the theoretical perspective). The purpose of these short arrows is to indicate that despite the research orientation and the dominance of the linear analytical research skills, a research study employing criteria-focused methodology exists in the literal-cultural space, and as such is also affected by intuitive research skills. With these skills, researchers may examine, albeit a small amount, of the components of the study, and if necessary make slight adaptations.

Study Design and Selected Population and/or Field Research

Various qualitative research methodological patterns have common elements of research and research phases: setting the theoretical framework, determining the research questions, assumptions or hypotheses, data collection, data analysis, and a summary of findings and conclusions. But each methodological pattern has its own emphases and characteristics which arrange these components in a unique way. As noted, the criteria-focused methodology pattern will be formatted in a linear-logical continuity, i.e., each stage leads directly to the next step. Researchers who focus on this research pattern can - and should - design all components of the study in advance.

Consequently, the research design is derived from a focus on research questions and the purpose of the study. "A research design describes a flexible set of guidelines that connect theoretical paradigms first to strategies of inquiry and second to methods for collecting empirical materials." (Denzin & Lincoln, 2005a, p. 25) Determining research questions and their hypotheses directs the investigators where to collect the data, how to collect it, and from whom. Accordingly, researchers have the task of finding the most appropriate research field. "The cases are expected to represent some population of cases." (Stake, 2005, p. 450) If this is an evaluation research, then the work of researchers will be based upon those who have previously determined where the project will be conducted. But generally, the work of selecting the participants is not easy at all. If the study does not deal with an organization or defined segment of society, researchers do not always require institutional approval, but in any case they need the approval of the participants, which is not always an easy task. However, if they need institutional agreement, researchers must determine not only the most suitable field for research, but also the one that's ready to be exposed to research. Research is perceived by many proposed participants, especially institutions, as an exam and a means towards control and criticism, thus in many cases they are reluctant to participate in the study or place very difficult conditions upon the researchers. Sometimes the permission of high-level administrations is required, such as government offices and the like. Researchers must also take into account the criteria of ethics, to ensure to avoid harming the participants' rights and privacy (Dushnik and Sabar–Ben Yehushua, 2001). Although qualitative research is not examined as clinical studies, the studies generally require approval and agreement in the spirit of the Helsinki Declaration of Research Ethics, which exists in many research institutions. (King, 2010)

In positivist-quantitative research, the research population is selected using a random sample of subjects, namely the participants are selected using a statistical tool aimed at ensuring the objectivity of choice. The assumption is that as the number of participants increases, so does the ability to safeguard the lack of intentionally-selected participants (Mason, 1996). In contrast, in qualitative research based on the verbal data of discourse, the number of participants cannot reach the extent of participants in quantitative research. Thus, conducting a random sample often loses its meaning. Moreover, qualitative research, whatever the preferred approach and methodology, is focused upon the narratives of individual participants. If it is the study of a group, institution or organization, the focus is on the cumulative individual narratives of individual participants. All participants in qualitative research are a unique "case" (Stake, 1995). The guiding principle of qualitative researchers is that the phenomenon under study has been learned from the specific experience of the individuals. Therefore, the purposeful sample might be better suited to qualitative research (Mason, 1996; Stake, 2005). Of course, there is nothing wrong in also employing statistical tools to ensure that the subjects will not reflect any hidden biases of the researchers.

However, in selecting subjects, it is important to remember that the participants will be exposed to both observations and interviews, asked questions and expressing their experiences by words. While quantitative research usually prefers closed questionnaires which do not demand the participants to invest much time, the time required for qualitative research interviews (even structured interviews) is much greater. This may force qualitative researchers to spend a relatively long time to locate the appropriate participants, which further reduces the possibility to load a random sample.

Data Collection Methods

After the stage of research design and establishing a field research and study population, the data collection phase may be activated. In criteria-focused methodology, the data collection phase is in direct linear continuation to the previous steps. Determining the theoretical framework in earlier steps basically involves designing the method of data collection. Since this is qualitative research whose primary characteristic is the use of the language of words, employing quantitative character tools such the closed questionnaire is indeed possible, but should be marginal and not overshadow the gathering of verbal data. Thus, research tools that match this methodological pattern should be those accepted by qualitative research in general, such as primarily initiated tools - interviews (individual or group) and observations - as well as reference into existing products, such as documents and objects. (Shkedi, 2005)

The main question is not what tools we use, but what form these tools should take. Can we use tools that give expression to human intuitive research capabilities, or do we prefer tools that reflect more control and limitation of intuitive capabilities (tools that appeal primarily to human analytic skills), or should we elect some kind of balance between the two? Any of the "interview" and "observation" tools includes a wide range of research practices. On one side of the spectrum are the interviews and observations that demonstrate the highly-structured model, and opposite them are open interviews and observations. It seems that the nature of the methodological pattern based on predetermined criteria will reflect the control and limitation of the researchers' intuitive skills. These would be relatively structured tools with fixed criteria (categories) that can ensure that researchers will be faithful to the theoretical perspective and criteria that guide the study, and that all interviews and observations will be loyal to the same theoretical perspective and the same categories. Therefore they are structured interviews. "In structured interviewing, the interviewer asks all respondents the same series of pre-established questions with a limited set of response categories." (Fontana & Frey, 2005, p. 701-702) The interview questions will be introduced in a logical order which is predetermined from the start for all the interviews, with no room for extensions toward issues that do not concern the specific study.

Observation, like the interview, will be criteria-focused (categories) set in advance. It will be closer to what is called "pure observation," and will be focused upon observation by its nature "where the researcher looks only at material that is pertinent to the issue at hand, often concentrating on well-defined categories ..." (Angrosino, 2005, p. 732). Therefore, it is an observation by an external, non-involved observer which viewed the phenomenon under investigation from the perspective of the categories. Such observers do not devote attention to events that are not in their research interests, which have not been planned in advance (Maykut & Morehouse, 1994; Guba & Lincoln, 1998; Shkedi, 2005).

We can also add the questionnaire to the research tools in this methodological pattern, where participants are asked to reply in writing to a series of questions (as opposed to a closed questionnaire, which requires only markings). The questionnaire can be seen as a form of a structured written interview. In many cases, researchers will not choose such a questionnaire, but prefer a structured interview, not because they do not appreciate the research contribution of the questionnaire, but in those cases where the subjects will not be willing to devote the necessary time and effort required to respond seriously to a written questionnaire.

We can describe the nature of the research tools used in criteria-focused methodology on the chart that appears in Figure 6C.

Figure 6C: The Nature of the Data Collection Tools in Criteria-focused Methodology

As illustrated in Figure 6C, the data collection tools in this methodological pattern are characterized by external criteria, and rely mostly upon analytical research skills. Note that these data collection tools are not situated definitively at the pole's end (as is typical of quantitative research). This is to stress that in every qualitative research represented by words and descriptions and even in such a structured methodology as this, there will always be a degree of responsiveness to the intuitive, unstructured and non-planned ways, though in this case with a relatively small degree of flexibility.

The data collected from interviews and observations must be preserved for documentation and analysis. An appropriate way to do this is by camera, video and/or tape recorder (or today's more sophisticated recording and photographic instruments), and then transcribing the data for presentation and analysis. Careful recording and transcription is important because they preserve the words exactly as expressed, ensuring the validity and reliability of the research (Greenwood & Levin, 2005; Guba & Lincoln, 2005). Unfortunately, the process of transcription is time-consuming and significantly increases the cost of the study. Researchers who choose the criteria-focused methodology aim to collect information that is relevant to the categories of the research as stated prior to the data collection, and therefore may choose to transcribe only the sections of the observations and interviews that are relevant to the research issues. (Pidgeon & Henwood, 1996)

The fact that in this methodology, researchers have targeted very specific categories leads some to conclude that the researcher can interview and simultaneously record directly by writing the interviewee's words, even if the participant's words are delivered at dictation speed. These researchers may also be seen as a blessing, since they would believe that limiting the rate of speech might allow the respondents to carefully consider their words. Although I believe that it is important to record and precisely transcribe every word and picture, there is no denying the fact that skipping the recording and transcribing process greatly shortens the time required, and certainly highly reduces the research cost. Indeed, in criteria-focused methodology, researchers can consider basing their studies on such a "shortcut." However, in other methodologies which will be described in upcoming chapters, this shortcut process would harm the principles of the research.

As noted above, researchers who adopt the criteria-focused methodology should try, while focused on receiving information related to predefined categories, to absorb information about the larger context of the phenomenon under investigation. Due to the structural nature of this methodology, we expect every qualitative research to represent the contextual picture, even limited, of the phenomenon under investigation. So the researchers are expected to also show the contextual picture in addition to the description and/or theoretical picture. To put it in terms of the qualitative research methodological principles presented in Part Two, we could say that researchers using this methodological pattern manifested mainly in the analytical research process should also give a certain measure of expression to the intuitive process with closeness, involvement and empathy, so as to absorb the data that might present the context of the phenomenon under investigation.

Research Categories

Categories are conceptual units that reflect the content or structure of the phenomenon under investigation. Each category is a concept linked to other concepts which are at the same level or situated above or below. As will be described in different chapters, the categorical division is common to all qualitative research methodological patterns, but there's a difference between various methodological patterns in terms of the status of the categories (fixed, changeable, etc.) and the timing they are set for (before coming to the study field, during the research process or after) (Polkinghorne, 1995). As mentioned above, in this methodological pattern the categories are determined before coming to the study field, and their status is relatively stable throughout the study. As defined, this methodology is characterized by focusing on external theory which constitutes the baseline guides of the research process. The means by which the researchers ensure that the theoretical point is saved are the research categories. (Anfara, Brown & Mangione, 2002; Charmaz, 2006)

Using categories as a tool for data collection and analysis can guide a definite direction of the data search and its analysis, and serve as a basis upon which to make sense of data and its classification. It is clear that the process of data collection and its sorting is an interpretive process of giving meaning to data (Gall et al., 1996; Ryan & Bernard, 2000). Categories are actually organizing tools that enable researchers to organize the data in accordance with relevant characteristics. Categories provide a means for explaining the behavior and discourse of the participants (Creswell, 1998; Schwandt, 1998). The meaning of the data is determined by the theory through which we view the data. Thus, different theoretical windows can supply various theoretical meanings to the same data (Fielding & Lee, 1998; Guba & Lincoln, 1998). The main purpose of the categories in criteria-focused methodology is therefore to serve as a means of data collection and analysis organization, based on an array of predetermined theoretical meanings.

To illustrate, Figure 6D illustrates a partial list of categories taken from a study conducted according to the criteria-focused methodology principles. The study deals with the roles of school principals and their functions, and its main category is "appropriate traits of principals." In this example, there are three upper categories, each of which is divided into several sub-categories. Researchers may use these categories to observe the principals as they work.

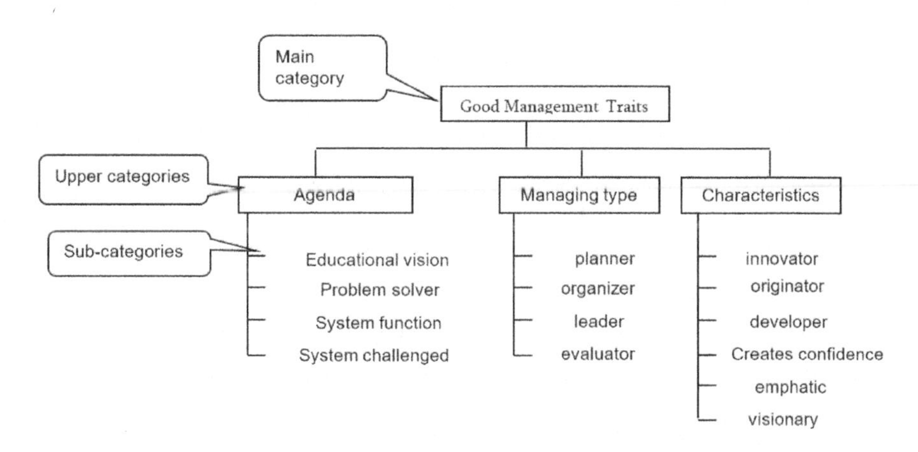

Figure 6D: Example of the Array of Categories Guiding the Data Collection and Analysis

As shown in Figure 6D, the category array has three levels. However, the option exists to create another level of sub-sub-categories, which reflects the details of each of the sub-categories. These categories in each level come from the theoretical assumptions of previous stages of research. The categories will accompany the research throughout, and will serve as the basis for researchers to prepare the guidelines for interviews, observations and other tools of data gathering.

Figure 6E illustrates an example of interview questions with the principal, prepared on the basis of the categories that appear in Figure 6D above. (Note that there is no direct link between a particular category and a particular question, but the questions do not extend beyond the scope of the categories and encompass almost all of them.)

1. **How do you understand your role as a principal?**

- **From the administrative perspective**
- **From the educational perspective**

2. **What are your priorities at work? How did you designate them?**
3. **What is your management concept?**
4. **What is the role of the teachers in the school?**
5. **What is your relationship with the students and the level of involvement with them?**
6. **What is your relationship with parents? What is their place in the school?**

Figure 6E: List of Interview Questions

This is, of course, only a partial demonstration of various questions prepared in advance for the interview. Researchers who adopt the criteria-focused methodology introduce all the questions to the participants, and even keep the order of questions and mind that all the respondents are presented with the same questions. Interviewers may, of course, explain their questions if necessary or intervene when only partial answers or no answer is given to the question asked. But after receiving a satisfactory answer, the researcher will move to the next question. However, as qualitative researchers, they can be attentive to comments and hints about the context of the phenomenon under study.

Data Analysis

The analysis process is a process of ordering and interpreting the data collected (Dey, 1993; Pidgeon, 1996). The analysis involves distributing data and organizing its parts into a new analytical order. The main task of the analysis is to connect pieces of data into a category. In the criteria-focused methodology, the data is collected according to predefined categories based upon theoretical concepts, research questions and hypotheses. Analyzing data and classifying it into categories is a process by which pieces of information are compared to each other to find similarities and differences. Those that are similar will be combined in the same category, and those which are different will be placed in different categories. The location of the categories will be determined in a way that reflects the connections between the categories, i.e., the degree of closeness or distance between one another. (Seidel & Kelle, 1995)

"By uniform instrumentation that has high reliability and validity... researchers can insert sufficient distance between themselves and their subjects to make formal knowledge claims on the basis of the scientific method." (Pinnegar & Daynes, 2007, p. 10) The analysis process in this methodological pattern flows directly from the data collection process, because of the same predefined categories used by the researchers in both processes of collecting data and analysis. Accordingly, we can suppose that the collected data was already collected and actually sorted according to predetermined categories. This last statement, as it is true, requires reservations and attention. It is true that researchers who adhere to a criteria-focused methodology observe and interview according to given categories. However, during an interview, for example, even when they are asked predefined questions according to certain categories, the participants may give answers which are not relevant to what was asked, but relevant to another, different question and category. As for observation, even if it is based on gathering focused information, more information may always be found which is unconnected to the predetermined questions but gives an important contextual picture.

Not only do the categories stem from a theoretical perspective of the research, their internal order is also drawn from the same source. The question of which of the sub-categories will be combined together under the central category arises from the theoretical perspective. One can imagine the category array as a kind of upside-down tree whose branches face downwards and trunk upwards. At the top are the more general categories, expressed in comprehensive conceptual (or even theoretical) terms. At the bottom, small branches and leaves are the categories with more descriptive wording. The categories thus offer the conceptual structure of the data (Richards & Richards, 1995), as shown in Figure 6F, whose categories were taken from research dealing with teachers' conceptions of their role in curriculum usage.

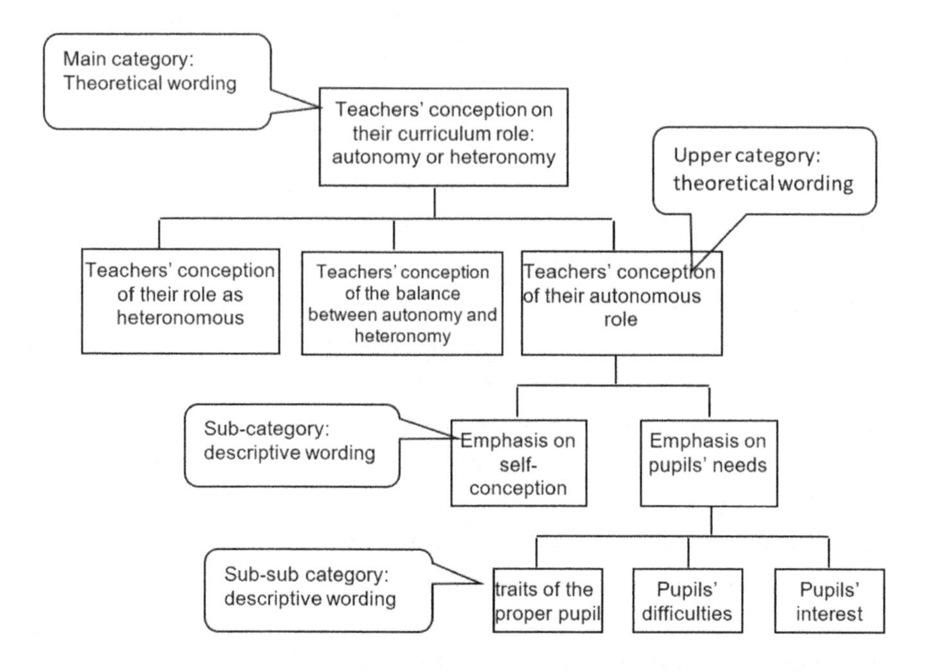

Figure 6F: Four Levels of Criteria-focused Methodology Categories

As shown in Figure 6F, the central category not only indicates the title of the research, but also gives the direction to the study drawn from theoretical literature: the tension between autonomy and heteronomy. Each of the three upper categories under the main category (of course, there can be many more upper categories) are formulated in terms drawn from the theoretical literature dealing with the role of teacher development and implementation of educational curriculum. Each of the upper categories could be split to sub-categories which are more descriptive than theoretical. Each of the sub-categories may be divided to sub-sub-categories, and so on.

As we shall see in the coming chapters, the analysis process in other methodological patterns will be more complex, and in most if not all cases, is a multi-step process. This is not the case when it comes to criteria-focused methodology. The fact that the criteria (categories) accompany the research process from the beginning makes the analysis process relatively simple. In the criteria-focused methodology, the data is located in a constant array of categories. This analysis process is called the "mapping stage of analysis". Basically, we can utilize this analysis to reach different levels of detail on the continuum, from description to theory (Strauss & Corbin, 1990). In many cases within the basic (lower) levels are the descriptive categories, and in the higher levels (of the "tree") are the theoretical categories (Strauss, 1987), as shown in Figure 6F above.

Although this is qualitative research that seeks to ultimately produce a verbal report rather than a quantitative-numeral report, it is possible to use numerical units. This does not mean that the researchers use sophisticated statistical systems, but they may give numeric expression to the phenomena under study. This is true for most, if not all of the qualitative methodological patterns, but is especially noticeable in the criteria-focused methodology which proposes high-level validity and reliability (Pinnegar & Daynes, 2007). According to this methodology, all subjects undergo a similar research process (for example, interviews with the same questions and observations that are focused on a similar category), thus researchers may conclude that it is important to note the frequency of occurrence of such expressions (Huberman & Miles, 1994; Miles & Huberman, 1994). Researchers may also use techniques of averages, frequencies and correlation (Shkedi, 1995). Using tools of descriptive statistics (but by no means sophisticated statistics) may well correspond to an intuitive perception of the research's readers and address the need to communicate with them.

In criteria-focused methodology, researchers are deductive in their thinking throughout the research process, i.e., the thinking of researchers during the study moves from general to specific, from theories and criteria to the data. Each unit of information acquires its meaning in association to the framework of clear criteria arising from the theory. Miles and Huberman (1994) suggest using several helper-tools to ensure a proper analysis of the data, including this partial list:

A. Sorting the data in several parallel columns

B. Creating a table of categories and inserting the data in each table cell

C. Creating a flow chart that reflects the relationship between the collected information units

D. Creating a table of frequencies of data units to categories and/or for different cases

E. Placing the information units on the chronological table reflecting the development over time, if time is a significant factor in the study

The Research Report

As with any quantitative or qualitative research, there are certain elements that appear in any final report: the display problem, literature review, research questions and assumptions or hypotheses, methodology, a description of the study population and the field, findings and analysis display, and discussion and conclusions (Crowley, 2010). The research report in criteria-focused methodology maintains all the above elements in a linear order from the theoretical perspective to the study's conclusion, to represent a logical order and a link between theory and the research process and its findings (Arksey & Knight, 1999; Jorgensen, 1989).

The unique element of criteria-focused methodology, which distinguishes it significantly from other qualitative research methodologies and adds justification for defining it as qualitative research, will be reflected prominently in the research report. As with all qualitative research methodology in any pattern, the findings will be mainly descriptive and narrative, with researchers seeking to design a verbal summary that communicates with the readers (Denzin & Lincoln, 2000) and reflects the logic of this methodology and its assumptions. The distinction between the paradigmatic and the narrative manners of expression (Bruner, 1985, 1996; Craig, 2001) may help describe the nature of a report in criteria-focused methodology. Quantitative research prefers a paradigmatic expression of concise numbers and formulas. The criteria-focused methodology pattern does not seek to reach a rich level of narrative phrases to express its messages, but chooses to do so in a verbal summary with an expression of focused-argument style. However, if the researchers have data that may form the basis of rich narrative descriptions, they may combine it in the finding, aside from the focused-arguments report.

Lieblich (2010) speaks in this context of tension between the "two contradictory poles, one which might be called 'traditional science' and the other 'artwork.'" (p. 381). According to Lieblich statement, we can say that the findings section in criteria-focused methodology would be closer to "traditional science" rather than to "artwork." Shor and Sabar-Ben Yehoshua (2010) distinguish between the writing style of "reporting" and that of "story." In "reporting," they mean writing in "thin description," and in "story" they mean "thick description." (Geertz, 1973) According to these concepts, data presentation in criteria-focused methodology will be closer to "reporting." These distinctions correspond to the methodological difference between the analytical and intuitive research skills, which places the criteria-focused methodology closer to analytical research skills. This is reflected in all processes of the study, as presented thus far, and also in the findings of the final report section, which is a more elemental-based style than story.

It seems that the most appropriate description of criteria-focused methodology is a "category-focused description" (Shkedi, 2003, 2005). This type of description displays the findings on the basis of predetermined categories defined in the early stages of the research. We could say that the categories are the building-stones of the description (Merriam, 1998). In this way, we can present each research case by itself (in a study of one case or several cases) or to present several cases with emphasis on the uniqueness and variation between each one. In a description of this type, each chapter or section is devoted to a specific issue reflecting one of the upper categories, a sub-category or several categories. The division into chapters, sections or sub-themes will more or less overlap the division into categories, and the order of the description will highlight the findings of the study and the message conveyed.

Technically, the writing process of such a description does not require a great level of creative writing which may be required in other methodological patterns described below. Although this is a literal rather than a quantitative description, the dominant verbal style is the argumentative style, and the order of presentation is very close to the predetermined categories. As mentioned above, there is nothing preventing combinations in the research description, although in a limited manner involving certain quantitative data corresponding to intuitive-human perception. It even seems that this often contributes to the clarity of the description. However, this description does not confine itself to presenting "the voice" of researchers, but also displays the authentic "voices" of the participants (Guba & Lincoln, 2005) by presenting quotations from their words.

An example of such a description is presented below in Figure 6G, which refers to research that examined the perceptions of teachers towards their students in the process of teaching ancient mythological literature. The arrangement of the described items matches the list of categories and research questions.

How the students were perceived by the teachers:

Generally, we found that the teachers indicated four patterns that express the students' attitude to the text:

A. **The students are not interested in mythological tales.**

B. **Students enjoy and benefit from studying mythological tales in specific contexts of time, place and content.**

C. **Students relate critically to mythological tales.**

D. **Students perceive mythological tales as important, legendary text, although not obligatory.**

Following is the description of all the properties:

A. **Students are not interested in the text.**

Lack of student interest in mythological tales is the most salient feature which the teachers used to define the students' attitude toward the text. Thirty-four teachers (out of 50) expressed this sentiment in one way or another. Most consider the lack of interest as the major element affecting the quality of the teaching process. Teachers use a variety of images to demonstrate the students' lack of interest, including using terms taken from other disciplines. For example...

Figure 6G: A Short Example of the Focused-Categorical Description

The description presented in Figure 6G was only an excerpt, of course, and went far beyond the text shown here. The discussion and conclusions of the findings presentation will also be focused around the research categories, which continue to be the common thread of the study from beginning to end.

Using a focused-categorical description style makes it easier for researchers to convey the theoretical picture. This kind of writing combines descriptions with the presentation of theoretical concepts (Merriam, 1998). The resulting picture should therefore reflect the theoretical perspective. The authors can incorporate theoretical arguments alongside the descriptive picture, place the theoretical perspective before the descriptive picture, summarize the descriptive picture with theoretical concepts, or create any other combination thereof that might correctly relate the picture with its theoretical meaning.

The final report of the criteria-focused methodology would be accompanied by a description of the research context. The researchers provide the reader a description of the physical and social environment of the case under study, as well as background information about the participants.

Research Quality Standards

The term "objective" is acceptable as a measure of the general quality of research in scientific research (Kirk & Miller, 1986), but also penetrates to the common lexicon in the social sciences and education. Considering the research as an objective process determines the basis for validity, reliability, and generalization. As will be shown in the upcoming chapters, most of the qualitative methodological approaches negate the "objective" stance, and suggest alternatives such as "subjective" or "perspective" (Lincoln & Guba, 2000; Schwandt, 2000; Smith & Hodkinson, 2005). As a result, researchers suggest alternative conceptual standards to express the quality of the research, such as "trustworthy" and other similar terms (Arksey & Knight, 1999; Guba & Lincoln, 1998, 2005), or alternatively use the terms "validity," "reliability" and "generalization" which emphasize the unique nature of qualitative research context (Greenwood & Levin, 2005; Shkedi, 2003, 2005; Stake, 1995).

It seems that when it comes to criteria-focused methodology, there is no reason not to use the conventional concepts terms of "validity," "reliability" and "generalization" as standards for the quality of the research. As mentioned, the criteria-focused methodology is based on a defined theoretical perspective, or at least on a set of criteria, concepts or categories which are determined in the first phases of the research and accompany the research process throughout. These categories are not just building-stones in the research process, but also a criterion for the quality of research. From this, it appears that regardless of whether researchers accept the positivist assumptions about the phenomena, the reliance upon a predetermined category creates the infrastructure to apply those concepts of quality accepted in quantitative research methodology in criteria-focused methodology as well.

The term "validity" is based on "the external criteria by which the validity of a text is then judged." (Bishop, 2005, p. 127) Or, in other words, the extent to which the research findings are compatible with the theoretical perspective of the study and interpreted correctly with it (or, the extent to which the research indeed studied what it intended to investigate) and can be applicable in criteria-focused methodology. Researchers as well as readers can examine the connection (or the lack of connection) between the theoretical perspective and its implication for categories and the research findings and interpretations. Because this is a qualitative research which presented its process in the language of words rather than the language of mathematics, researchers are required not only to keep the suitable balance between the data and a theoretical perspective, but also to expose them to a continuous consideration both in regard to the research itself, to colleagues, external judges, and also to the readers of the study.

Researchers need to present the database (interviews, observations, documents, etc.) as a chain of evidence for all stages of the study, in order to examine if it is faithful to the stated and formulated theoretical perspective (Huberman & Miles, 1994; Mason, 1996). As will be described in the upcoming chapters, determining external judges is a problematic process in most of the qualitative methodologies patterns, but not in this case. The fact that this methodology is targeted upon predetermined and external criteria allows for judges to examine the validity and reliability of the research (Merrick, 1999). We can appoint an odd number of judges (three or five), show them the process of analysis, the research findings and the research theoretical perspective, and request their verdict. We use an odd number of judges to reach a majority conclusion, lest there be divisions in opinion.

Just as the term "validity" is relevant to criteria-focused methodology, so the term "reliability" "is founded on the realist conception that what we choose to study can be thought of as having an independent object-like existence with no intrinsic meaning." (Pinnegar & Daynes, 2007, p. 29) Reliability is the possibility of repeating the same process of research and getting the same results, even if the research is carried out by other researchers (Yin, 1994). In most of the qualitative methodology patterns, it is difficult to expect that other researchers would achieve similar results, even in identical situations (Merrick, 1999; Schofield, 1989). However, in the criteria-focused methodology, which is based on predetermined research criteria and categories, it could be expected that researchers who examine the same phenomenon and use the same predetermined categories will come to a similar result, namely, that reliability will be achieved in the traditional sense. The way in which the researchers claim reliability is similar to the way they claim validity. The researchers present a database, reveal the processes of analysis, and present a detailed description of the final report findings (Mason, 1996; Pidgeon, 1996). This indicates that the researchers themselves, external judges, colleagues and readers can examine the reliability of the study (Dey, 1999; Peshkin, 2000).

One of the criticisms against qualitative research is that the ability to generalize the findings in comparison to other cases (external validity) is almost impossible. One reason for this is that qualitative research focuses on a relatively small number of investigated cases, and sometimes on even a single case study. Thus, qualitative researchers suggest alternative concepts to "generalize," which highlight the uniqueness of qualitative research. Steak (1995, 2005) speaks of "naturalistic generalization," and Guba & Lincoln (2005) suggest the phrases "transferability," "comparability," and "verisimilitude." But it seems that in the case of criteria-focused methodology, we are very close to generalization in the traditional sense, for at least two reasons. One is that throughout the study, this methodology relies on a relatively stable theoretical perspective, bringing us close to the concept of "objective," which is a key concept to the traditional argument for the possibility to generalize. Another reason is that the language of words used in this methodology pattern to describe the findings, is essentially a thin descriptive argument language and a less thick description. (Geertz, 1973) Such description is more limited and is based on more focused data, allowing a relatively short process of data collection to be gathered from a relatively large population. Of the three types of qualitative research generalization-- case-to-case, analytic, and generalization to the population (Firestone, 1993), the latter two are the most relevant in the criteria-focused methodology pattern. Case-to-case generalization is less relevant to this methodology. A reference to each of the studied cases in this methodology pattern is expected to be relatively narrow, and therefore does not warrant a rich, extensive descriptive picture of each of the investigated cases, and thus cannot supply a proper basis for comparison with new cases in order to reach generalizations. In contrast, the analytic generalization, which is based on theory, may be very relevant, because this methodology is based on the theoretical concepts, which guide the research from start to end. In this way, it can be argued that the research findings can also be valid in other cases, which are based on and explained by the same theoretical concepts (Dey, 1993; Firestone, 1993; Marshall & Rossman, 1989). Generalization to population, from a sample population under study to other populations, may be possible in situations where the sample population is relatively large, which is possible in the case of the criteria-focused methodology pattern as emphasized above (Arksey & Knight, 1999).

CHAPTER 7

PARTIAL CRITERIA-FOCUSED METHODOLOGY (ALMOST POST POSITIVIST RESEARCH APPROACH)

The partial criteria-focused methodology which close to post-positive research approach, presented in this chapter, is situated, by definition, closer to pole of analytical skills, but maintains some extent of affinity to the pole of the researcher's intuitive skills. In fact, many of those who employ this methodology perceive it as being in the center of the spectrum (see figure 7A. Although this methodology is based on a clear theoretical system and criteria, it is characterized by being highly flexible with respect to these systems, unlike the criteria-focused methodology, which is soundly based throughout the research process upon a guiding theoretical system and clear criteria.

Partial criteria-focused methodology draws its logic mainly from the post-positivistic paradigm, characterized by a much more flexible approach with respect to a theoretical system and the criteria arising from it. While in the chapter dealing with criteria-focused methodology, we expressed reservations about the possibility of finding a complete correlation between the methodology and a positivistic paradigm, it seems that in the case of partial criteria-focused methodology, there is a greater correlation - though not identical – between the post-positivistic paradigm and this methodology. However, it actually seems that a good number of adherents to the partial criteria-focused methodology may hold, at least in part, the assumptions of the constructivist approach.

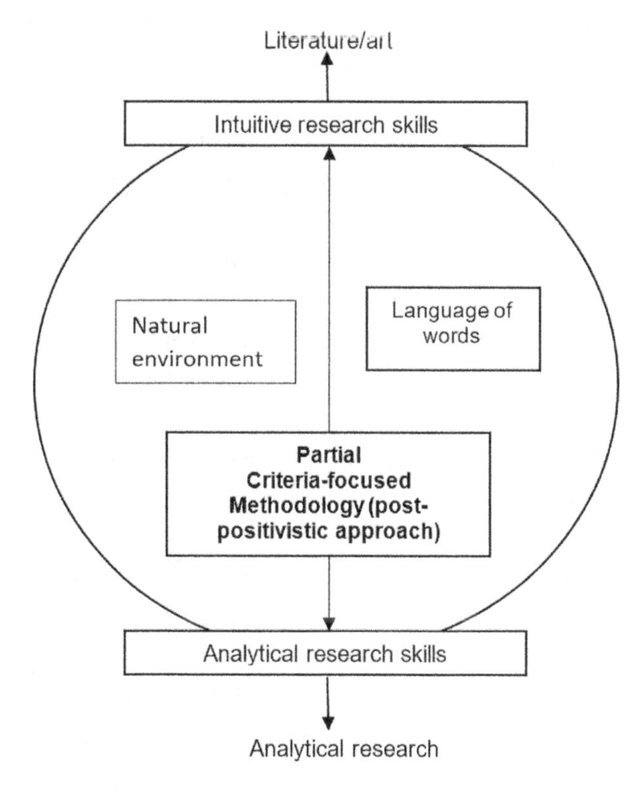

Figure 7A: The Location of Partial Criteria-focused Methodology in the "Space" of Qualitative Research

The Role of the Post-Positivistic Approach in this Methodology

In order to understand post-positivism, it is advisable to compare it to the positivist approach. Positivism is a foundationalism philosophy which believes that knowledge is organized in a stable and orderly fashion, and disclosed to the people by a scientific-empiricist cumulative process. Positivism argues for the presence of absolute truth, not only in the natural world, but also in the social world. According to this approach, the role of science, both natural sciences and social sciences, is to describe the world accurately and objectively (Guba & Lincoln, 2005; Philips & Burbules, 2000; Sullivan, 2010). This knowledge can be formulated in a clear, theoretical manner which aims to reflect the truth and is expressed by using quantitative and statistical methods (Bruner, 1985, 1996). The positivist approach pervaded the research in science and social science for 400 years, but lost its seniority (though it has not disappeared) during the twentieth century (Guba & Lincoln, 1998).

"Post-positivism is a non-foundationalist approach to human knowledge that rejects the view that knowledge is erected on absolutely secure foundations" (Philips & Burbules, 2000, p. 29). Post-positivism assumes that facts and assumptions are not independent entities and they can only be perceived through a "window" of defined theory (Guba & Lincoln, 1998, 2005). Consequently, while positivism is based on verifying research hypotheses and assumes that we can achieve complete verification, post-positivism is based on the falsification of the hypotheses (Clandinin & Rosiek, 2007). "Not only are facts determined by the theory window through which one looks for them, but different theory windows might be equally well-supported by the same set of facts." (Guba & Lincoln, 1998, p.199) Hence, any one theoretical framework would provide a different picture than another. Post-positivism is primarily associated with the philosopher Karl Popper (1959), who claimed that people can never reach absolute truth, but only theory falsification. Popper (1902-1994) established the principle of falsification: a theory can be considered as a science only if we can think of a way to disprove it, if it is falsifiable (Dawkins, 2006).

One of the additional elements distinguishing between positivism and post-positivism is the role of the researcher's values in the study. While positivism speaks of absolute objectivity and rejects the possibility that researchers' values will take a role in the research process (Sullivan, 2010), the concept of post-positivism, which assumes that any phenomenon can view various theoretical perspectives, holds that the values of the researchers can be part of the process. Consequently, what the researchers view is affected by their values world. This does not mean that there is no value- free objective reality, but what the researchers see is what is grasped by their perspective, which is filled with values (Tashakkory & Teddlie, 1998).

Post-positivism is identical to positivism in its ontological assumptions about the nature of reality, expressed in the claim that there is an "objective reality" (Guba & Lincoln, 1989; Maykut & Morehouse, 1994). According the post-positivistic approach, It should be obvious that "we cannot always tell if a statement or theory or description of or about a situation or area of interest is true; but this does not mean that there is no truth of the matter in such a context" (Philips & Burbules, 2000, p. 37).

Moreover, what motivates the research and directs its aim is the desire to test hypotheses and arrive at the truth. For example, in the 1940's and 50's, biologists believed that the number of chromosomes in a normal human cell is 48. This knowledge was achieved by controlling the scientific process and was perceived as the objective truth. Years later, it was found that this knowledge is incorrect. Even so, it should be emphasized that the scientists were fully justified to refer to this number as the objective truth, since this was the best possible hypothesis at that time, based on valid theories and observations. Later it was determined that the number of chromosomes in the human cell is 46, and this number became perceived as "factual truth." Based on post-positivistic assumptions, the fact that what is perceived as true in one research context was later found to be erroneous does not eliminate the desire to seek truth, even if we assume that it is never really certain and absolute.

Positivism offers a linear, well-defined methodology for scientific discovery based on previously validated knowledge. Post-positivism offers a methodological process based on a permanent process of checking how hypotheses survive the test of refutation. The researcher constantly examines the data and processes, and decides which hypotheses to keep and which to omit. Post-positivism is identical to positivism in its preference for an ideal formal system – i.e., favoring the precise, concise expression of the research findings, and the resulting mathematical-statistical expression as the worthy goal of the study (Philips & Burbules, 2000). However, since this is an inherently skeptical approach which indeed seeks the "truth," but assumes that any statement is not absolute but subject to falsifying, the verbal expression, more obscure by nature, is worth accepting, although through focused and verbally-expressed arguments more than open-story expression (Woods, 1996).

Between Post-Positivism and Constructivism

As mentioned at the beginning of this chapter, many researchers using the partial criteria-focused methodology may be closer to the constructivist approach than to the post-positivist approach. Therefore, it is important to clarify some methodological distinctions implied by the comparison between these two approaches. "Whereas post-positivists seek a description of a reality that stands outside human experience," the constructivist "seeks a knowledge of human experience that remains within the stream of human lives" (Clandinin & Rosiek, 2007, p. 44). Thus two different positions are implied regarding the involvement or remoteness of the researchers from the phenomenon under study. Researchers employing the constructivist approach will tend to be very much enmeshed in the phenomenon being studied, involving their personal values and regarding this as an important research advantage. In contrast, researchers employing the post-positivistic approach endeavor as much as possible to separate their personal values from the research process, still believing that they cannot achieve absolute detachment (Guba & Lincoln, 2005; Tashakkory & Teddlie, 1998).

There is a significant difference with regard to the research context. The post-positivist approach accepts the influence of the environment upon the phenomenon under study, but strives to limit its place in the context of the study as much as possible and to reduce its impact upon the study's results. In research adhering to the constructivist approach, the place of the context is most prominent (Clandinin & Connelly, 2000). Another difference between the two approaches is the extent of clarity that the researchers achieve. While post-positivism expresses a degree of absoluteness, the constructivist position assumes a lack of clarity in understanding the phenomena under study. From this, researchers employing a post-positivistic approach criticize their constructivist colleagues for their uncertainly in presenting the phenomena under study, while constructivist researchers may criticize their post-positivistic colleagues on their simplistic perceptions about the nature of human experience and the behavior of people and society (Clandinin & Rosiek, 2007).

Those who adopt the partial criteria-focused methodology as a qualitative research methodology choose the language of words, with its complexity and non-clarity, as the language of the study over the language of mathematics, marked by great precision, which is the preferred language of post–positivistic researchers. Therefore, it appears to be difficult to identify qualitative researchers adopting the partial criteria-focused methodology as being pure post-positivists. It seems that we can talk about an actual symbiosis of the post-positivistic approach and the constructivist approach, which differs from researcher to researcher in its extent.

Much of the literature of recent years dealing with qualitative research does not pay sufficient attention to qualitative research adopting the post-positivist approach and the partial criteria-focused methodology. However, it seems that many researchers employing the qualitative research methodology have opted for this methodology, even unconsciously. It is assumed that these researchers will find special interest in this chapter.

The Study's Theoretical Perspective

Partial criteria-focused methodology, like criteria-focused methodology, is based on the pre-formulation of the theoretical perspective before any other research activities. With respect to the tensions between the external point of view and the internal perspective, between the emphasis on human intuitive research capabilities and the focus upon analytical processes, distance and control, the scale leans more toward the external perspective and distance and control, as reflected in Figure 7A above. The assumptions of this methodology invite the researcher to clearly formulate the theoretical framework before entering the field study (Creswell, 1998). Since this methodology is not absolutely dependent on predetermined criteria but is open to changes and additions, the theoretical framework may be flexible and changed to some extent during the study, in light of new insights that arise at the encounter between the theoretical and the data systems.

Although there are particular similarities between this methodology and the criteria-focused methodology, there is a clear difference between them. Criteria-focused methodology is characterized by a tendency to be highly responsive to the researcher's analytic abilities and to external standards, while in the partial criteria-focused methodology, there is a greater expression of the researcher's intuitive skill and internal criteria, albeit the emphasis is still on analytical, distance and control characteristics and predetermined external criteria.

This methodology assumes that "Every inquirer must adopt a framework or perspective or point of view [before entering the field study]. Given this framework or perspective, he or she may see phenomena differently from the way other investigators see them. But it does not follow from this that there is no fact of the matter or that "anything goes" – relativity of perspective does not necessarily lead to subjectivity, and relativity does not always warrant the charge of being biased." (Philips & Burbules, 2000, p. 46)

The need to adopt an a-priori theoretical perspective is also due to practical reasons. At the foundation of this methodology is the assumption that research cannot be conducted properly without a guiding theory or model. The theoretical perspective enables the researcher to conduct the study properly and in control along the whole process (Fetterman, 1989).

The study will start, of course, with a review of literature which defines the theoretical perspective (Shaw, 2010a). This perspective, as in criteria- focused methodology, will become the anchor for the entire research process. This literature review will be complete, clear and coherent, to give direction to the research. The nature of this methodology, and the fact that it is not absolutely dependent on predetermined criteria, may lead to certain changes in the literature review, especially additions arising from the development of the research process. However, because the criteria guides the study from the beginning (as opposed to participant-focused methodology), we cannot expect to achieve dramatic changes in the literature review.

Formulation of Research Questions and Assumptions

The partial criteria-focused methodology, similar to the pattern upon which criteria-focused methodology is focused, asks not only to describe the phenomenon under investigation, but also to explain it (Guba & Lincoln, 1998). According to Miles & Huberman (1994), researchers that can be identified largely with the partial pattern-focused methodology differentiate between research processes seeking to describe the phenomenon and those seeking to explain it. It is important to recognize that even the criteria-focused methodology, in part or in full, does not necessarily lead to the present explanation of the phenomenon under study, but may in many cases be satisfied with describing based on predetermined theory and criteria. Researchers using partial criteria-focused methodology are faced with several options, ranging from theoretical explanations and accompanying descriptions to studies seeking to offer causal relations between elements of the phenomenon under study. "Qualitative researchers should be interested in exploring the causal relationship, a demanding objective that remains uncommon in contemporary qualitative research." (Fielding & Lee, 1998, p. 44). It seems that researchers who choose this methodology will see their aim as being to provide theoretical explanations for the phenomenon being studied, and not be satisfied with providing a description.

Research Questions

Research questions will be derived from the theoretical framework guiding the study. Unlike the participant-focused methodology reviewed in the previous chapter, where the study begins with open-ended questions and becomes more focused as the research progresses and the extent to which participants' "voices" are heard, in partial criteria-focused methodology, the research questions are focused and predetermined. This methodology articulates research questions at two levels: descriptive questions and explanatory questions. Descriptive questions are characterized by such interrogative words such as "what," "how," "how to," and the like, and explanatory questions are characterized by the word ""why," with all its parallel and similar question words (Holstein & Gubrium, 2005). Here's an example of descriptive research questions dealing with research on teachers' implementation of the curriculum (Shkedi, 1995):

1. **What are the teachers' attitudes towards the possibility of using an external curriculum?**
2. **What are the teachers' attitudes towards each of the external curriculum components?**
3. **What are the teachers' perceptions of the external curriculum contents?**

The answers to these descriptive questions will be provided directly from the discourse of the participants. Understanding this discourse will be based on the course of the theoretical perspective of the research.

The explanatory questions may be formulated as follows:

1. **What is the relationship between the teachers' years of experience on the job and the use of an external curriculum?**
2. **What is the connection between the ways teachers perceive the curriculum and their pedagogical beliefs?**
3. **Why are teachers interested in receiving an external curriculum?**

While the answers to the descriptive questions will be a description, even if they are based on a theoretical perspective, the answer to the explanatory question will be theoretical-conceptual in nature. Although they are based on descriptive data, they present the theoretical aspects of the phenomenon.

The Research Hypotheses

Based on the theoretical perspective defined at the beginning of the study and the research questions, the researcher seeks to predict the answer to the research questions, and derives hypotheses using components (variables) relevant to a given theoretical framework. Hypotheses will be tested by a clear set of processes, with a theoretical background that supports these assumptions (Travers, 2001). However, contrary to the criteria-focused methodology which poses hypotheses in order to verify them, in partial criteria-focused methodology the hypotheses, if established, are not for verification, but to examine if the findings refute them (Guba & Lincoln, 2005). Hypotheses in the context of descriptive questions aim to formulate the theoretical-conceptual framework and define the relevant variables for examining the phenomenon under investigation. The place of hypotheses will be more significant in the context of explanatory questions. The following are examples of research hypotheses connected to the three explanatory questions presented above:

1. **"The greater a teacher's seniority, the less his/her use of an external curriculum."** This research hypothesis is based upon theoretical and research material indicating that as teaching experience is gained, teachers' confidence and independence grows.
2. "**Teachers will comprehend the curriculum in accordance with their personal pedagogical beliefs.**" This research hypothesis is based on theoretical material dealing with the issue of reading comprehension, suggesting that different people who read the same text will interpret it differently based on their personal perceptions.
3. "**Teachers are interested in receiving an external curriculum in order to receive teaching ideas, not to use it verbatim.**" This research hypothesis is based on studies conducted on the issue.

It is clear that these hypotheses are rooted in the review of theoretical literature for the study. Clearly, these hypotheses, like any other research hypotheses, can be refuted, and the researcher may suggest other sub-hypotheses on the basis of different theoretical materials in accordance to the data gathered, or at least to suggest changes in the theoretical framework of the study. But in any case, the hypotheses are the anchor of the research process, a reference that drives the research and gives it the theoretical-conceptual framework. Figure 7B sums up (from left to right) the connections between the research components of the partial criteria-focused methodology presented so far.

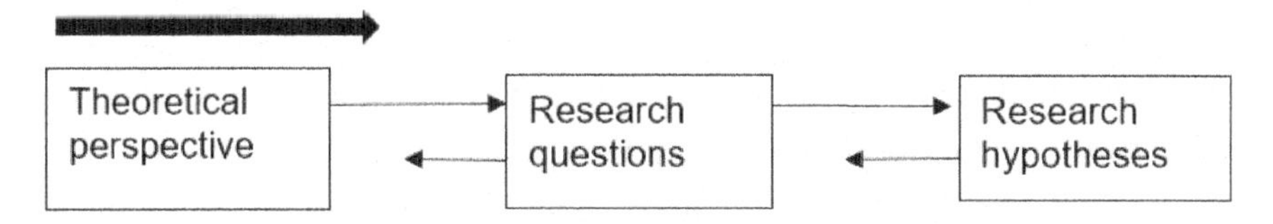

Figure 7B: The Relationships Between the Research Components of Partial Criteria-focused Methodology

The solid arrows from left to right indicate the dependence of each component on the previous components. However, the broken arrows, from right to left, indicate a certain two-way direction of the research process. We can see a certain degree of returning to and examining the previous steps, in light of the picture that emerges in later stages.

The Study Design and Population Selection and/or Field Research

The process of selecting a field study and its population in partial criteria-focused methodology is similar to all qualitative research methodologies, and assumes no special character. It should perhaps be taken into account that the nature of the partial-criteria focused methodology will require the participants a greater degree of exposure than the criteria-focused methodology. The cooperation of the participants and their sense of trust become more significant in this methodology (Silverman, 2006).

The participants will be required to devote more of their time, both in observations and especially interviews, than required in criteria-focused methodology. Researchers will have to find respondents willing to devote the required time involved. If possible, two rounds of selection should be carried out. The first will be based on a random sampling. Then, following negotiations with potential participants, a purposeful sample based on the original random population sampling should be created. (Mason, 1996)

As we shall see in the next section, the methods of data collection in this methodological pattern are relatively structured tools. This means that the participants are required to address questions and comments arising from an external perspective, rather than those arising out of their own world or from dialogue with the researchers. This type of research tool often creates the sense of an external examination. Researchers need to be attentive to this potential difficulty. This augments the need for a purposeful sample reaching the right people who would be willing to cooperate, and even to replace participants should it become necessary. Obviously investigators need an adequate level of ethics to ensure avoiding harm to the participants' rights and privacy, in the spirit of the Helsinki Agreement (Dushnik & Sabar - Ben-Yehoshua, 2001; King, 2010).

Data Collection Methods

Although the partial criteria-focused methodology is relatively flexible compared to the rigidity of the criteria-focused methodology, its fundamental principle is that the data collection must be based on a predetermined theoretical perspective, questions and hypotheses. Miles and Huberman (1994) suggest that before embarking into the research field, the researchers should prepare a list of categories. Moreover, they suggest adopting a process they call "data reduction," where data is collected selectively according to a list of categories which can be changed during the research process (Fielding & Lee, 1998).

Some data will be collected using initiatory tools such as observations, interviews (of individuals or groups) and field notes that the researchers recorded during their stay in the field of the study. There is also "naturally occurring data" collected, such as documents and objects (Silverman, 2006). It is difficult to see a significant utilization of quantitative research tools such as structured questionnaires, unless within a mixed methodology that combines the research with quantitative tools (see Chapter 10). "Open" questionnaires which invite the participants to respond to written questions will be less acceptable, but optional.

The nature of the research tools will reflect the character of the partial criteria-focused methodology, namely, being close to the criteria array while being flexible during the study. Therefore, a structured interview or a structured observation of the kind mentioned in criteria-focused methodology will not be accepted here, because this methodology provides a more significant expression to the intuitive inquiry skills of the researchers, besides their analytical skills. However, the research tools will be based on guiding criteria in a way that assures at the prescribed theoretical perspective will guide the study and be implemented in a similar manner for all participants. However, these standard tools will be flexible and allowed to be changed during the study. On the continuum between structure and openness, between predetermined criteria (categories) and creating criteria during the meeting with the participants, it appears that the tendency is towards predetermined criteria, but with openness to the voices coming from the study field. Figure 7C illustrates the difference between the methodological research tools of the three methodology patterns dealt with so far.

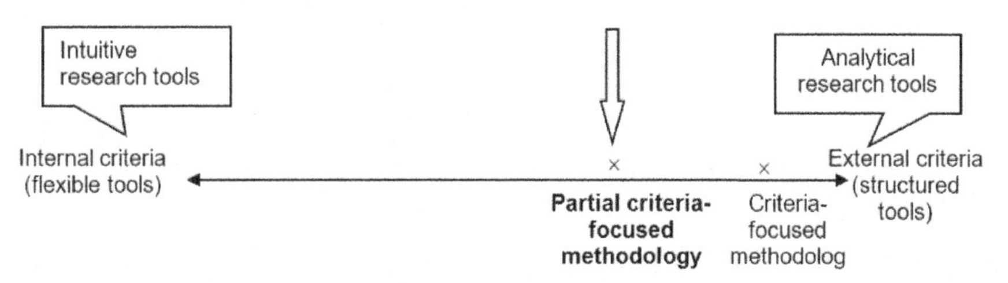

Figure 7C: Characteristics of the Data Collection Tools of Partial Criteria-focused Methodology

It is important to note that the "location" of the research tools in partial criteria-focused methodology on the continuum in Figure 7C gives a more meaningful expression to the external criteria determined before the departure to collect data, tools which are relatively structured. However, the chart also indicates a particular expression for internal criteria that was not planned for in advance.

On one hand, the interviews will be of the semi-structured variety, as opposed to structured interviews, in-depth interviews, or open interviews (Silverman, 2006). Semi-structured interviews are based on a list of prepared questions (Sparadley, 1979). The list of questions is repeated in all the interviews, thus ensuring that all participants are introduced to the same questions. Compared to structured interviews, semi-structured interviews "can provide greater breadth to the phenomenon under study than do other types." (Fontana & Frey, 2005, p. 705) Unlike the structured interview where researchers focus only on questions prepared in advance, the semi-structured interview allows and even encourages researchers and participants to add spontaneous questions to enrich the research picture. However, the prepared questions still remain the focal point of the interview. Figure 7D describes the characteristics of the semi-structured interview, compared to other types of interviews.

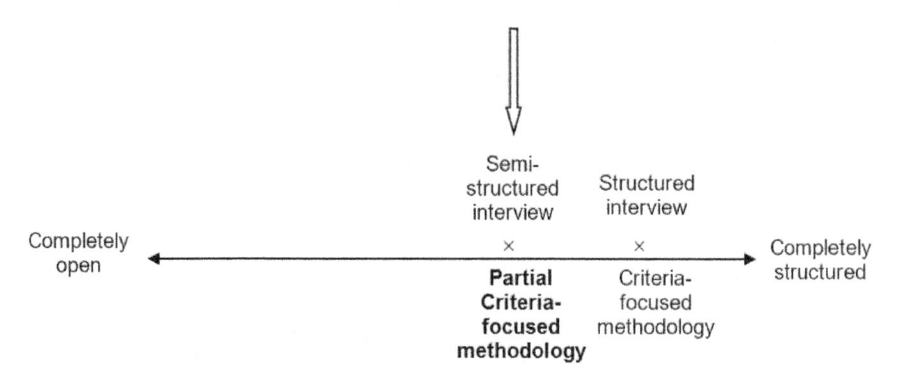

Figure 7D: Characteristics of the Interviews in Partial Criteria-focused Methodology

Observation in partial criteria-focused methodology is not of an extremely structured format. Investigators maintain some kind of external position towards their research, being careful not to be overly involved, and using their intuitive inquiry skills in a moderate way. However, although they arrive at the field with a list of criteria prepared in advance, they are open to unexpected events, and if necessary to expand the circle of the observation (Jorgensen, 1989; Marshall & Rossman, 1989; Tedlock, 2005). This type of observation can be called a "semi-structured observation." Figure 7E describes the "location" of semi-structured observation across the continuum between pure observation and pure participation.

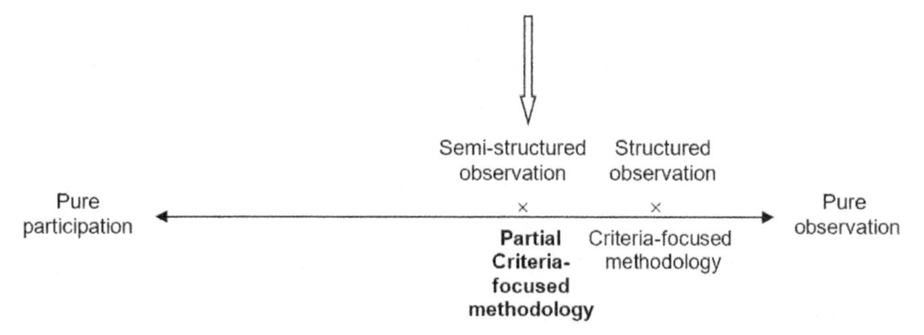

Figure 7E: Characteristics of Observation in Partial Criteria-focused Methodology

Although semi-structured interviews and observations are typical to partial criteria-focused methodology, it is possible that researchers who adopt this methodology will prefer using less-structured interviews and/or observations. The interviews and observations will utilize recording devices such as a tape recorder, video camera and other electronic tools to transcribe the events, so that the material may be used by researchers for adequate analysis. Recording and/or photography and transcribing will also ensure that the discourse and events are preserved exactly as they occurred, so researchers can vouchsafe for the validity and reliability of the study (Clandinin & Connelly, 2000; Shaw, 2010a). Researchers employing this methodology may choose to record only those parts of the observations and interviews that are considered relevant issues according to their research conception (Pidgeon & Henwood, 1996).

Research Categories / Themes

As emphasized in previous sections, the categories in partial criteria-focused methodology are determined before entering into the research field, and flow directly from the theoretical perspective, the research questions and the hypotheses (Polkinghorne, 1995). In this sense, the picture is the same as that presented in the criteria-focused methodology. But the difference is that in the partial criteria-focused methodology, the categories may change during the study, although to a limited extent: a few categories may be omitted, others may be added, but primarily the array of categories may expand.

There are at least two reasons that the categories are determined in advance. One is the implicit assumption of this methodology, which considers a solid theoretical system to be the cement that bonds the phenomena under study (Guba & Lincoln, 2005; Philips & Burbules, 2000). Basically, as noted in previous sections, in terms of the tension between the external and internal aspects, the balance tilts towards the external aspects. Thus, this methodology is based on an a-priori array of categories. The other reason to create an array of categories before entering into the research field is essentially practical. The assumption is that in the absence of such a system, researchers may lose the direction of the study. The role of the categories is to direct the process of collecting data and prevent researchers from losing their way. A clear list of categories helps the researchers focus on what is perceived as the principle theoretical perspective. However, once researchers enter the field study, they are open to pay attention to other voices (Philips & Burbules, 2000). Within a given theoretical perspective, researchers are open to further refine the existing categories according to other voices they hear.

The example in Figure 7F deals with observations on school principals. The researchers came to the research field with a list of categories, seeking to observe the principals' characteristics. The categories of the observation can be the following: "innovative versus conservative," "original versus conventional," "develops new ways versus keeps safe ways," "creates trust versus mistrust," "concentrates on people versus concentrating on tasks," " visionary versus technocrat," and the like. The researchers observe the school principals and characterize them according to these categories by determining their position on each of the categories. During the process of semi-structured observation, the researchers may identify more or consider other characteristics of the principals and add them to the category list, such as "dealing with conflict versus ignoring conflicts" or "focused on organization and order versus focused on people," etc. During the observation process, the researchers can also characterize sets of sub-categories for some or each of the above categories.

Figure 7F illustrates a partial picture of an array of categories. The darker frames are those of the categories added during the data collection process. The other categories are those that the researchers prepared before departing to collect data.

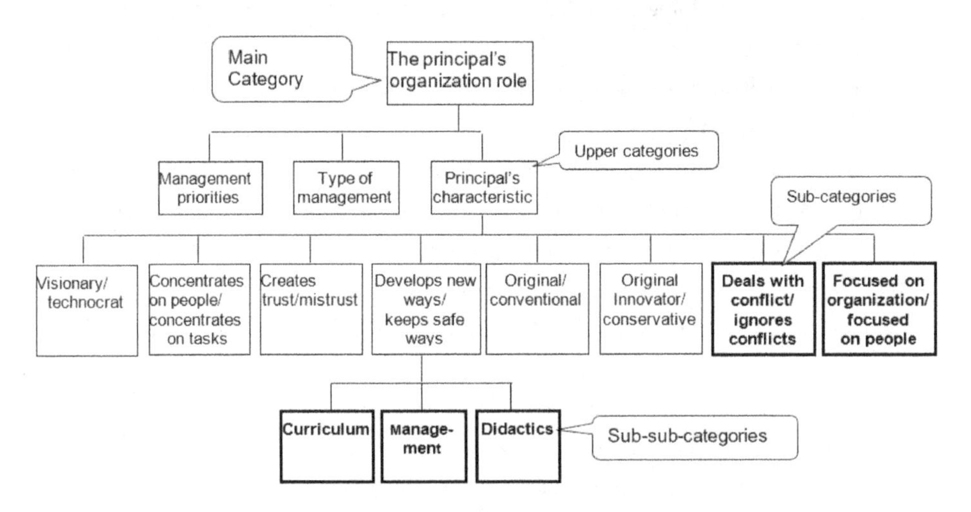

Figure 7F: An Example of a Developing Category Array of Partial Criteria-focused Methodology

There is a significant difference between the interviews and the observations. In partial criteria-focused methodology, the observer-researchers are external viewers, avoiding involvement as much as possible in the research field. Certainly, they are not the participant observers and do not want to influence the events in the research field. However, by its nature the interview is based on the direct interaction between researchers and participants, resulting in a mutual influence of researchers and interviewees upon each other. The interviewers in this methodology conduct their interview from a pre-prepared list of questions based on the research categories. Throughout the interview, researchers seek to be connected to the category list (Silverman, 2006), but they are open to the expansion and honing of these categories. The interview questions focus on the categories, but do not prevent the interviewees from expanding their responses.

Following is a demonstration of this process during an interview in the context of a question the principals were asked:

1. How do you understand your function as a school principal?

A. **From an administrative aspect**
B. **From an educational aspect**

During the interview, the interviewees were asked the above questions, based on the pre-set categories. The interviewers were open to extending certain questions, and willing to depart, albeit to a limited extent, from the predetermined set of questions. In response to this question, if the principal interviewees spoke, for example, about their communal and political role, the researchers extended the sub–questions and categories for those elements that they had not originally expected. Following is the set of questions modified according to the new insight:

2. How do you understand your function as a school principal?

A. **From an administrative aspect**
B. **From an educational aspect**
C. **From a communal aspect**
D. **From a political aspect**

However, the overall categorical picture will not change, and the researchers continue to adhere to the predetermined array of categories.

Data Analysis

As emphasized above, in partial criteria-focused methodology the data is analyzed in accordance with predefined categories, through a process called the "mapping analysis". The analysis process itself is, as stated, the process by which raw data (interviews, observations, etc.) is connected with the categories in a manner that gives meaning to the data. During the analysis process, bits of data are compared to each other to find similarities and differences. Those that are seen to be similar will be placed together under the same category, and those seen as dissimilar will be placed in different categories. The degree of closeness or distance between the categories will determine their location and emphasize the connections between them (Seidel & Kelle, 1995). Technically, the researchers place the data and the list of categories before them, and arrange the data into categories according to the content suitability (Gordon-Finlayson, 2010; Charmaz, 2005).

Since the data collection process is based on predetermined categories (even if they were modified), it seems that there is a certain correlation between the interview questions and the specific categories or between the observation view and the categories. However, as noted in the section dealing with collecting data, during the data collection researchers are open to receive more data than what was targeted within the given categories, resulting in a richer picture than what was supposed. Consequently, the analysis process will start by placing a predetermined category array, including changes made in the categories during the data collection phase.

Throughout the analysis process, researchers continue the process of changing and updating the category array. Researchers should examine the data and those pieces of data which are not suitable for existing categories, but seem relevant, placing them into new categories created during the analysis process. Figure 7G contains an example of adding categories during the analysis phase:

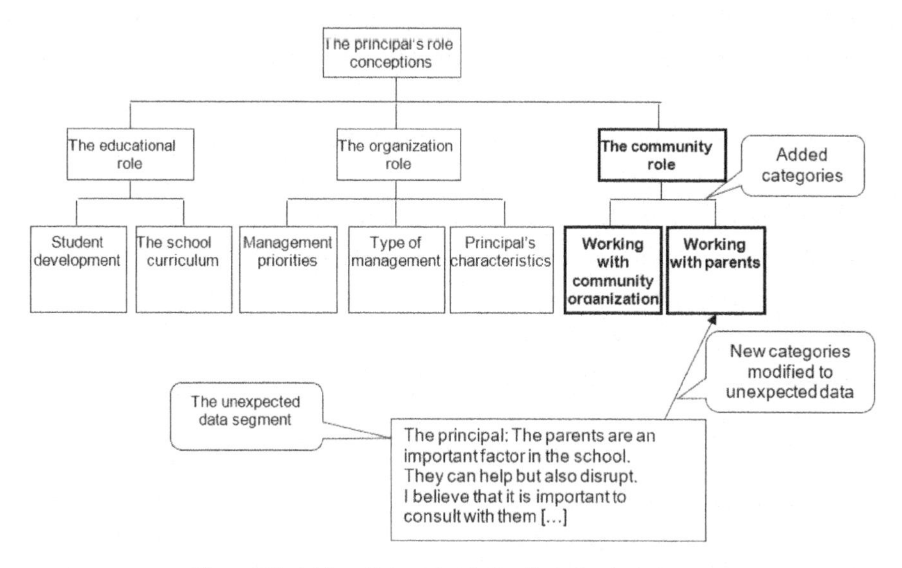

Figure 7G: Adding Categories in the Data Analysis Stage

The theoretical framework established in the early stages of the research continues to accompany the researcher at all subsequent stages, and also becomes the basis for a theoretical presentation of research findings (unlike the participant-focused methodology, where the theoretical framework emerged from the data). However, following the data analysis process, researchers may go back and examine the theoretical perspective and the research hypotheses and questions, making adjustments in light of new insights. In the partial criteria-focused methodology, there will probably be only minor changes, with the predetermined theoretical framework basically preserved to a certain focus, refinement and clarification. The structure of the research components will essentially be retained, but there may be some degree of flexibility in this structure.

The Research Report

The research report in partial criteria-focused methodology includes all the elements accepted in research reports, and generally takes the "standard" form of order: beginning with a literature survey, continuing to research questions, hypotheses or assumptions, methodology, presentation of the population and the field studied, findings, discussion and conclusions (Crowley, 2010). In this respect, most of what is written about the research report in Chapter 6 dealing with criteria-focused methodology is also relevant here. As in any qualitative research report, the emphasis will be on a presentation in the language of words. Unlike the criteria-focused methodology, the partial criteria-focused methodology used in the report can be more narrative, based on the fact that the researchers have not limited themselves to the predetermined theoretical perspective and categories and may deviate to unpredicted data which makes the research picture more vivid and rich.

As in any qualitative research report, the most challenging part is the description of the findings. We can say that an appropriate description of this methodology will be a combination of the "focused-categorical description" with "focused-narrative description" (Shkedi, 2003, 2005), where categories will be the backbone of the description (Merriam, 1998), but the writing style will deviate from the argumentative style into richer narrative expressions. In order not to burden the readers and to prevent the report from being too lengthy, most of the report will be in the focused-categorical description style, while a select number of episodes or instances will be presented more in-depth in the narrative style. In this way, the report emphasizes the description according to the categorical and theoretical logic, but at the same time allows the reader to look deeper into the phenomenon, at least in the context of a number of episodes or incidents.

One example of a focused-narrative description accompanied by a theoretical explanation is taken from a study that deals with students' perceptions of the Bible (Shkedi, 2001). The description includes a theoretical explanation attached to the data description. Figure 7H presents part of the description:

<u>Students' Perception of Alienation from the Bible</u>

Four students from a population of 52 respondents expressed their overall attitudes toward the Bible which can be defined as alienation. This relatively small number of negative responses quite surprised the teachers and did not fully comply with their expectations. A typical alienated response was expressed in the words of Rachel, a 16-year-old high school student who defines herself as secular.

"The Bible belongs to me? No. No. No. And I'll tell you why. It is something that's connected to the God. If people are bad in the eyes of God, He is angry. If it's good for God, then He is not angry. Do what is best for God.

Everything we learn in Bible classes is about sin and punishment. I do not believe it. It seems inappropriate to reality today and I do not believe that it was ever true."

Figure 7H: Example of Focused-narrative Descriptions

The description above combines theoretical language ("the alienation attitude") with a description in the language of the participants, giving expression to the character of this methodology which focuses in part on categories guided by a theoretical perspective and at the same time bringing the reader to a validated picture of the research by a direct quote from the participants. By employing this description style, researchers attempt to communicate with the readers and at the same time to maintain the quality of the research (Dey, 1993; Arksey & Knight, 1999; Denzin, 2000).

Research Quality Standards

Research employing partial criteria-focused methodology criteria, adhering post-positivistic research approach, can use conventional concepts of research quality like validity, reliability, generalization and objectivity as a benchmark for the quality of the research (Schwandt, 2000; Lincoln & Guba, 2000; Smith & Hodkinson, 2005). As defined, this methodology is based on a theoretical perspective, or at least on a system of conceptual categories, predetermined at the beginning of the research and accompanying the investigators throughout. These categories are used not only as building blocks for research at all stages, but also as the criteria for examining the quality of the research. They are a compass that guides the research process. The dependence upon an ever-present guiding structure of the categories provides a basis for using the conventional quality research standards typical of quantitative-positivistic methodology in partial criteria-focused methodology as well.

All references to the concepts of validity and reliability dealing with criteria-focused methodology that were presented in Chapter 4 are also valid in the case of partial criteria-focused methodology. Regarding the possibility of generalizing the findings of this research methodology, it seems that the three types of qualitative research generalizations, case-to-case generalization, analytic generalization and generalization to the population (Firestone, 1993) may be relevant. It is important to note that in this methodology, unlike criteria-focused methodology, generalization from case to case may be also relevant. The fact that we portray a relatively deep and broad description may provide enough richness to give the readers a basis for comparison and to examine the extent of other cases taken from their daily lives or from the research literature they encounter. In addition, the generalization to the population will be relevant as the population sample becomes larger, which is common to this methodology (Arksey & Knight, 1999). Analytic generalization may be also relevant, since this methodology is based on the theoretical concepts guiding the study from start to finish (Marshal & Roseman, 1989; Dey, 1993; Firestone, 1993).

Implementing research quality-checks for reliability, validity and generalization can be bolstered when researchers adopt the process of triangulation, namely the use of a variety of information sources and research methods (Fontana & Frey, 2005; Stake, 2000; Denzin & Lincoln, 2000). It seems that in fact the partial criteria-focused methodology researcher may use many information sources, such as interviews, observations, documents and more. To ensure the claim of validity, reliability and generalization by using triangulation, researchers must make certain to keep the database open and exposed, record all the various data analysis procedures, and preserve them as evidence of quality research. As it is qualitative research based on words and narrative descriptions, there are no statistical formulas like in quantitative research to vouch for research quality, but only the clear presentation of the final report with a transparent database.

CHAPTER 8

USING THE NARRALIZER SOFTWARE FOR CRITERIA AND PARTIAL-CRITERIA ANALYSIS (POSITIVE AND POST-POSITIVE APPROACHES)

When the stage of the reading and arranging of the data is completed (see Chapter 5), the actual analysis process is starting by splits into two different tracks, one for adherents of the methodological patterns that begin the process of research and analysis with predetermined categories, and the other for those affiliated with the methodological patterns whose categories are determined during the research process itself (see chapters 11-12). The Criteria-focused methodology and the partial criteria-focused methodology, (positivist and post positivist approaches) all generally begin with a-priori theories and categories. This chapters offer detailed guidance in qualitative research data analysis. Thus, after the phase of reading and arranging the data, we move directly to the phase of determining categories. priory categories, determining the categories during the analysis process itself.

The Mapping Analysis category array

The category array guides the researchers in the data collection and data analysis. This category array can take the graphic form of tree (see Figure 8A), the branches face downward and upward helping to express the relationship and the hierarchy between the categories.

At the top of the category tree is the "main category" ("Teachers' conceptions of their students"). One level below are the "upper categories." In Figure 8A, there are seven upper categories. Each of the upper categories can split into several "sub-categories." In our example, only two upper categories divided into sub-categories, but of course, all the others can also be divided. In the lower-level subdivisions, "sub-sub-categories" are created, and so on. Basically, all categories at all levels can be divided, and the number of vertical axis division levels can exceed five (the number in Figure 8A). The division on the horizontal axis can also reach a situation where each category is divided into a large (but reasonable) number of sub-categories.

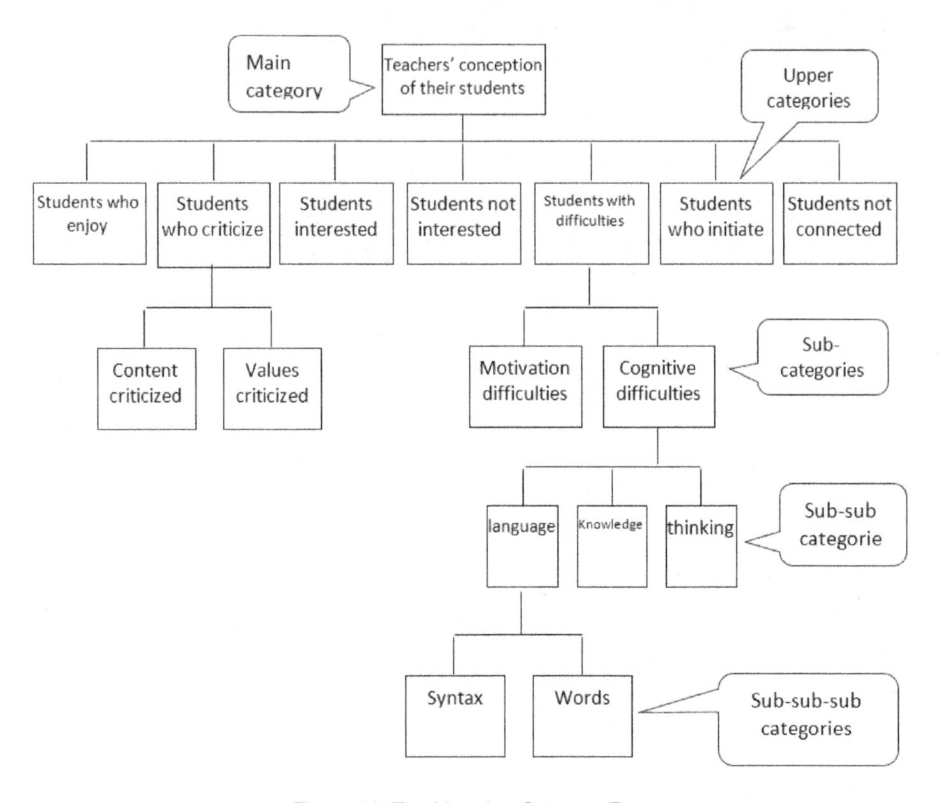

Figure 8A: The Mapping Category Tree

Similarly, we can create a vertical array of categories, as shown in Figure 8B. This category system is completely identical in its characteristics to the category tree system. The vertical array is more convenient to use in word processed documents and in the **Narralizer** software. This array is a very convenient base to attach relevant data segments belonging to the categories (as will be explained in the upcoming sections).

Teachers' conceptions of their students

1.Students not connected

2. Students who initiate

3. Students with difficulties

3A. Cognitive difficulties

3A1. Thinking difficulties

3A2. Knowledge difficulties

3A3. Language difficulties

3A3.A Words

3A3.B Syntax

3B. Motivation difficulties

4. Students not interested

5. Students interested

6. Students who criticize

6A. Values criticized

6B. Content criticized

7. Student who enjoy

Figure 8B: Category Vertical Array (Category Tree)

In the next step, after the creation of a category array, the researcher will place the data in the vertical array, as explained below.

Creating the Category Array in the Narralizer

1. Open the **Narralizer**. On the **working screen,** the **research document** appears as the default option.
2. When opening the **Narralizer,** the category **Narralizer 1** is displayed in the **Categories pane** (when opening subsequent documents, the names **Narralizer** 2, 3, etc. will be displayed). This is just a temporary name and may be changed by the command **Rename** to insert the same name as the main research category. The name of the main category is also the name of the Research Document.
3. Highlight the temporary name (Narralizer 1) and choose the **"Rename"** command. Alternatively, use the right mouse button to select "Rename" as shown in Figure 8C

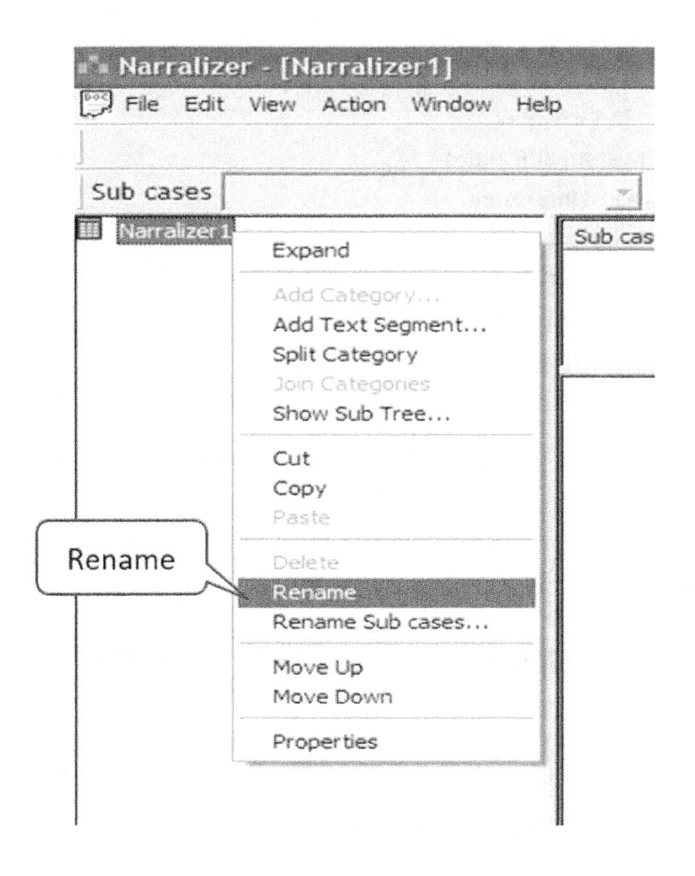

Figure 8C: Selecting a Name for the Research Document

4. With the selection of **"Rename,"** a "**Save As**" box will open. At the top of the box, under the title "**Save in,**" specify the directory in which to place the document. The new document can be placed in an existing directory or a new one (Figure 8C).

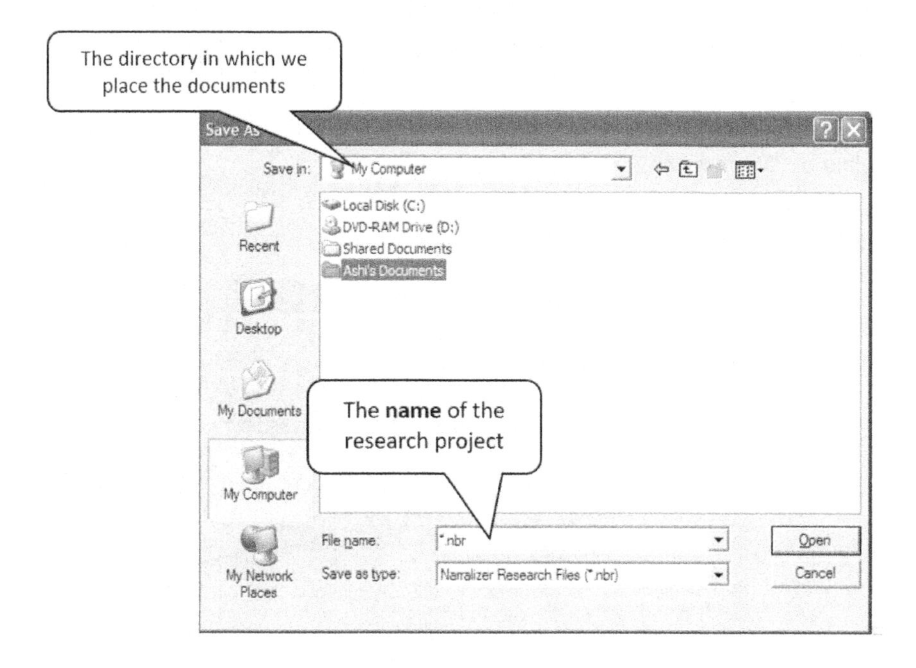

Figure 8D: Placing and Naming the New Source Document

5. At the bottom of the box, in "**File**" name, write the name of the research project (just as is customary in Windows™). In the example presented here, the selected research project is "Teachers' approaches" (Figure 8E).

   Bear in mind: One of the advantages of the **Narralizer** is the ability to maintain the research documents of each of the phases and sub-phases of the analysis process. Thus, we can go back and examine the process at any time, revise and amend if necessary, in a way that lets us claim the validity and reliability of the analysis process. To preserve a research document that reflects a certain stage of the analysis, use the command **Rename** to create a new document. The "old" document will be kept in the research documents directory (library), and is available for review at any time. To facilitate the tracking of the documents, it is recommended to add the date (e.g. 01/01/2011) or any other identification (for example, "mapping analysis") to the main category of each document stored.

6. Point to the main category to split (in order to build the category array). **Right click** on the mouse reveals the **Split Category** command (see Figure 8E).

You may also split the main category by clicking the **Split Category** icon on the toolbar or Action menu.

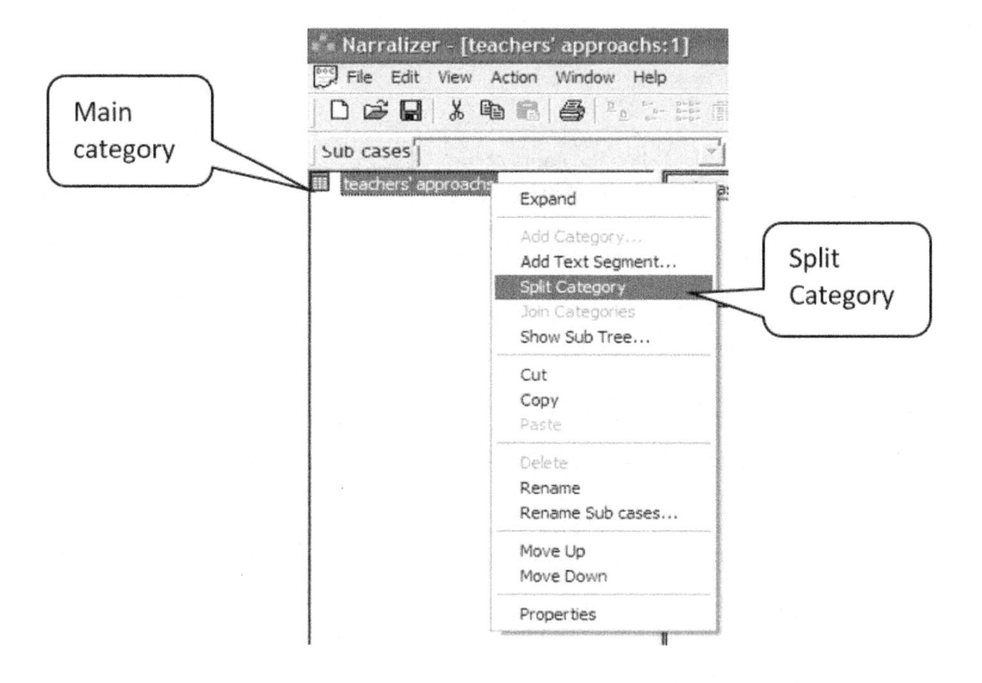

Figure 8E: Split Category Command

7. Selecting the **Split Category** command causes a dialog box to open. Type the name of the new category in the upper text box labeled **Name** (in the following example, Figure 8F, the sub-category is "student attitudes").

If comments are required, note the large text in the dialog box labeled **Comments.**

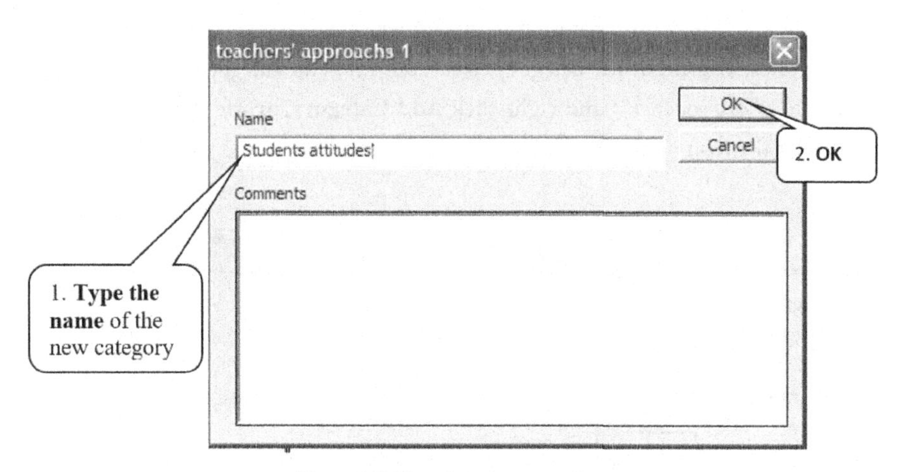

Figure 8F: First Sub-Category Box

8. Clicking **OK** automatically opens a similar dialog box to the previous one. Now, type the name of the second new category (in Figure 8G: "students' difficulties") and click **OK**. The dialog box will close and two upper categories will appear in the Categories pane under the main category.

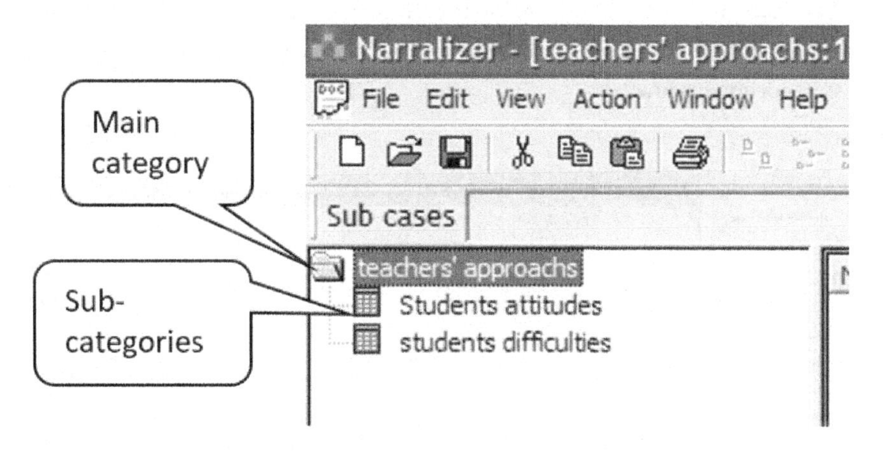

Figure 8G: The Category Array Following the Split Category Command

<u>**Bear in mind**</u>: When we split categories, there must be at least two branched categories. The **Split Category** command therefore automatically opens the dialog box twice. If we fail to create two categories, the program will create a second category and assign it a temporary name, which can be changed any time.

9. Having split the main category into two upper-categories, you may now add sub-categories. **Highlight** the category to be split into sub-categories (in Figure 8H: "teachers' approaches") and right click **Add Category, or** use the Action menu, or click the icon on the Toolbar.

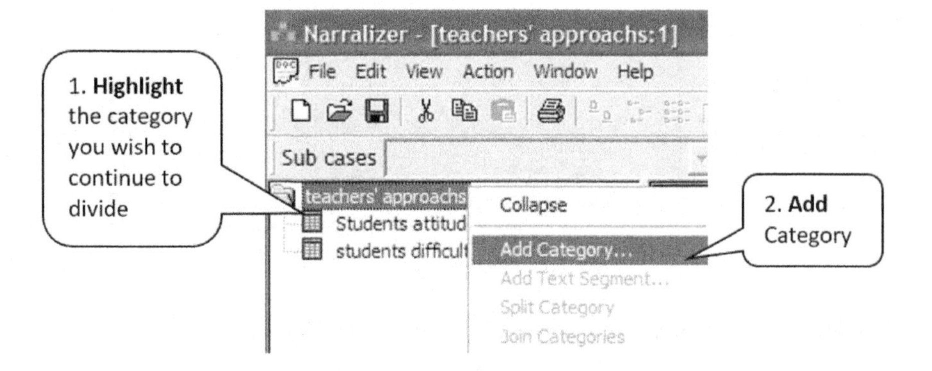

Figure 8H: Add Category Command

10. **Add Category** opens a new dialog box. Type the name of the new category in the upper text box **Name**. Click **OK** and the new category will be added to the **Categories window**.

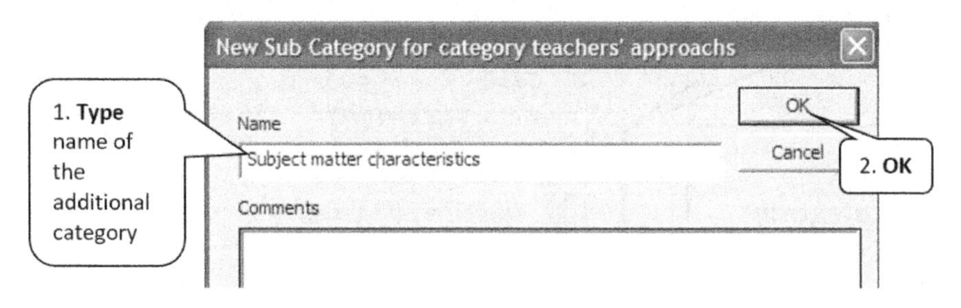

Figure 8-I: Typing the Name of the Added Category

<u>**Bear in mind**</u>: When a category is split, its icon changes from a hatched square to a folder. The new category icon is now a hatched square (see Figure 8J).

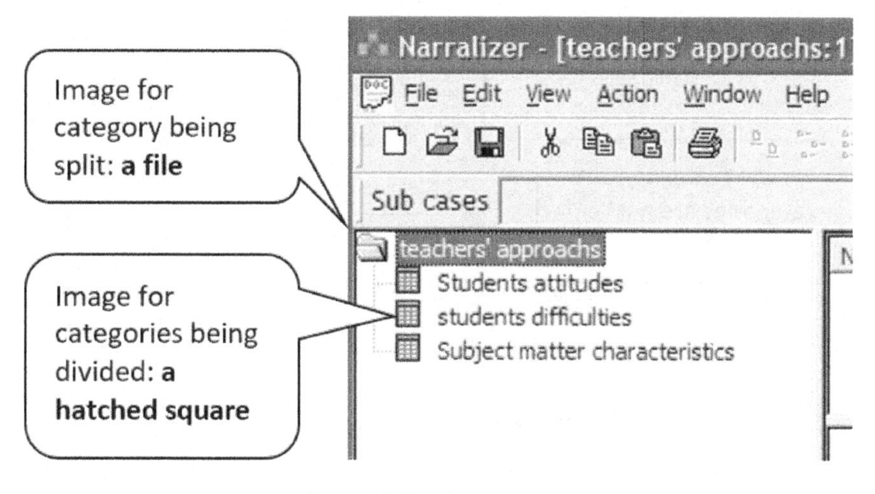

Figure 8J: The Category Icon

11. We can thus add (split) as many categories as needed from the main category. In the **Categories pane,** the **main category** represents the first level of the **Category array**, and in the second level are the **upper-categories**. **Narralizer** allows us to split categories into third, fourth, and fifth levels, etc., according to the analysis requirements.

Constructing the Complete Category Array

We can continue the process of constructing the category array. In the **Categories pane**, each upper-category (second level) can be split into sub-categories (third level). This is identical to the procedure for splitting the main category described above.

1. **Highlight** the category to be split using the Split Category command. Create two new sub-categories and type their names. The icon for the category that was split will be changed in the **Categories pane** (in the example shown: **students' attitudes**), and show a folder instead of a hatched square. The icon in the **Categories pane** shows a closed folder, and the sub-categories that split from it are inside and do not appear. The + (plus) sign to the left of the folder indicates that the folder is closed, as illustrated in Figure 8-K.

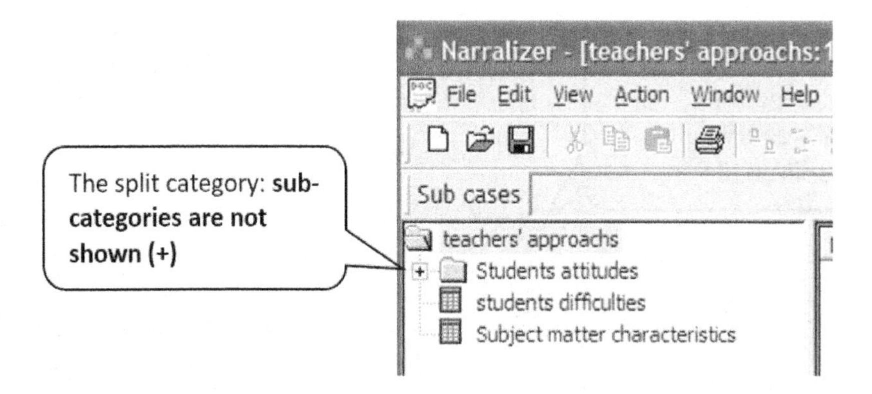

Figure 8K: Closed Icon When the Categories Are Not Shown

2. To view the sub-categories in the **Categories pane**, click the square with the + (plus) sign. This will reveal the sub-categories and the + sign will be replaced with a – (minus) sign, as illustrated in Figure 8L, below.

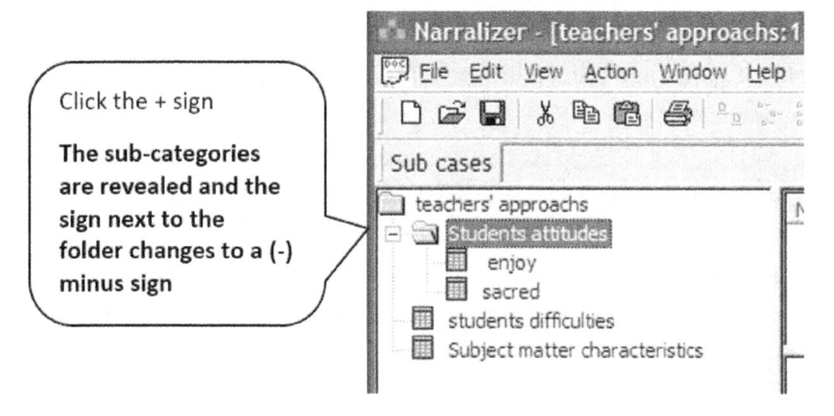

Figure 8L: The Icon is Opened and the Sub-Categories are Revealed

Similarly, clicking the square with the (-) minus sign hides the categories and replaces the (-) sign with a (+) sign.

Another way to display the sub-categories is to point the cursor at the category whose sub-categories we wish to display, and right click the mouse to reveal a pane with the command **Expand.** Left clicking **Expand** causes all the sub-categories related to the category above them to appear, as shown in Figure 8M.

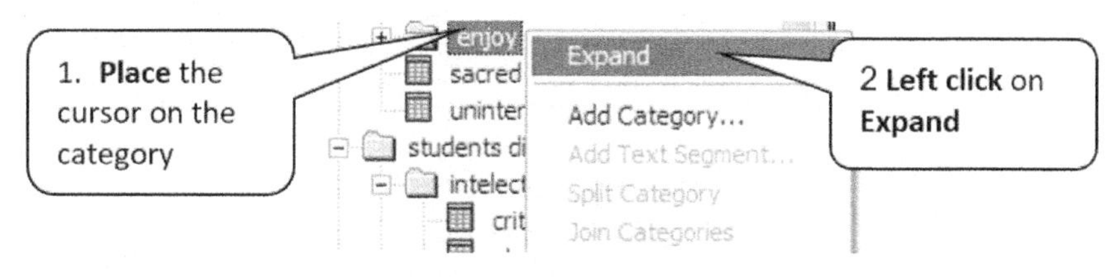

Figure 8M: Expand Command

The same method is used to **collapse** the sub-categories. Place the cursor on the category; right click with the mouse to open a pane labeled **Collapse**. Left click on Collapse to collapse all the sub-categories into the category above them.

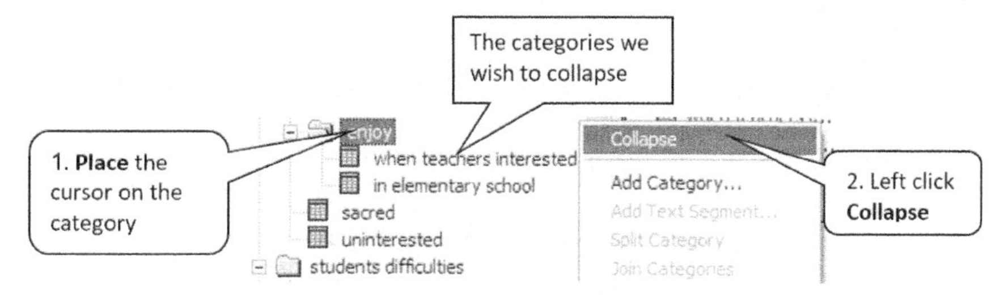

Figure 8N: Collapse Command

3. To add new sub-categories to the category that was split, use the **Add Category** function described above. Use **Split Category** and **Add Category** to split all other categories in the **Categories pane.**

In this way we are operating the **Categories pane** on three (or more) levels of categories:

Level 1: Main category
Level 2: Upper category split from the main category
Level 3: Sub-categories split from each of the upper categories

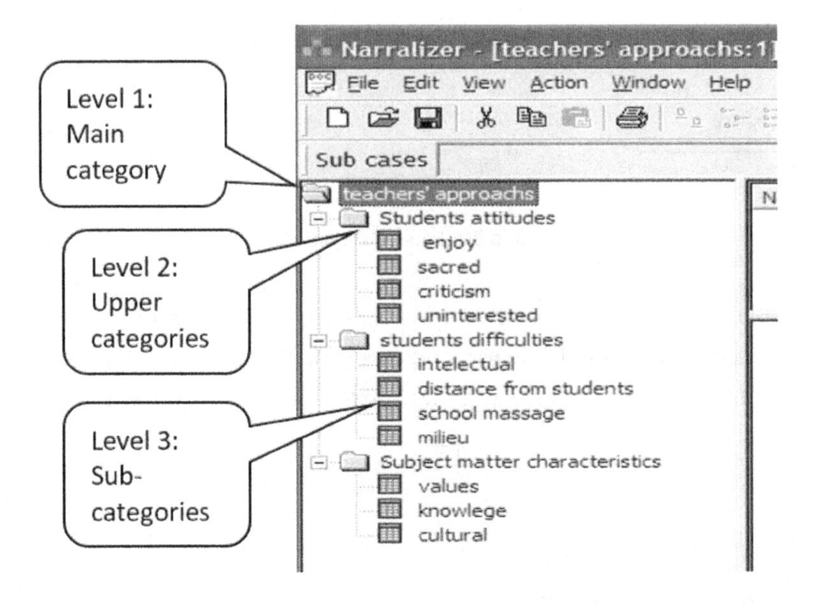

Figure 8-O: Conducting Three Levels of Categories

Use the **Add Category** function to add as many categories as needed on both Level B and Level C, and/or split categories to create more category levels, as will be explained below.

In the same way, we can continue splitting sub-categories into fourth, fifth, and more levels of categories, depending on how much detail the research analysis requires. After splitting the categories into a fourth level, the categories will appear as shown in Figure 8P:

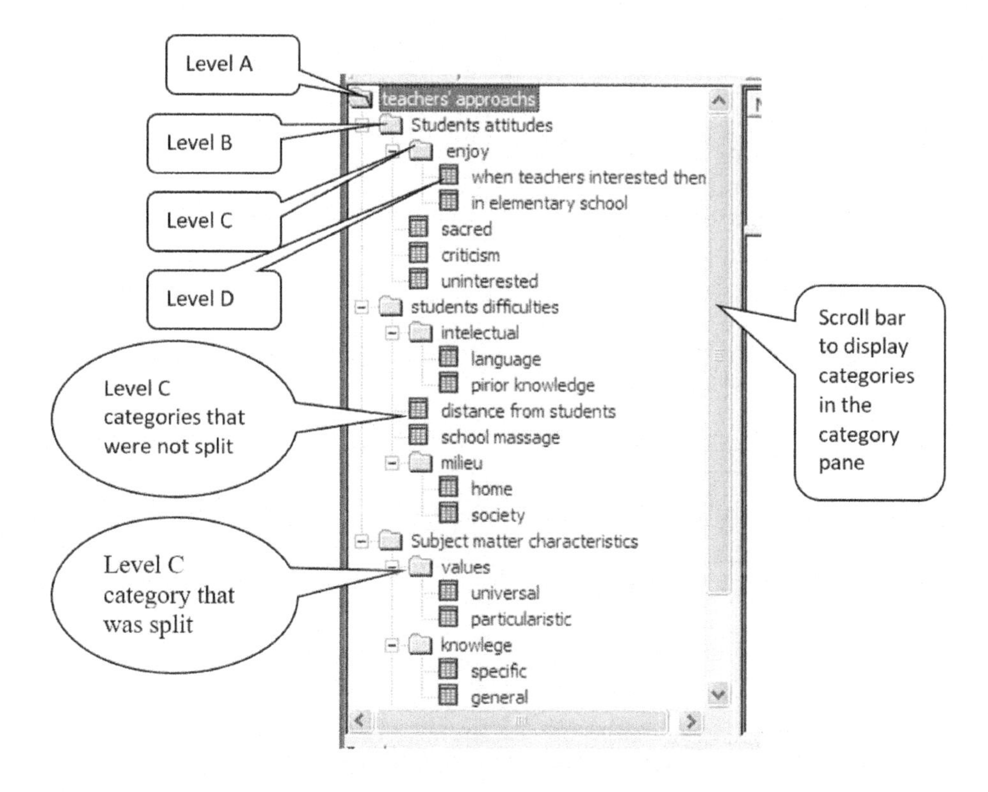

Figure 8P: Category Array with Four Category Levels

Bear in mind: Eventually the category list will be larger than the **Categories pane**. When this happens, use the horizontal and vertical scroll bars to view all categories.

As mentioned earlier, the software can create as many categories as required. For example, we may split certain sub-categories as far as Level E, and leave other categories at Level C, as shown in Figure 8Q.

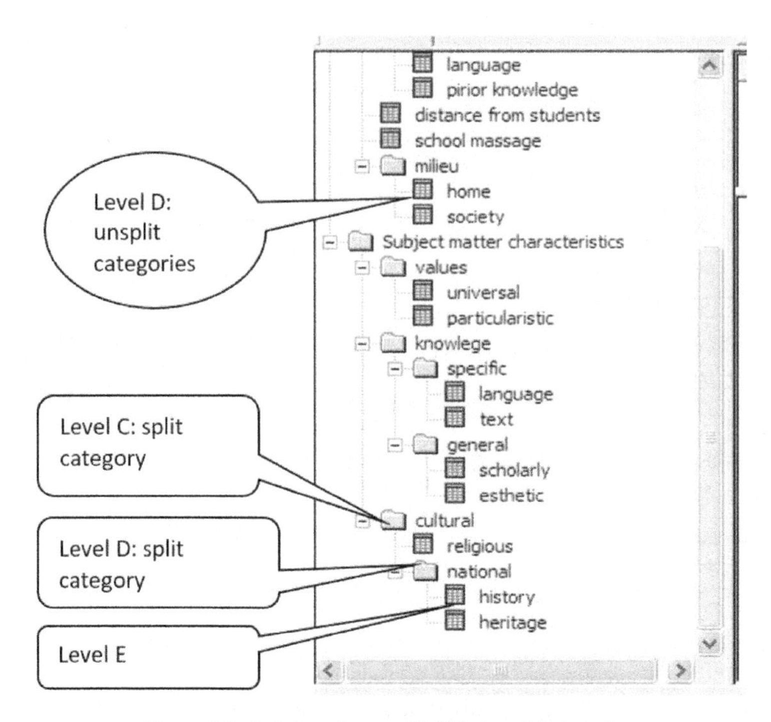

Figure 8Q: Category Array with Fifth-Level Categories

Distinction between Indication Categories and Content Category

As seen in the previous sections, the entire range of the categories is located on different levels that reflect the relationships between categories, from higher levels of categories (main categories) to upper categories, sub-categories, sub-sub-categories and so on. Some categories only specify the properties of the data, and some also contain the data segments. Accordingly, we can distinguish between "indication categories" and "content categories." Content categories are categories which include content segments. Indication categories, however, do not include data segments, but include the categories beneath them. The indication categories characterize the properties and relationships between categories and are located in the upper levels of the categories array. Only the categories that are not split and are not branched into other categories are content categories that contain the data segments. Each content category can be changed to an indication category if you continue and split it into at least two categories. In other words, each indication category was originally a content category before being split. Figure 13R shows an indication category ("curriculum") which is branched into content categories associated with data segments.

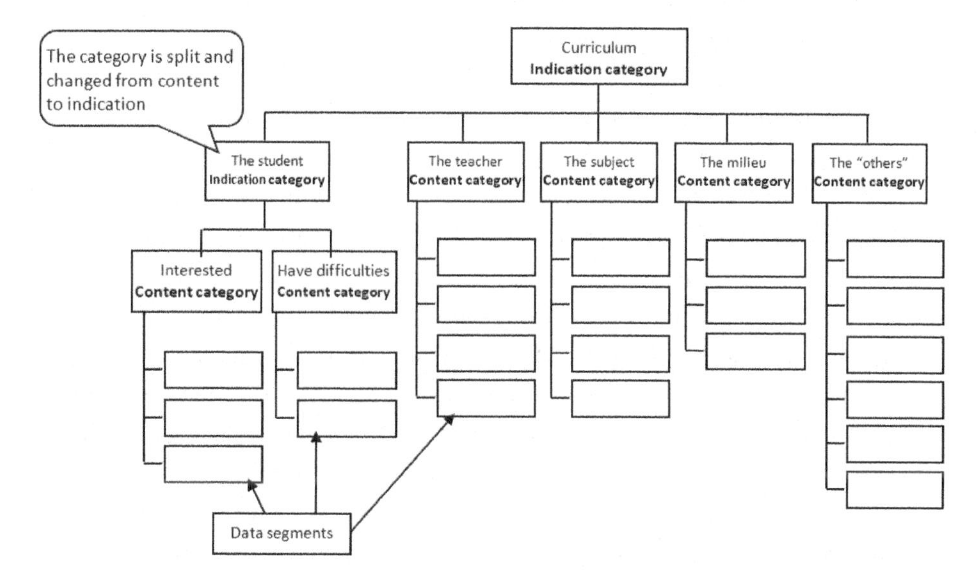

Figure 8R: Indication Categories, Content Categories and Data Segments

When the content category splits into two content categories and becomes an indication category, the content (data segments) moves from the category that split into the newly-created content categories. The category "student," originally a content category (see Figure 8R), is now divided into two categories to become an indication category. The two new categories are now content categories, and the content segments move respectively to the new categories.

Note that while the category "student" split and became an indication category, the same-level categories ("teacher," "curriculum," and so on), which are not divided, remain content categories even though they all remain on the same level. As seen, the content categories, which carry the data segments, are always the categories at the end of the "tree branches."

Inserting the Data Segments into the Categories

Categorization is based on data classification. It is conducted by distinguishing and separating the original data (of the interview, observation, document, etc.) into data segments and re-combining these data segments to a new location and a new order to assign meaning to the data. Categorization consists of two elements: first, the process of dividing the raw data into separate segments, and second, associating the segments into different groups (categories). (Charmaz, 2005; Strauss & Corbin, 1990; Arksey & Knight, 1999). When reorganizing the data according to categorizations, all data segments which are associated with a certain category are attached together. The categorization is carried out by comparing different segments of the data to find similarities, differences and connections between them (Seidel & Kelle, 1995).

The categorization process that characterizes the methodologies based on the predetermined categories is conducted as follows: First, determine the category array. Then, insert the relevant segments to the appropriate content category. In practice, the raw document data to be analyzed should be placed alongside the category array to which the segments will be inserted. Carefully read the data, identify the segments associated with each content category, copy the segment, and move the pieces of relevant data into the appropriate content categories. Figure 8S shows an example of a category array before inserting the data segments.

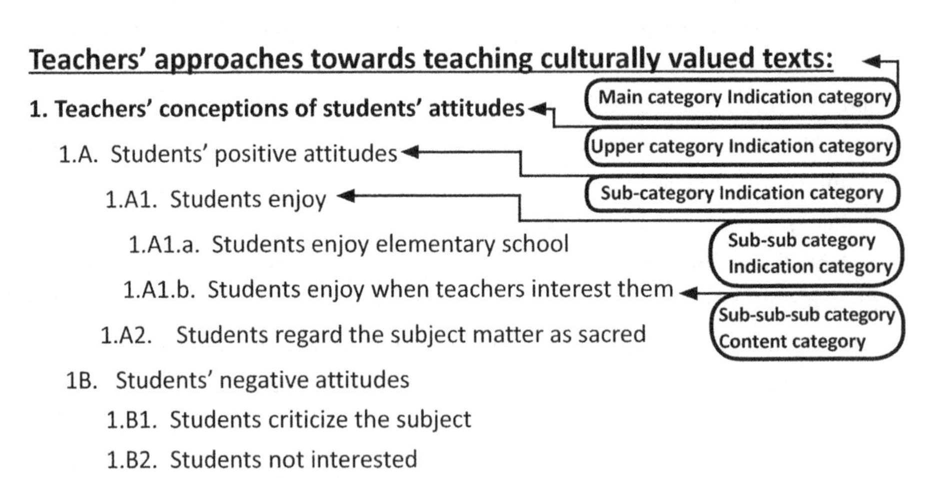

2.A1.a. Students' language difficulties
2.A1.b. Students' difficulties due to deficient prior knowledge
2.A2. The subject matter is distant from the student.
2.B. Students' contextual difficulties
2.B1. Students' difficulties in light of school messages
2.B2. Students' difficulties in light of milieu messages
2.B2.a. Messages from students' homes
2.B2.b. Messages coming from society

3. Teachers' conception of the subject matter characteristics

3.A. The values characteristic of the subject matter
3.A1. The subject matter contains universal values
3.A2. The subject matter contains particularistic values
3.B. The knowledge characteristics of the subject matter
3.B1. The subject matter contains specific knowledge
3.B1.a. The subject matter contains language knowledge
3.B1.b. The subject matter contains special types of texts
3.B2. The subject matter contains general knowledge
3.B2.a. The subject matter contains scholarly knowledge
3.B2.b. The subject matter contains esthetic issues
3.C. The cultural characteristics of the subject matter
3.C1. The subject matter contains religious issues
3.C2. The subject matter contains national issues
3.C2.a. The subject matter contains historical knowledge
3.C2.b. The subject matter contains knowledge about heritage

Figure 8S: The Category Array before Inserting the Data Segments

The process of inserting segments from the raw document data into the category array may be performed by using the basic Copy and Paste operations of the word processor. Immediately after pasting the section, attach the identification marks of the segment (name of the sub-case or any other necessary identifying mark). If the researcher needs to keep the original place of the segment, be sure to write the serial number of the passage (see appendix 1). It is important to remember that the data segments can be inserted only into the content categories. Figure 8T illustrates the placement of data segments in the content categories (the same category array that appears in Figure 8S above)

Teachers' approaches towards teaching culturally valued texts:

1. Teachers' conceptions of students' attitudes

1.A. Students' positive attitudes

1.A1. Students enjoy

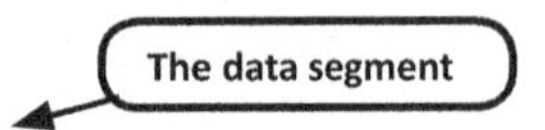

1.A1.a. Students enjoy elementary school

"In the third grade, it's a story, it's enjoyable, it's like a novel. In the third or fourth grade, the kids like Bible... There's excitement in the classroom. When we get into other things... that are not particularly plot-oriented stories, that's when the difficulties begin." (David, 2)

1.A1.b. Students enjoy when teachers interest them

"When you take an awkward text that looks strange, and you analyze it and suddenly they understand something about it, and there are all kinds of ideas, they like it..." (Sharon, 1)

1.A2. Students regard the subject matter as sacred

"If a kid's Bible falls down, you'll often see the child kiss it... They feel it's something more exalted, different... He knows he has a connection with it, even subconsciously." (Jim, 1)

1B. Students' negative attitudes

1.B1. Students criticize the subject

"It might be a disrespectful dismissal of the divine power: `Who is this God? And if He exists, why didn't He save the Jews from the Holocaust?'..." (Rachel, 2)

1.B2. Students not interested

"Not interested, irrelevant, despite what I do, for the most part the text will interest some of the students, for most of the children it is not important... they make sure to shout out their lack of interest; usually: 'What do we have to do with this text?'" (Jim, 2)

Figure 8T: Inserting the Data Segments into the Category Array

Inserting the Data Segment by Narralizer

1. Examine the data text following its order alongside the category array of the analysis document (see explanation above) and look for segments of text that match the categories in the Categories pane.

2. Find a text segment that relates to the existing categories. Highlight the segment in the data document and give the **Copy** command.
3. After copying the data segment, turn to the analysis document. Position the cursor on the category relating to the data segment (in our example in Figure 8U, the category is "distance from students") and **highlight** the category. **Right click** on the selected category (see Figure 13V). Position the cursor on the **Add Text Segment** and **click.**

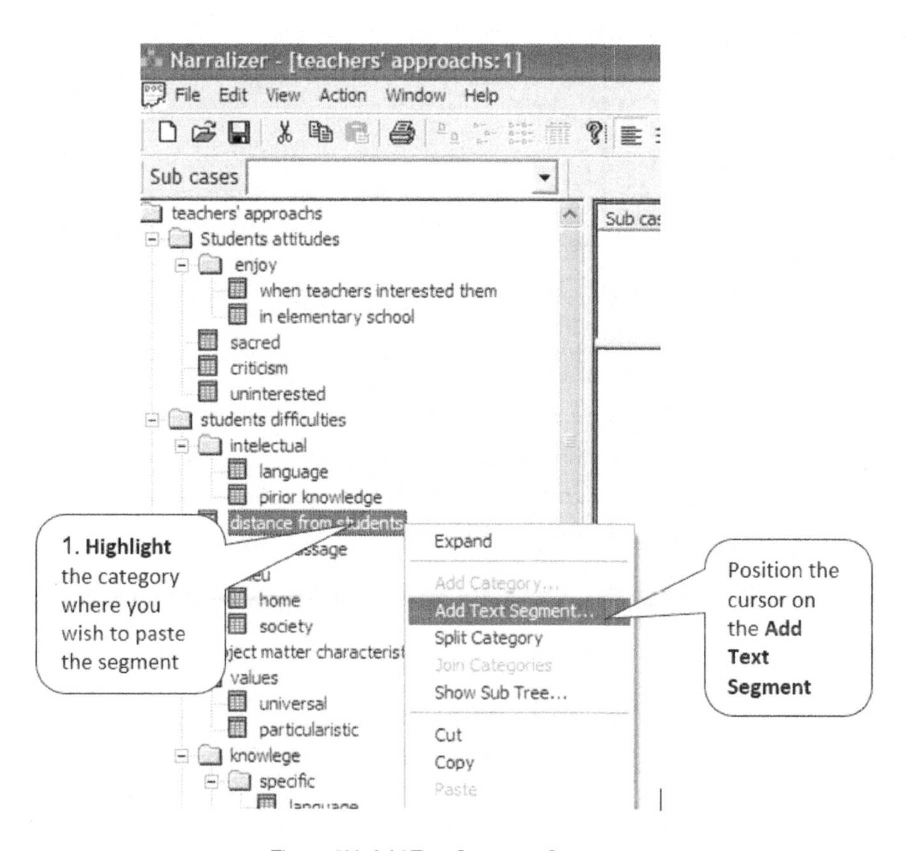

Figure 8U: Add Text Segment Command

4. After clicking the **Add Text Segment** command, a dialog box is opened. The box will automatically display the text copied from the data document. (There is a quicker method for opening the box, using the **Add Text Segment icon** on the Toolbar.) Write the sub-case name in the top pane of the dialog box. If the sub-case name is already assigned, follow instruction #5 below.

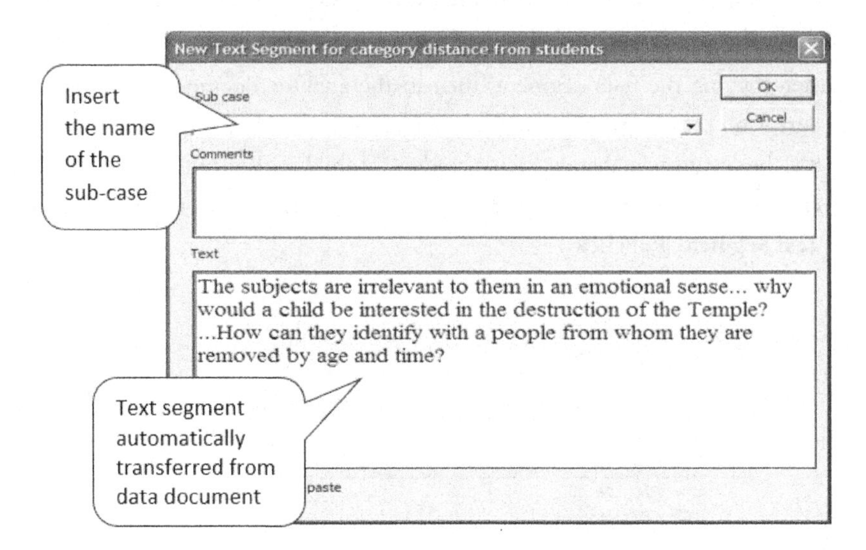

Figure 8V: Add Text Segment dialog box

5. If the name of the particular sub-case already exists as part of the sub-cases list, click the right arrow of the sub-cases pane to display the complete list of sub-cases. Left click on the name of the relevant sub-case and release in order to place the name in the pane. Click **OK** to close the box, and the text is placed in the appropriate location on the analysis document.

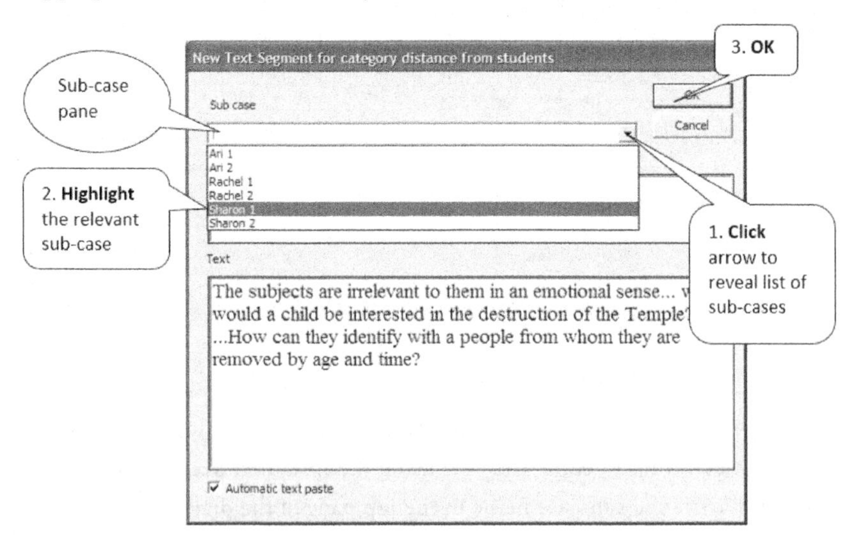

Figure 8W: Choosing Sub-Case Name From the Sub- Cases List

6. After clicking **OK,** the segment is added to the data pool and the name of the sub-case will be shown in the **Content pane** (the pane at the top of the Work Screen) under the Sub-Case column (in our example in Figure 8X: "Sharon 1"). A green hatched square will appear next to the name of the sub-case. This is the text segment icon. The initial words of the data segment can be seen in the text column in the **Content pane**, allowing us to identify the segment.

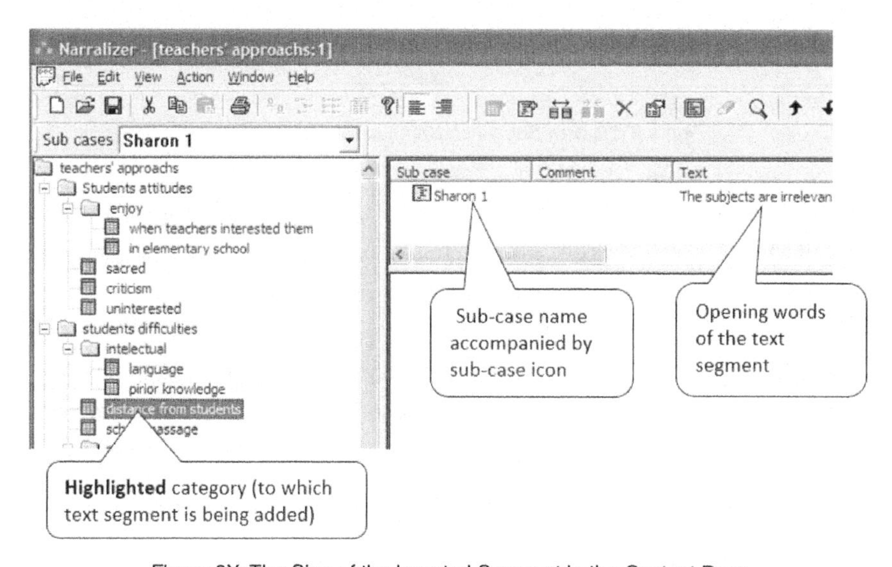

Figure 8X: The Sign of the Inserted Segment in the Content Pane

7. Use the same method to assign additional text segments to categories and insert them in their correct location, one after the other. A new Text Segment box reopens each time, showing the data segment copied from the original data document. The name of the sub-case placed in the box automatically appears in the sub-cases pane. This makes it much easier to work, since researchers generally work continuously on the data for the same source document (sub-case). However, it is important to remember that when changing to data from another source document (sub-case), we must change the sub-case name by displaying the list of sub-case names and clicking the relevant name. After clicking OK, the icon indicating a new segment will appear in the **Content pane**, displaying the new segment name (in Figure 8Y, "Rachel 1").

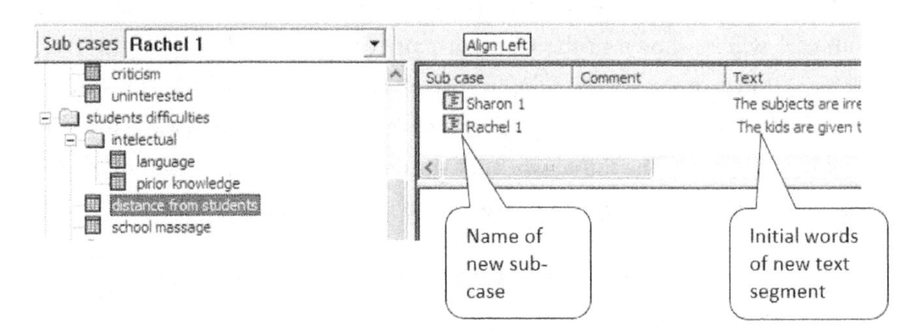

Figure 8Y: A New Sub-Case Name in the Content Pane

Note that a text segment (or part of one) can be assigned to several categories at once, according to the researcher's decisions. In this case, the researcher must perform the same **Add text segment** action for each category individually.

Bear in mind: All new Text Segment boxes have a Comments pane to record comments regarding the data text: memos, segment abstract, and researcher's notes.

CHAPTER 9

EXPANDING THE MAPPING CATEGORY ARRAY

Chapters 8 presented the process of creating a mapping category array and inserting data into the categories. At the end of chapter 8, the data segments were inserted in their respective categories. This chapter will address the expansion that researchers can make in the category array during and after the process of entering the data segments. These processes involve changes in the category array, which can sometimes be quite significant.

Changing the Order of Categories (in the Horizontal Dimension)

In mapping analysis documents, the categories are ordered on the vertical dimension and the horizontal dimension (in reference to the category tree). On the vertical dimension, the categories are ordered on category levels, starting with the main category, passing to the upper categories and continuing to the sub-categories, sub-sub-categories and so on. During the analysis process, attention was paid to the vertical order of the categories. Less attention, if any, was paid to the horizontal dimension category order. For example, the upper categories located on a horizontal line under the main category are not necessarily arranged from left to right (or right to left) in a purposeful and significant order, and in many cases the order is completely random (in Figure 9A, the three upper categories are arranged randomly right to left: "Cultural messages," "Knowledge," and "Values"). The same incidental arrangement on the horizontal dimension level may also be in the sub-categories below the upper-categories, as well as in the subsequent category levels. After the category array is completed (see Chapter 8), researchers may seek to change the horizontal arrangement of the categories.

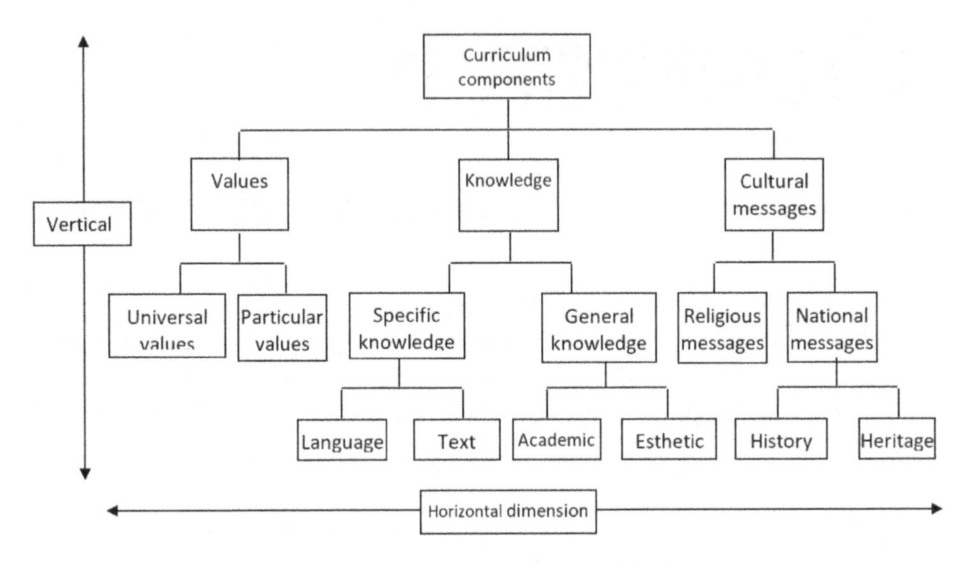

Figure 9A: Vertical and Horizontal Dimensions of the Category array

Figure 9B illustrates the category tree in a word processor arrangement, so that the categories are displayed one under the other and the data associated with them (shown symbolically by a rectangle with the word "data") is arranged under each content category. The category order is represented on a vertical dimension by numbering, and the different level of the categories is noted by the place of each category level. In the example in Figure 9B, the second upper category "Knowledge" is moved above to the top of the list. This transfer is done by Cutting & Pasting the upper category, including all of its sub-categories and all the attached data segments (in a "real" study, there are many pages). With the change of the category order, the numbering should change accordingly.

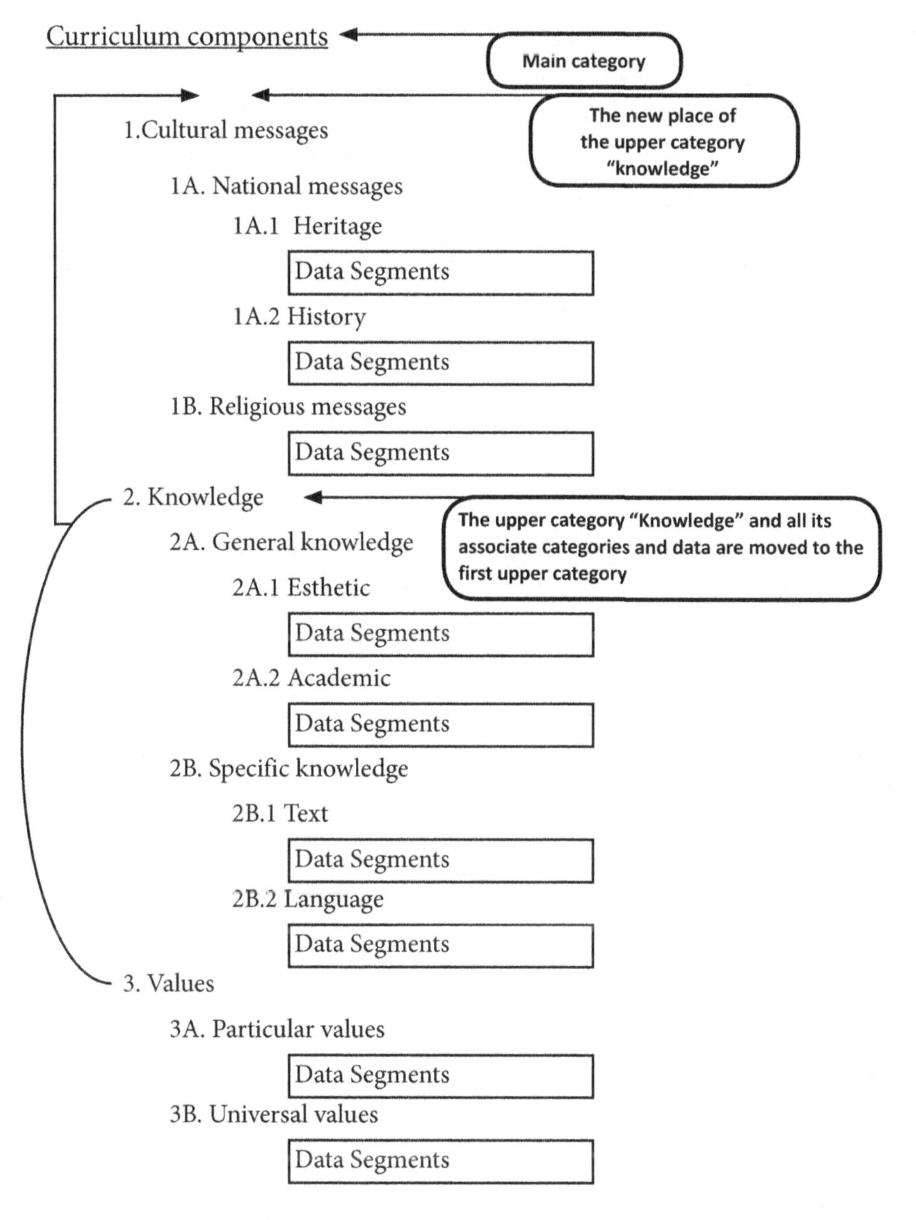

Figure 9B: Changing the Category Order in the Horizontal Dimension

When moving an indication category to a new location, make certain that this action will include all categories and data segments attached to the category. In the example in Figure 9B, when moving the indication category "Knowledge" (an upper category), we transfer all the categories (either indication or content types) with all the data segments attached to this category. Thus, changing the horizontal order does not involve a change in the nature of the categories, their characteristics or the relationships between categories. It has only changed the significance of each category within the category array.

Change the Order of the Categories (in the Horizontal Dimension) Using Narralizer

The categories are arranged in the Category pane in the same order as they were entered. At any time, the horizontal arrangement of categories in the category pane (namely, the order in the category pane) can be changed in a **Narralizer** document.

> **Bear in mind**: "horizontal" is a phrase borrowed from the category tree, and requires a degree of abstraction. By "horizontal," we mean all the categories that are at the same level under the same category.

The order can be changed, as described in Figure 9C:

1. Point the cursor at the category in the **Categories pane** (indication or content category) that is to be relocated.
2. Click **Move Up/Move Down** icons (or the Action menu /right click the mouse) to move the category, until the categories are listed in the right order.

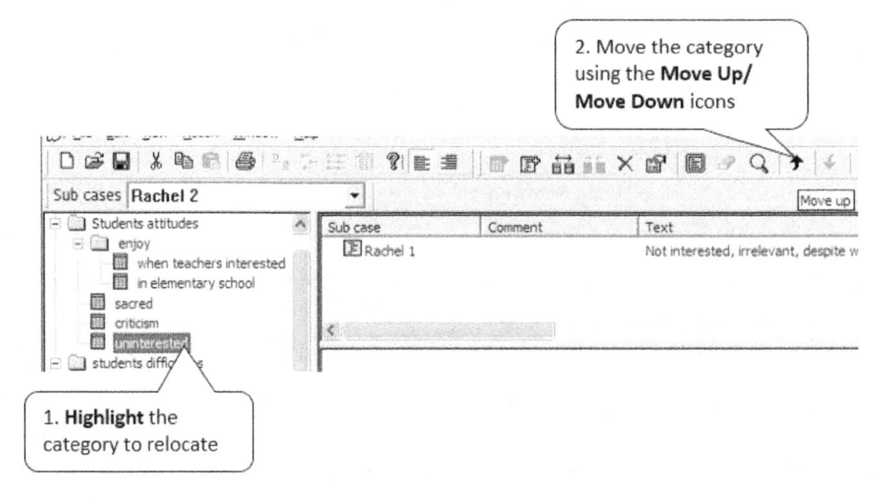

Figure 9C: Change the Order of the Categories

Bear in mind: that when we move an indication category (a category with a folder icon) in the **Narralizer**'s category pane, all the categories (indication and/or content) attached to it also move automatically.

The order in which the data segments are listed in the **Content pane** can be rearranged in the same way.

Moving Categories from Place to Place (in the Vertical Dimension)

Sometimes researchers decide to move a category from under one particular category to go under a different category. This is a vertical change- - moving to another place in the category array that changes the nature and characteristics of the category and its relations with other categories. This operation is illustrated in Figures 9D – 9E.

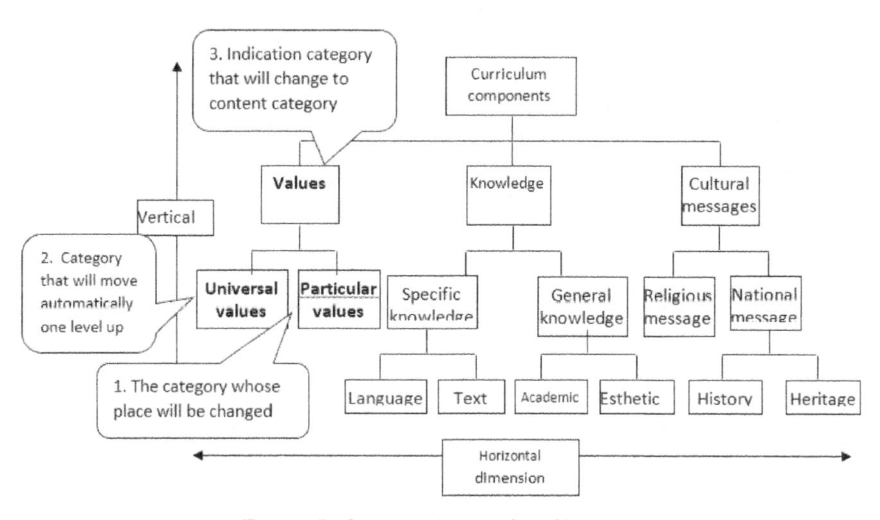

Figure 9D: Category Array before Changing

Technically, moving a category on a word processor document analysis is relatively simple. Researchers identify the category that they wish to move, use "Cut & Paste" to move it together with its attached categories and data segments, and insert it in the new place. Researchers should be aware of the significance of the change and the new order it has created. The change does not apply only to the category that was moved, but also to the categories remaining in their original place. Moving the category may change the characteristics of the categories that were above and below it (as each category is characterized by the categories surrounding it). Following this operation, the name of the category should change accordingly.

In Figure 9E, we can see that the category "particular values" has changed its place under the category "cultural messages," which changed its name to "particular cultural messages." As a result of the changes, the category "universal values" remains without a parallel content category, and thus became a content category and moved up one level.

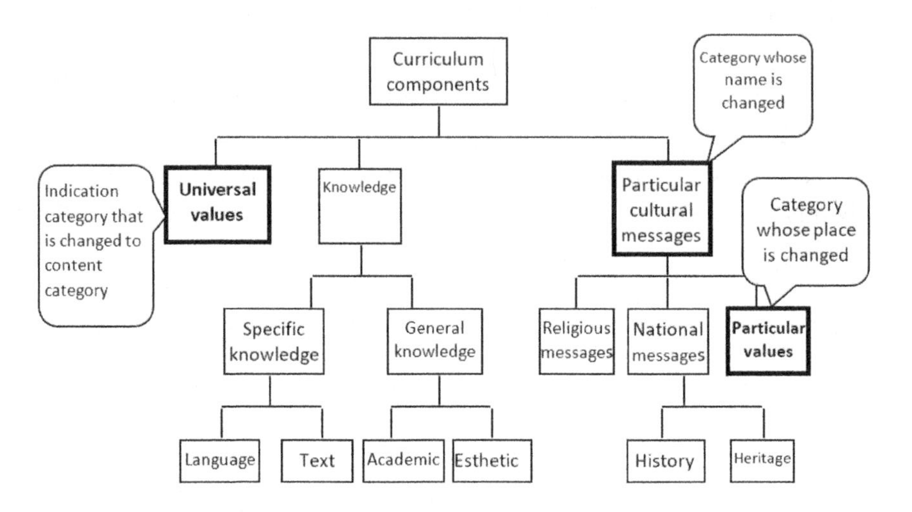

Figure 9E: The Category Array after Moving Categories from Place to Place

Moving Categories from Place to Place Using the Narralizer

1. In the example in Figure 9F, we will move the content category "criticism." To do this, **highlight the indication category one level above it** (in this case "student attitudes"), which causes the category one level below to be displayed in the **Content pane**.
2. **Highlight** the category in the **Content pane** (upper right hand pane) to be moved.

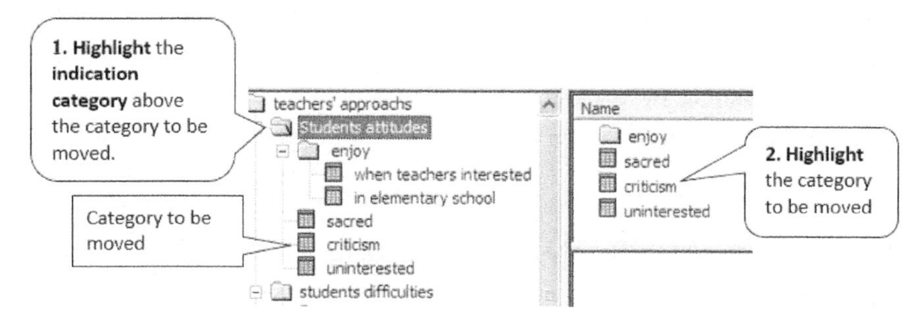

Figure 9F: Highlighting the Category to be Moved

3. **Drag** the category from the **Content pane** to the **Categories pane** using the left mouse button, and place it in the **indication category under which** it is to be relocated. Alternatively, use **Cut & Paste**.

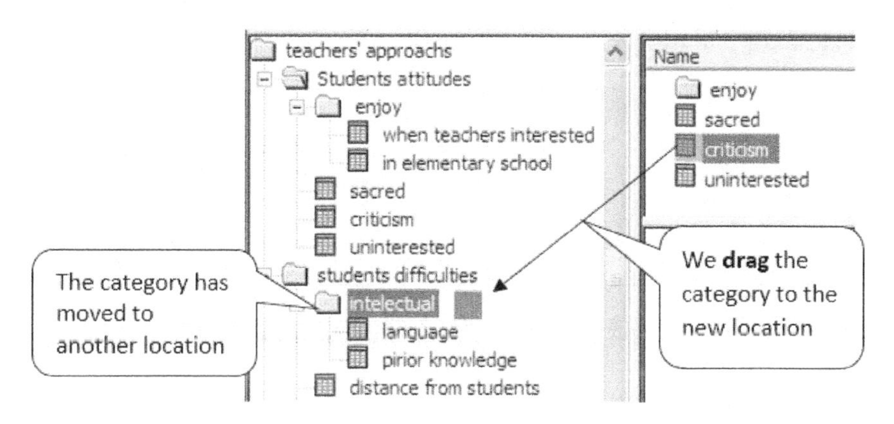

Figure 9G: Dragging the Category to its New Place

<u>**Bear in mind**</u>: Be sure to drag the category to the indication category above the target place. If inaccurate, the category may be transferred to an unfavorable place, even without the researcher being aware.

To relocate a category and also keep it in its original location, use the above steps. However, instead of dragging or cutting, use **Copy and Paste**. When two categories have the same content, it is recommended to give them different names reflecting their different locations.

Bear in mind: The software does not ever allow the existence of only one sub-category under a higher category, and automatically corrects the situation. Splitting, by its nature, must divide the category into at least two sub-categories. Otherwise there is no significance to the fragmentation.

Creating New Indication Categories of Existing Categories

The process of constructing a category array also includes insights that lead to creating new indication categories from existing categories located below them. During the process of constructing the category array, the researcher may identify certain characteristics that are shared by several categories all located under the same category, and decide to give expression to the common characteristic. Figure 9H illustrates four categories under an indication category.

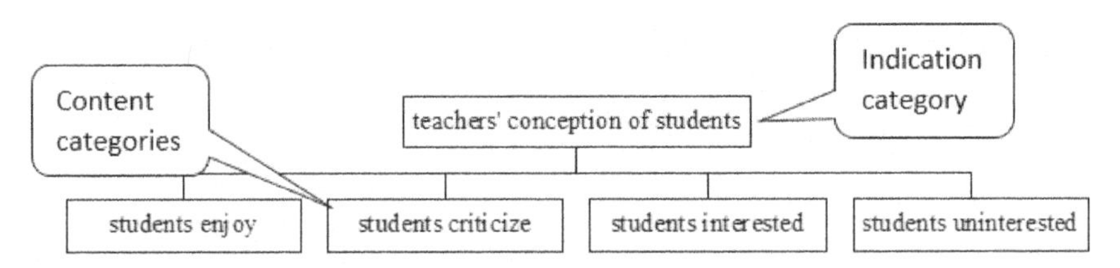

Figure 9H: Indication Category with Four Categories Below

The researchers noticed that the four categories can be divided into two groups under new categories. In practical terms, this is expressed by adding two intermediate indication categories under the top category, each of which contains two categories, as shown in Figure 9-I.

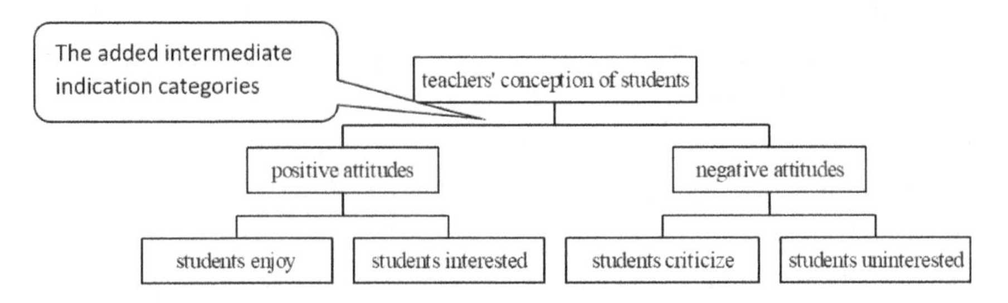

Figure 9-I: Creating Intermediate Indication Categories between Existing Categories

Creating New Indication Categories for Existing Categories using Narralizer

1. The researcher concludes that existing categories shown in **categories pane** share common properties

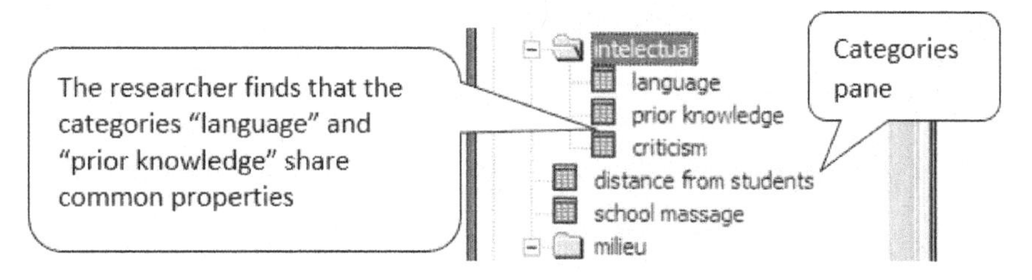

Figure 9J: Identifying Common Properties in Existing Categories

2. **Highlight** the indication category in the **Categories pane** containing the categories to be placed in a new indication category. The categories are now displayed in the **Content pane**.
3. In the **Content pane**, **highlight** the categories to be grouped under the new indication category. Highlight two or more categories using the **Ctrl key**.

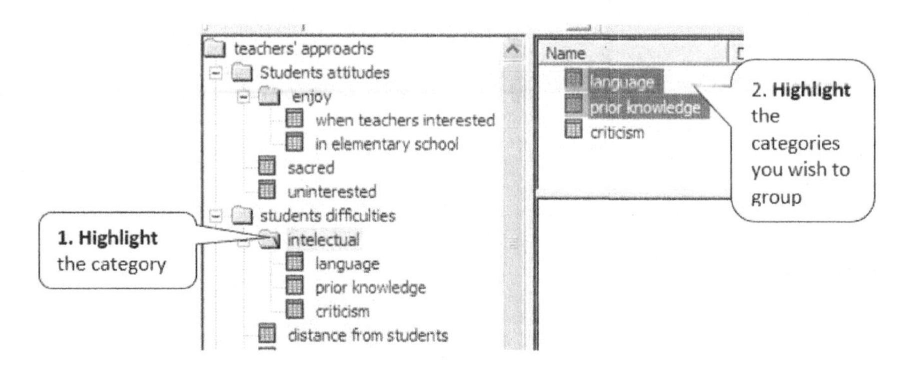

Figure 9K: Highlighting the Categories to be Grouped

4. After highlighting the categories in the content pane, right click the mouse to select **Join Category**/click the **Join Category** Command Toolbar icon/select **Join Category** in the Action menu.

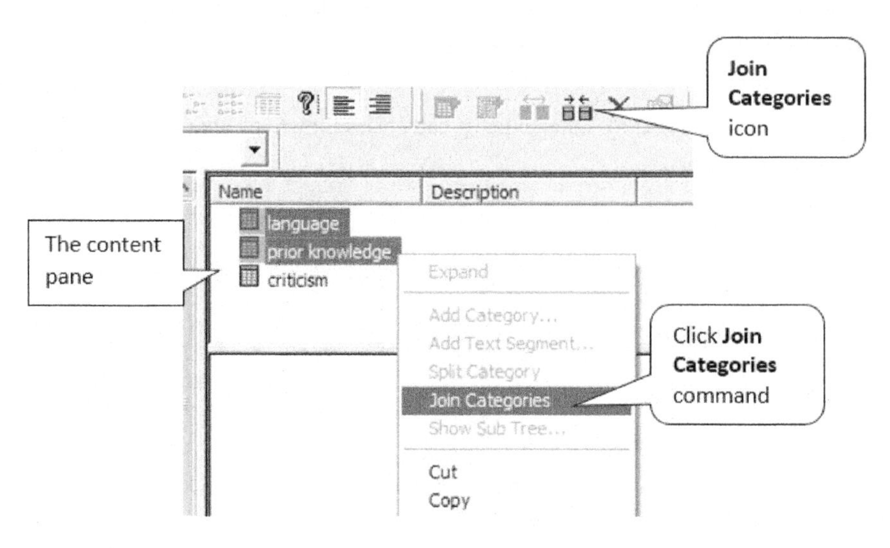

Figure 9L: Join Categories Command

5. **Click** the **Join Categories** command to open the **Join Categories box**. In the upper pane of the box ("Name"), **type the name** of the new indication category. See Figure 9M.
6. Click **OK** to complete the procedure.

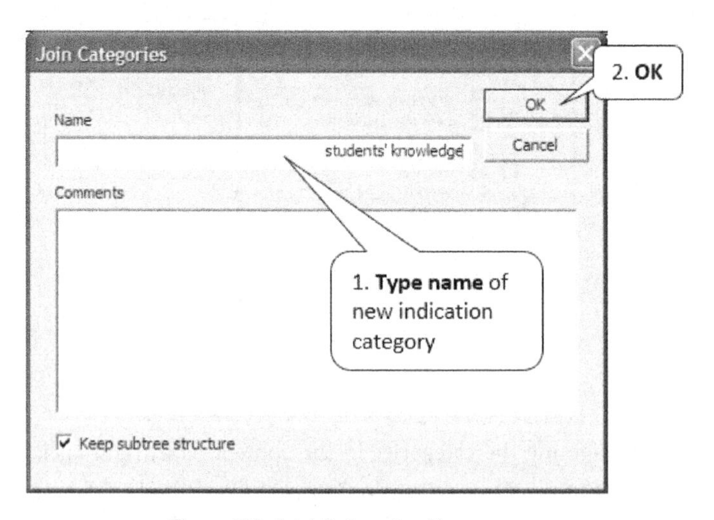

Figure 9M: Joint Categories Box

The highlighted categories are now grouped under a new indication category in the **Categories pane** (see Figure 9N). The content categories are under the new category and may be revealed by clicking the + square at the right of the prime category or by clicking the Expand command (which changes the + sign in the square to a – sign).

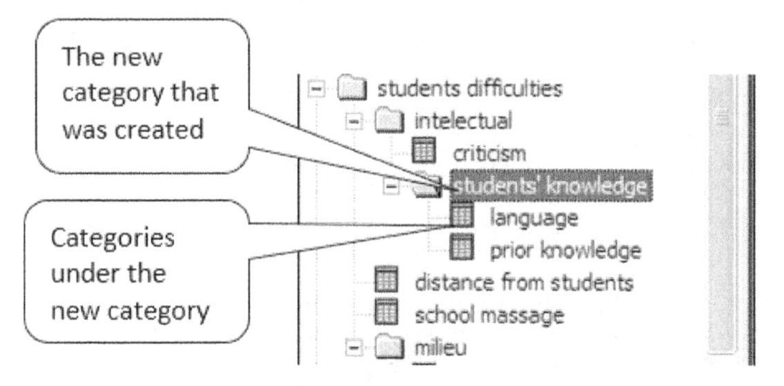

Figure 9N: The Changed Categories Array

Join Several Categories to One Category

Sometimes researchers decide to join two or more categories into one. This is a situation where researchers conclude that there is no justification for two or more categories to exist separately, whether because they each contain too small a number of data segments, or because the differences between them do not seem significant, or because one of them is empty of data segments. In display 9-O, the researcher has decided to join two content categories which seem identical.

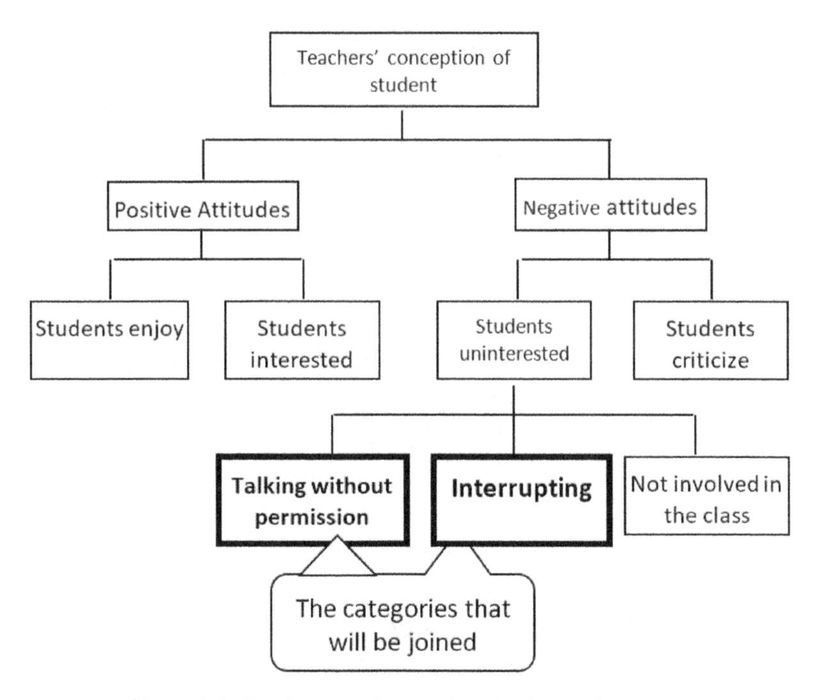

Figure 9-O: The Category Array before the Joined Categories

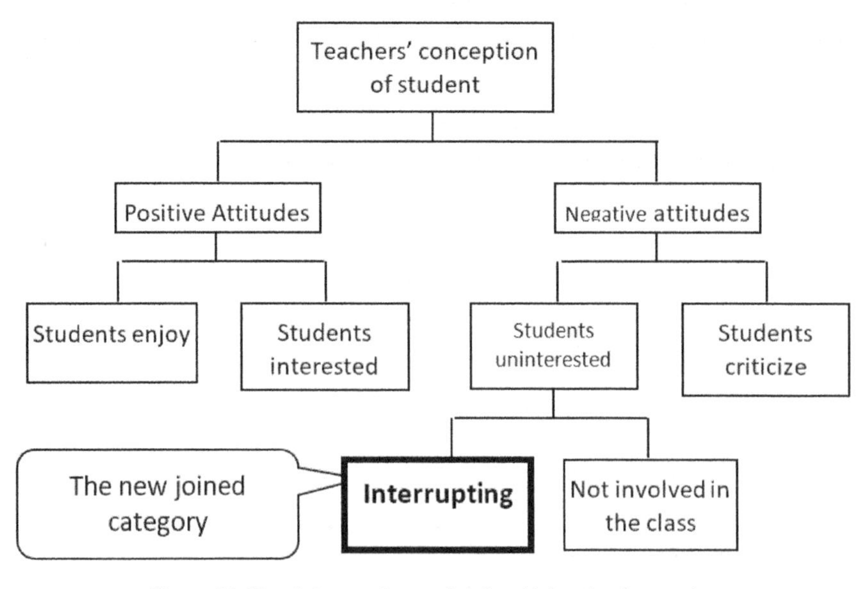

Figure 9P: The Category Array after Combining the Categories

In Figure 9P, we can see the category array after joining data segments of two categories under one category. In this case, the joined category received the name of one of the original categories ("interrupting"). The example above deals with joined content categories. The case of uniting indication categories is very similar, and the categories under the two separated similar categories will combine under the new joined category.

Join several categories to one category using the Narralizer

1. In the **Categories pane, highlight** the indication category, which contains the content categories to be combined. The content categories are displayed in the **Content pane** (see Figure 9Q).
2. In the **Content pane, highlight** the content categories to be combined (see Figure 9Q)

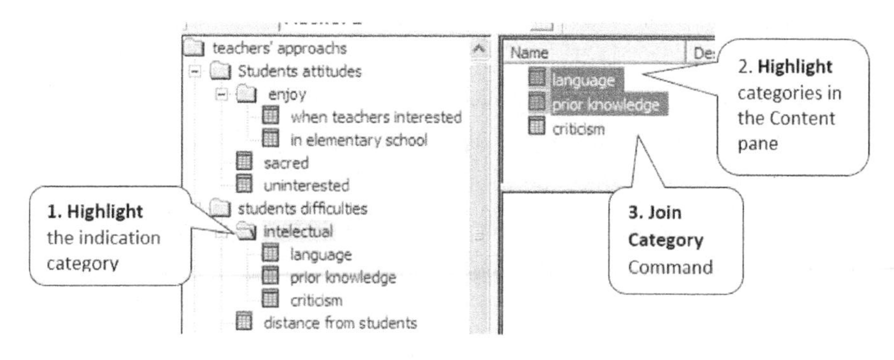

Figure 9Q: The Join Category Command

3. Clicking the **Join Categories** icon in the Command Ruler/right clicking the mouse/using the Action menu will open the **Join Categories box** (same procedure as in the previous section). At the bottom of the box is a small pane, labeled **keep sub-tree structure**, which contains a ✓. **Click on the label to cancel the ✓.** (See Figure 9R)

   Bear in Mind: It is important to cancel the ✓. If it is not cancelled, the content categories will not be merged into one content category, but stay disconnected as separate content categories under a joint indication category. (There may be times when we want this to happen – see previous section for a procedure explanation.)

4. Type the name of the new content category in the upper pane labeled **Name.**

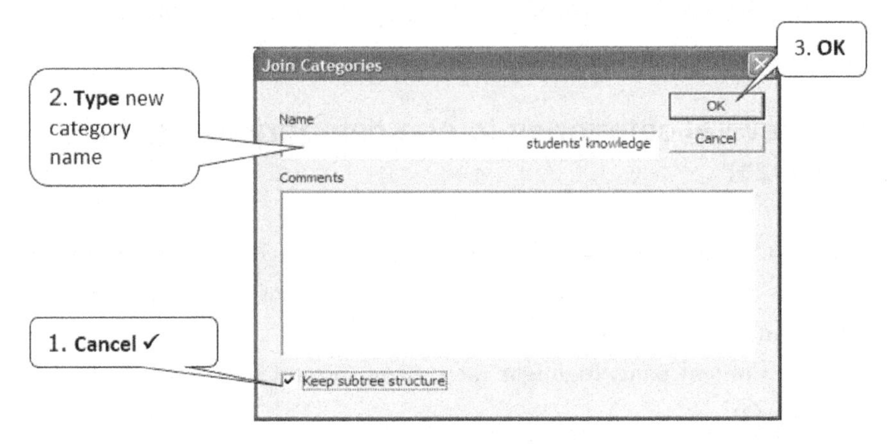

Figure 9R: The Join Categories Box

5. Click **OK** to assign the new content category to the Category array of the **Categories pane** which replaces the separated content categories. See two illustrations in Figure 9S.

I. Categories before merger (three categories beneath the main category "intellectual")

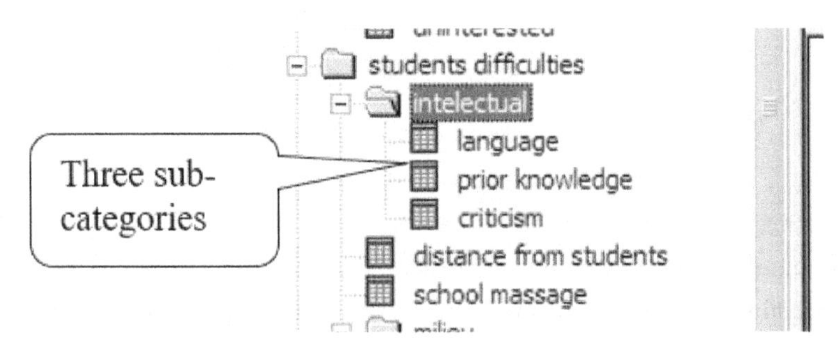

II. Categories after merger (two categories merged; now there are two categories beneath the main category "intellectual")

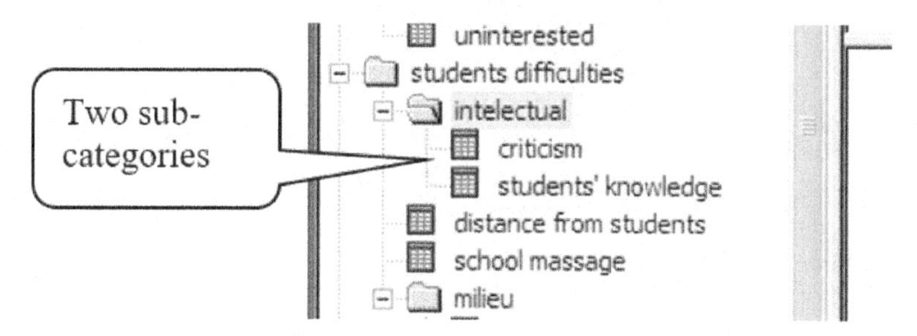

Figure 9S: The Category Array Before and After the Join Category Command

Copy/Paste Data Segments within the Analysis Document

Sometimes, while re-reading the data segments that are already attached to the categories, researchers conclude that a particular segment, as a whole or in part, is not appropriate for its category, or is also inserted in another category. Researchers may cut or copy these sections to transfer them to another place in the same category array. This operation is, of course, performed by using the Cut & Paste command in a word processor. It is important to move the piece with its identification marks (the name of the sub-case) to its new place. Otherwise the data segment will lose its research value.

Copy/Paste Data Segments within the Analysis Document Using Narralizer

1. When reviewing the data displayed in the **Display pane,** the researcher may decide that a particular data segment or part of it relates to another category as well. **Highlight** the segment (or part of it) in the **Display pane** and **Copy** (see Figure 9T).
2. With the cursor, **highlight** the category in the **Categories pane** to which the segment or part of it is to be assigned.
3. Give the **Add Text Segment** command

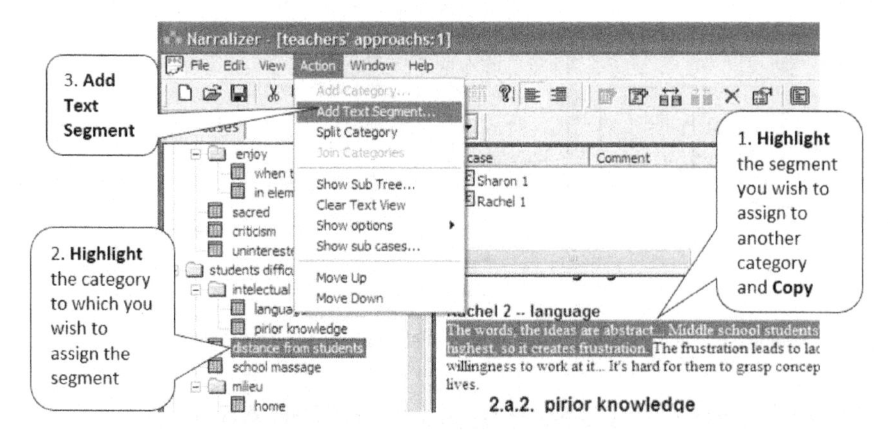

Figure 9T: Marking the Data Segments to be Copy/Pasted

4. A **New Text Segment** box opens containing the highlighted text. Now, verify that the **name of the Sub-case** shown at the top of the segment highlighted in the **Display pane** appears in the right place in the box window. To do so, use the arrow on the right side of the pane to locate the name of the Sub-case (see Figure 9U).

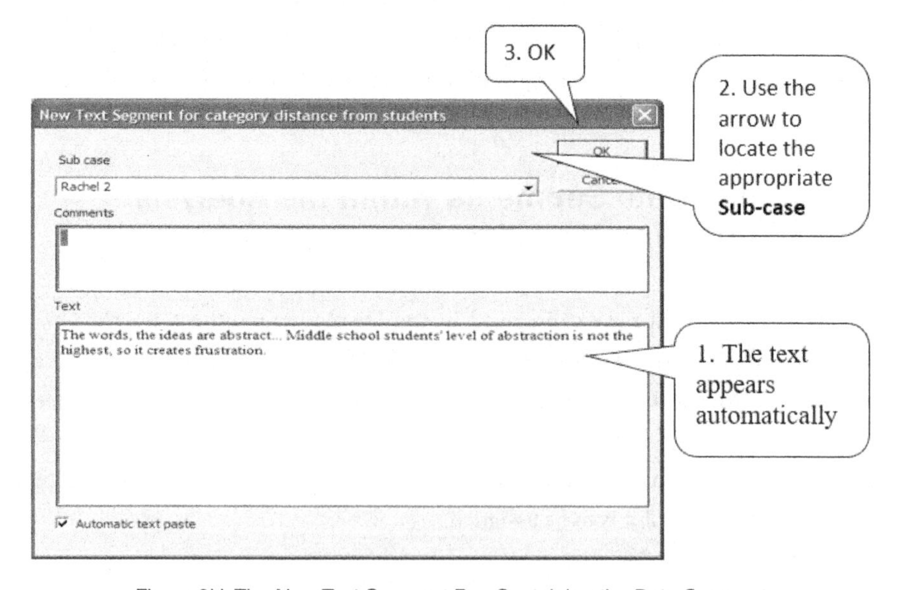

Figure 9U: The New Text Segment Box Containing the Data Segment

5. Click **OK** to add the segment to its new location.

Bear in mind that the change is not shown in the **Display pane** immediately. It will appear the next time you use the **Show sub-tree** command.

Search and Count in the Data Text

Throughout the analysis process, researchers frequently need to search for a specific word or expression. Using the word processor, researchers utilize the command "find." However, in many cases researchers are not satisfied with just searching, but also want to get the numerical dimension, i.e., the number of times the phrase appears in the document. A quantitative search operation may be relevant to all methodologies, but specifically required in the structured-focused methodology. "Word™" software allows us to search and obtain a quantitative summary. But the task becomes more complex, yet possible, when researchers wish to obtain a quantitative picture only in part of the document regarding certain categories or even one category. Moreover, the data document includes both the identification marks of each segment and the name of each category. Therefore, to accurately count the words in the pure data document, we must ensure that these external components will not be included in the count.

Search and Count the Data Text Using Narralizer

The **Narralizer** can conduct complex search functions and search for up to three expressions at a time. The researcher can search via the **Categories pane** or the **Display pane** when the cursor is in one of these panes. To search using the **Display pane**, first display the data to be searched. Thus, before searching the **Display pane**, click **Show sub-tree.**

1. Point the cursor at the **main category** to highlight it. Right click or use the File menu or icon on the icon toolbar to highlight the **Show sub-tree** command.

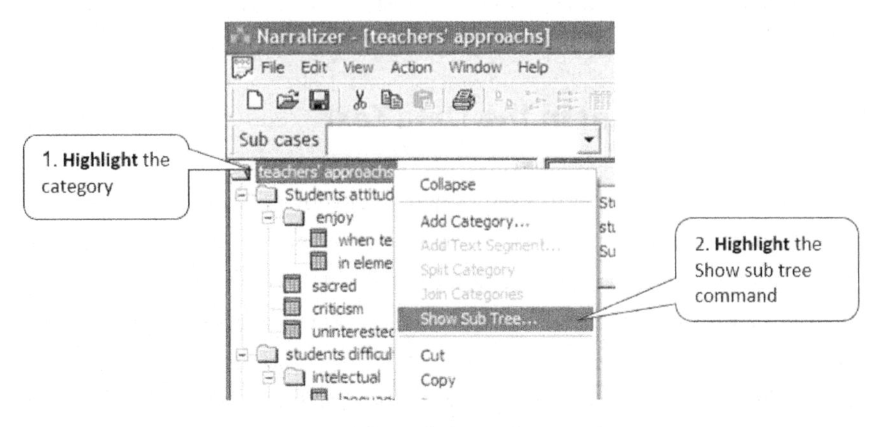

Figure 9V: Show Sub-tree Command

2. Click on the icon or on the **Show Sub-tree** command to open the **Select Sub-cases** box. A list of all sub-cases relating to the analysis document is displayed on the left, and above it, a list of all the sub-case groups in the analysis document. To view the complete analysis array for all categories, click **Select all** (Figure 9W).

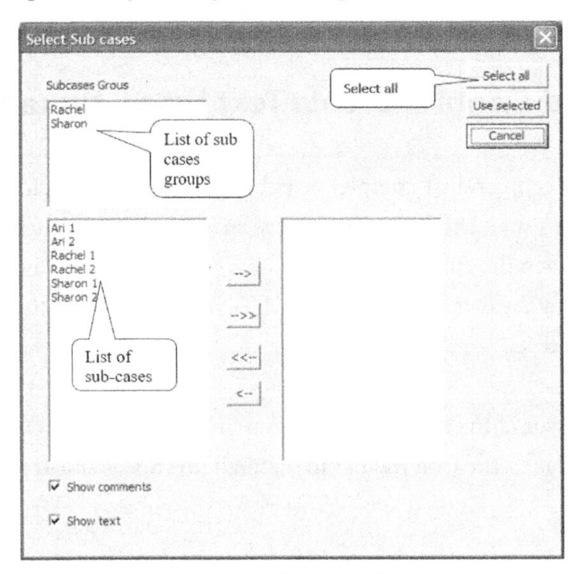

9W: Select Sub-cases Box

3. Click **Select all** to display all of the text segments according to their categories (as shown in Figure 9W).

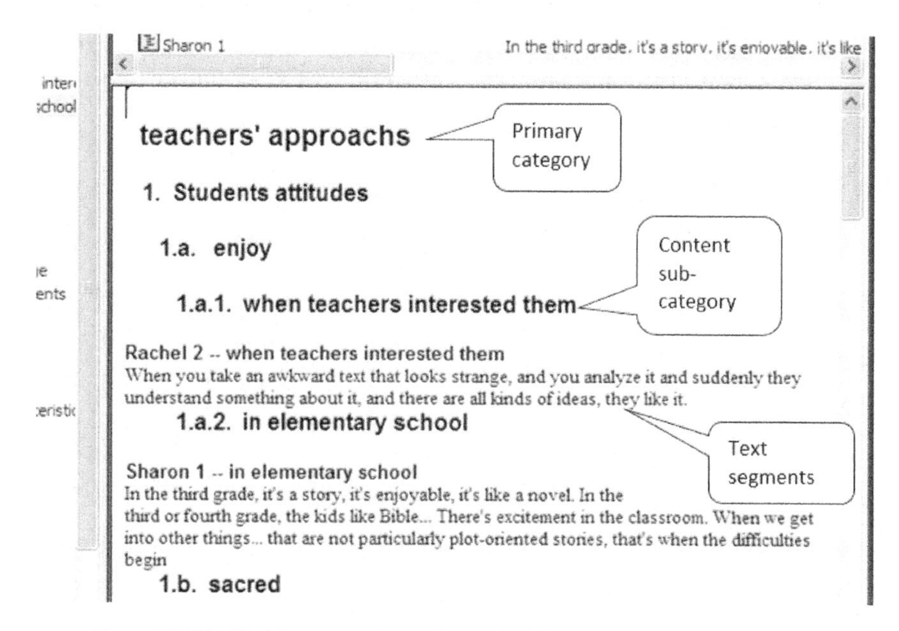

Figure 9W: The Text Segments According to their Categories in the Display Pane

4. To display the data array for just one or more sub-case, open the **Select Sub-Cases Box** as described above. In the left hand pane in the box, **highlight** the sub-case using the mouse.

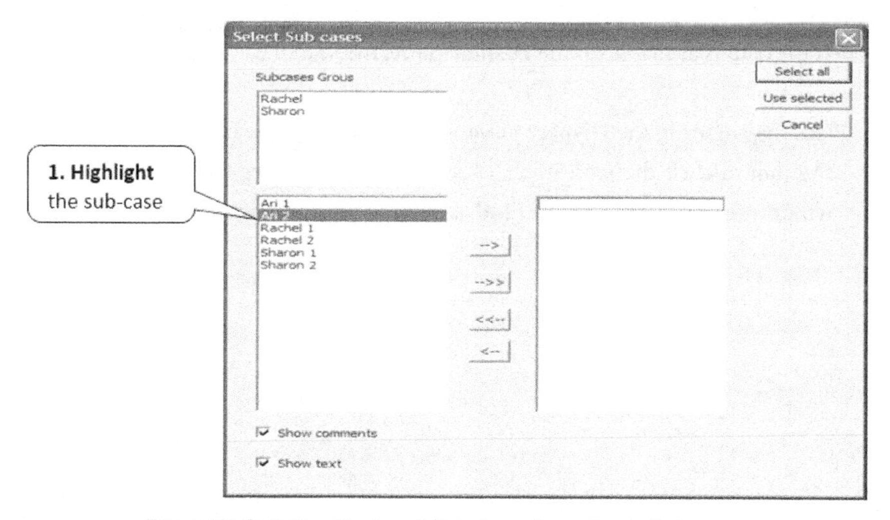

Figure 9X: Selective Display of Data Array According to Sub-cases

5. **Click** the top arrow in the middle of the box to transfer the name of the **sub-case** (one or more) to the right hand pane using the top **arrow** in the middle of the box. **Click** on the **Use selected command** to display all data segments of the sub-case highlighted in the **Display pane**.

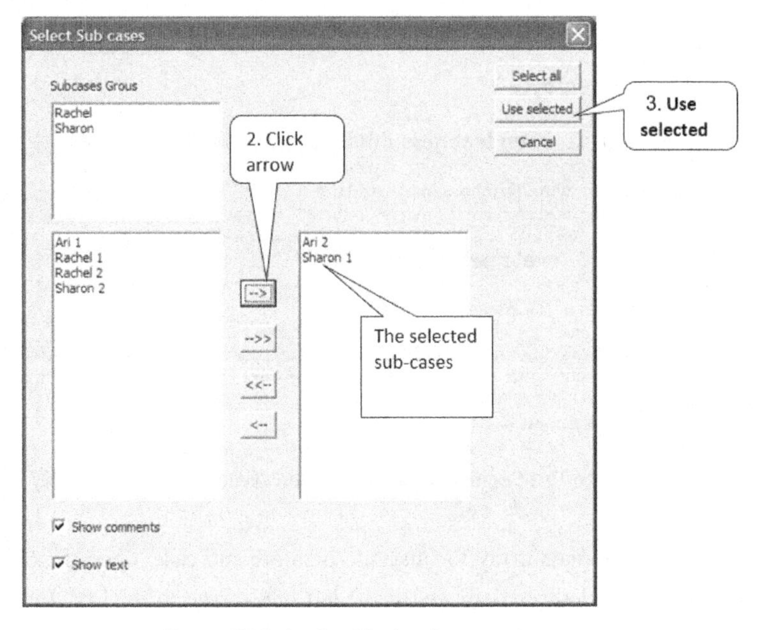

Figure 9Y: Selective Display Command

When the data is available on the **Display pane,** the search can begin.

1. Place the cursor in the **Display pane** and click **Find** in the **File** menu/right click the mouse/click the icon in the Activities toolbar/enter **Ctrl + F**. When the window opens, **highlight** the **Find** action as shown in Figure 9Z.

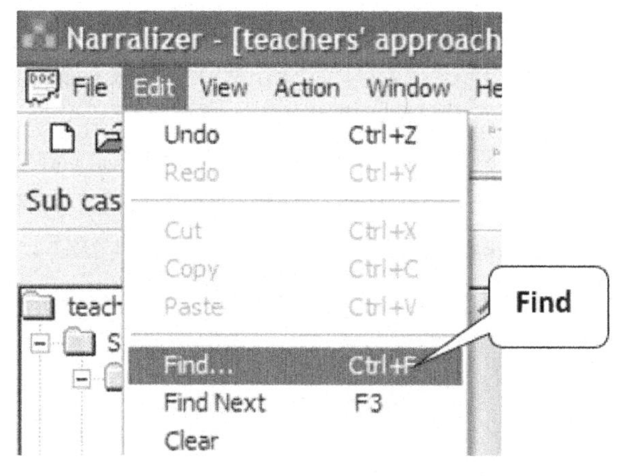

Figure 9Z: The Find Command

2. After clicking Find, the **Find box** appears. Now type the word or expression (in Figure 9-1: "kids") you wish to find through the Find command. You can search for one word or a multi-word expression. Words and expressions from the text may be copied and pasted in the box by using the **Copy and Paste** function.

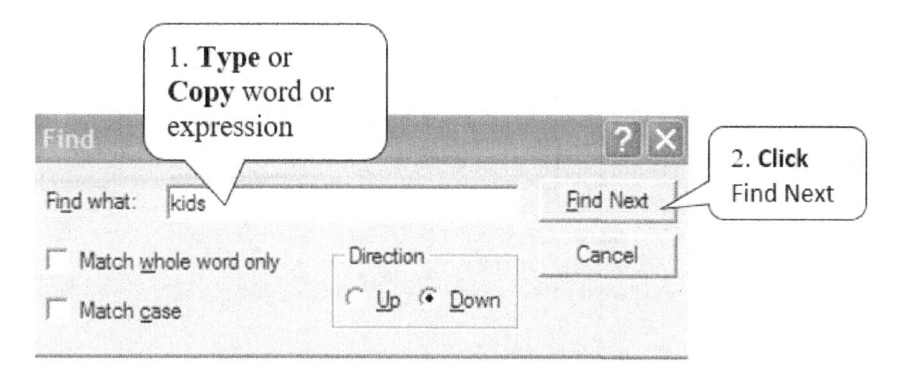

Figure 9-1: Write or Copy the Word/Expression to Be Found

3. Click **Find Next** and the requested word will be highlighted in color in every place it appears throughout the document. You may use **Find Next** to identify every occurrence of the word or expression you seek in the document.

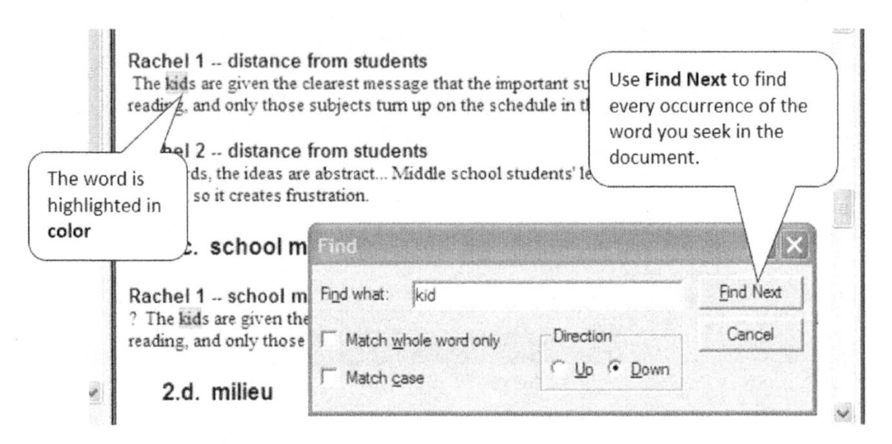

Figure 9-2: Every Occurrence of the Word You Seek is Highlighted in Color

When using **Find Next**, the number of occurrences of the word or expression in the document is displayed at the bottom left of the screen. In Figure 9-3, the word "kid" occurred five times.

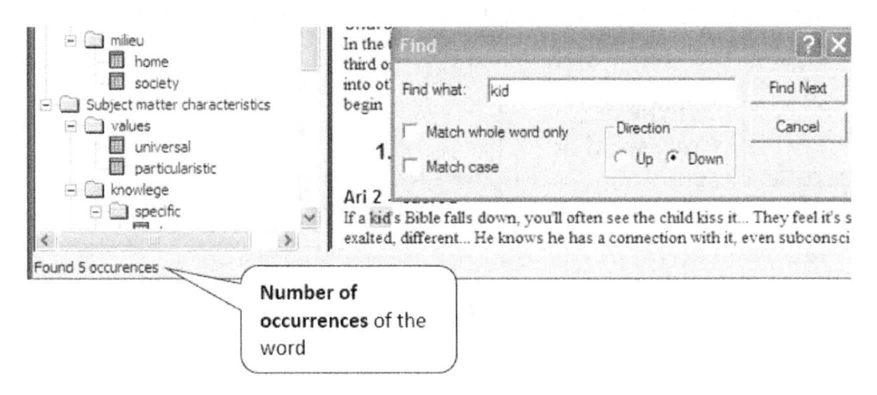

Figure 9-3: Number of Occurrences of the Word

Remember that each time you search, the number of occurrences of the last word or expression appears on the bottom left of the screen.

4. Continue searching for other words and expressions, up to a maximum of three words or expressions. Each desired word or expression will be **highlighted in a different color**. Thus, three different words or expressions will appear on the screen in three different colors for easy detection. In Figure 9-4, the second

search marked the word "student" and the third search marked the word "child" (the first search marked the word "kid").

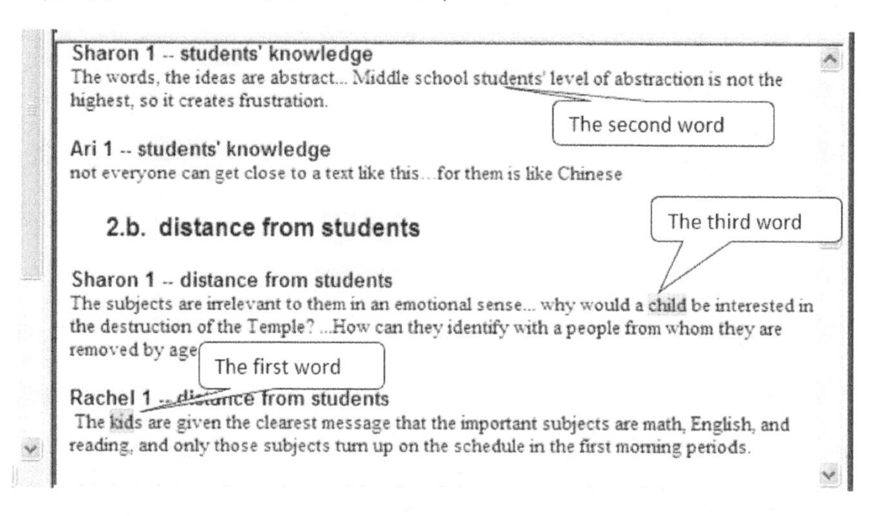

Figure 9-4: The Three Different Words Appear on the Display Pane

<u>Bear in mind</u>: **Find** not only informs the researcher how many times a word or expression appears in a document, but also where to find combinations of specific words or expressions (if they exist). Researchers can use this function and its options to locate the properties of investigated data and make decisions relating to categories.

5. In **Narralizer** analysis documents, all the identification marks of each segment as well as the name of each category appear automatically after the command "show sub-tree." Therefore, to accurately count the words in pure data segments, we must verify that these external components will not be accounted. **Narralizer** offers a special operation:

   I. After clicking **Select Sub-tree** to bring up the document in the display pane, the default of the opened **Select Sub-Cases box** presents the document text with its identification marks. At the bottom of the dialog box, "Show comments" and "Show text" appear automatically with the V sign next to each of them (see Figure 16-1). Clicking on a V sign cancels it.

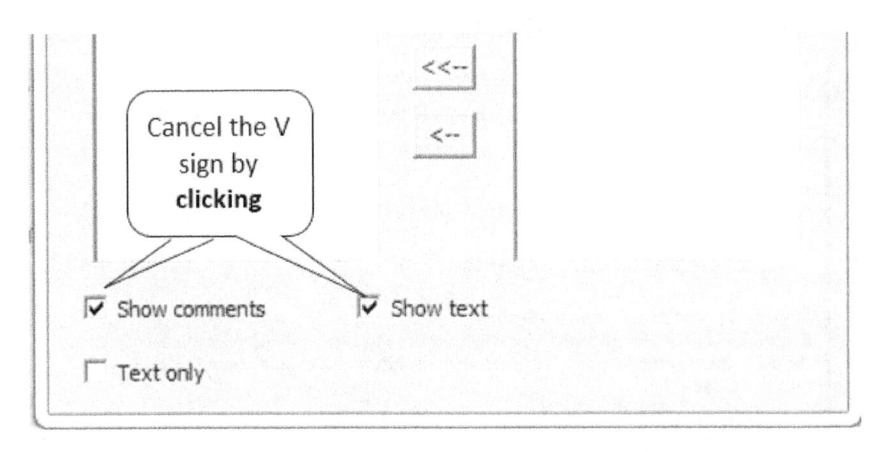

Figure 9-5: Canceling "Show Comments" and "Show Text"

II. **Click "V"** beside the sign ""Text only" (see Figure 9.6) and after the command "select all" for all the sub-cases to be displayed, or use "select" to display one or part of the sub-cases (see explanation in Section III below).

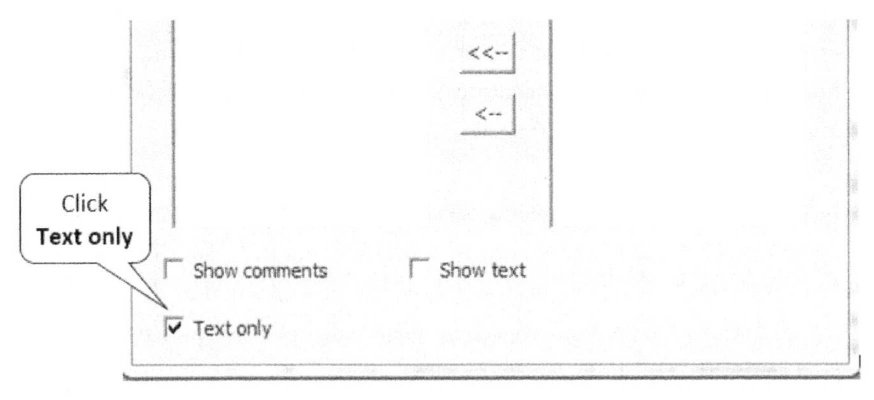

Figure 9.6: Clicking "Text Only"

After displaying the text without the additional remarks, perform the "find" command to receive the pure number of occurrences in the document.

Bear in mind: The **Narralizer** software version accompanying this book is a limited version (learning version) in terms of the amount of data and operations, but not in capabilities. At the base of the **Narralizer** working screen in the Learning version, information regularly appears about the degree of utilization of the software. Thus, it will be impossible to see the information about the number of times the phrase appears in the text. This is the only function of **Narralizer** that can't be applied using the Learning version.

To obtain a data array for certain categories but not others, mark the desired categories on the category pane, and use the Show sub-tree function described above. When the box opens, click Select all to obtain a selective data array. Clicking an indication category (folder icon) will summon all data contained in the categories below it, as shown in Figure 9-7:

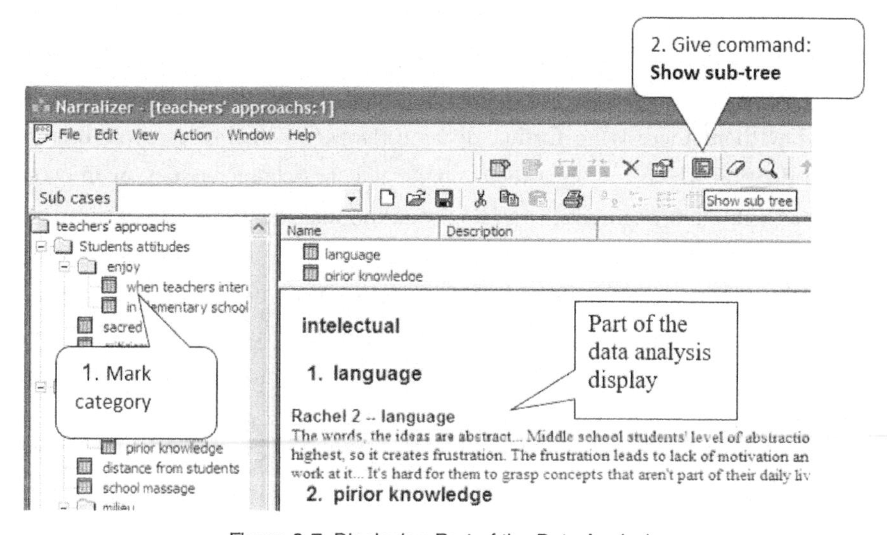

Figure 9-7: Displaying Part of the Data Analysis

CHAPTER 10

EXTENDING THE MAPPING ANALYSIS

This chapter is a direct continuation of Chapter 9, and extends the scope of the mapping analysis to more complex situations faced by researchers in their study.

The Dimensions of the Content Categories

Until now, the content categories have been presented at the lowest level of each branch of the category tree, i.e., categories to which the data segments are adjacent, and below them there is no further division. Now, we shall offer an additional level of division which can hone the content category's characteristics, or in other words, sort the data segments according to additional identifying characteristics. These dimensions are the smallest units of information in the analysis process. To specify the dimensions of the categories, researchers take the properties of content categories and place them along a continuum to view the possibilities of the category properties from one end to another (Creswell, 1998). The need to address different dimensions of the same content category increases as we enlarge the number of content categories under the same indication category aiming to create a distinction between them, and when we wish to compare different instances of content categories in different cases or in the same case over a long duration.

There are at least two kinds of dimensions, one referring to frequency, and the other to the degree of prominence. Dimensions of frequency are used in such content categories as "teachers' difficulties in math classes," for example. To specify the frequency of this category's property, we refer to the dimensions "always," "often," "sometimes," "rarely," and "never." Figure 10A illustrates the sequence characteristics of this content category.

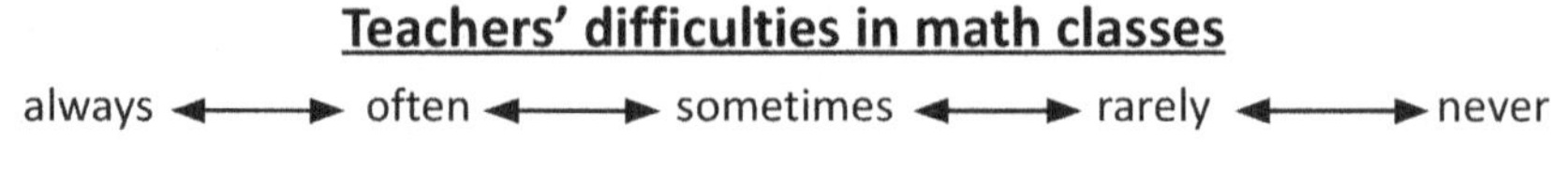

Figure 10A: Dimensions of Frequency in the Content Category

Referring to the dimensions of frequency of categories lets us distinguish between several sub-cases. In one sub-case, for example, the teacher regularly encounters difficulties in mathematics classrooms, while in another sub-case the teacher only rarely has such difficulties. In this way we can identify the differences between the sub-cases of an identical participant in separate situations, or between several participants, or to identify changes over time and/or in different contexts of place or social setting.

Another type of reference to the dimensions of content category is the degree of prominence. Consider, for example, the category "the contribution of mathematics to the development of students." This category indicates the participants' attitude that mathematics promotes the development of students. Sometimes it is not enough to describe this property, but it is also important to note the degree of its prominence: Is this a significant feature of the participants' attitudes, or not? This type of content dimension can also be organized along a continuum. Figure 10B shows the division of category dimensions according to their prominence and frequency.

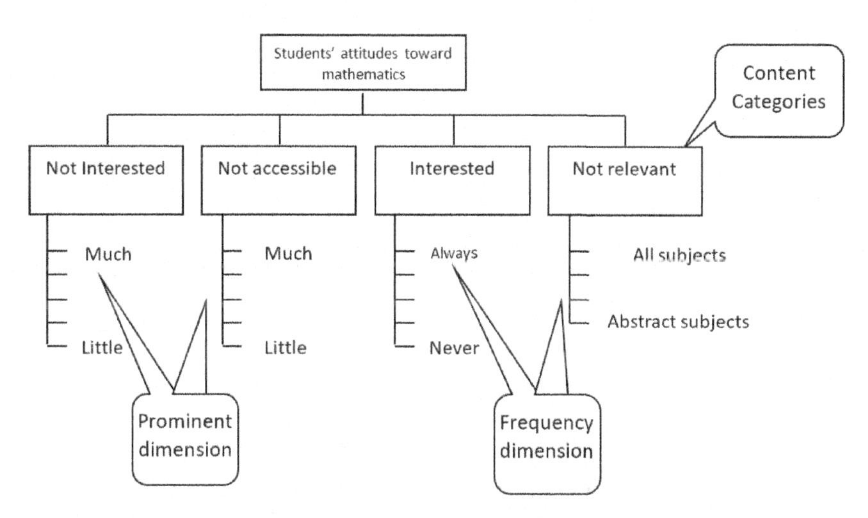

Figure 10B: The Dimensions of Content Categories According to their Prominence and Frequency

In practice, the classification of the content into several dimensions is similar to dividing the category into sub-categories located in the sequence set. But it is not necessary to make every content category into a new content category dimension (and as such, the original content category will become an indication category). It is possible to keep the original content category and to mark its content dimension beside each data segment. This identification will be added to the regular sign of each data segment: the sub-case name.

Specifying the Dimensions of the Content Categories Using Narralizer

1. **Double clicking** on the **data segment icon** in the **content pane** (or command **Properties**) opens the "Text Segment Properties" box, which automatically displays the data segment identification details.
2. In the Comments pane of the box, **record the dimensions** of the specific segment, as shown in Figure 10C.

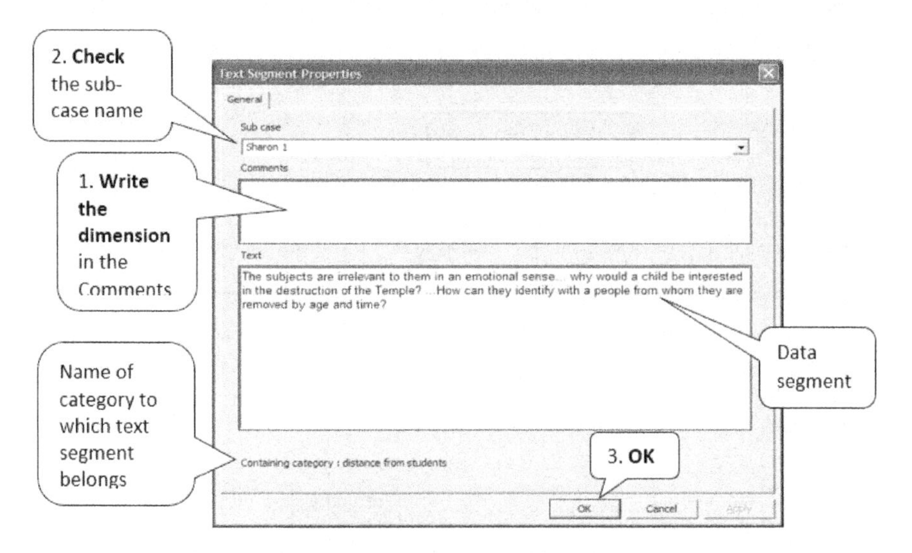

Figure 10C: Recording the Dimensions

3. After clicking **OK**, the box is closed and the properties of this segment will appear in the content pane.

Multi Sub-Case Mapping Analysis

The qualitative research project can focus on one, two, or many sub-cases. Accordingly, we can perform an analysis process for each sub-case separately or for all of the sub-cases together. Basically, the analysis principles of one sub-case and many sub-cases are identical. In the process of the mapping analysis, it is essential to decide whether to create a uniform category array for all sub-cases, or to conduct a unique category array for each sub-case. However, it is advisable to construct a common category array and to take advantage of the flexibility of this array to meet the similarities and differences between each of the sub-cases.

The benefits of creating a common category array lie in its possibility to give researchers uniform criteria to represent all the sub-cases together and each one separately, and to consider what is common and unique to each sub-case. Even if a uniform category array is used, this does not mean that every category will necessarily be expressed in all of the sub-cases. The difference between the sub-cases may be expressed differently in some of the content categories for each of the sub-cases. In some sub-cases, there will be a marginal or lack of expression for several content categories, while in other sub-cases there will be a marginal expression for other categories. The content categories will express the variety among the sub-cases, and the similarities and differences between them.

Figure 10D illustrates a category array with three sub-cases, A, B and C. The existence of each sub-case is mentioned under the name of each category.

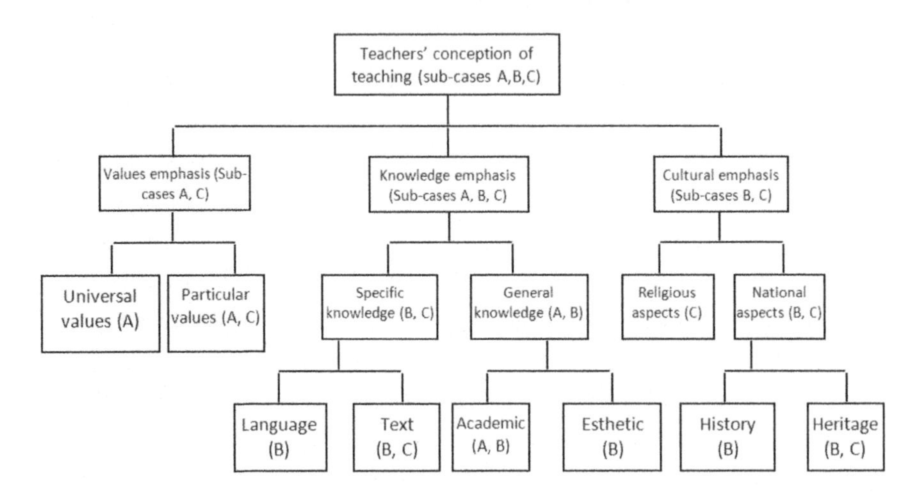

Figure 10D: Category Tree Emphasizing the Difference between the Three Sub-Cases

We can see that the use of a uniform category array allows one to see both the uniqueness of each sub-case and what they share in common.

Researchers using a word processor for the analysis process will prepare several copies of the category array (one copy for each sub-case) and insert the data segments which belong to the specific sub-case into each category array document. At the end of the analysis process, there will be several categorical analysis documents to compare, each belonging to a specific sub-case. There will probably be some "empty" categories (without data segments) in each of the sub-case's analysis documents. These empty categories will not delete. In this way, you may discern the similarities and differences between the sub-cases.

Multi Sub-Case Mapping Analysis Using Narralizer

The recommended procedure is to use one **Narralizer** research document with a common category array for analysis of all the sub-cases. In this way, you can easily compare between the sub-cases to find similarities and differences. The **Narralizer** can always be used to divide the sub-cases into several different research documents (this procedure will be explained in Chapter 20). It can also analyze each sub-case in the individual research document but maintain that all the parallel analysis documents of all sub-cases will use the same category array. Thus, the different documents can always be merged into one unified analysis document, as will be explained in Chapter 20. The process of inserting the data segments in multi sub-case research documents is identical to that described in previous chapters. Be certain to ascertain that each segment is accompanied by its identifying marks.

Copying an Existing Category Array to a New Document

Frequently, after inserting the data segments into the analysis document, the researchers wish to create a new analysis document with the same category array devoid of data. In other words, they want to duplicate an existing category array. This need may arise during multi sub-case research and/or in a longitudinal study over some period of time, and/or when several researchers work together in the same study. In such situations, it is sometimes preferable that each sub-case and/or each researcher and/or any period of time will be analyzed in a separate document. However, it is desirable that all sub-cases in any period of time will be analyzed according to a uniform category array. To perform the analysis using a word processor, simply create a new document with the same category array.

Copying an Existing Category Array to a New Document Using Narralizer

Narralizer allows us to open an existing analysis, copy the array without the data categories it contained, and paste the category set into new empty document analysis. This is done as follows:

1. Open the document whose **Category array** is to be duplicated.
2. Click **Divide** in the **File** menu, as shown in Figure 10E.

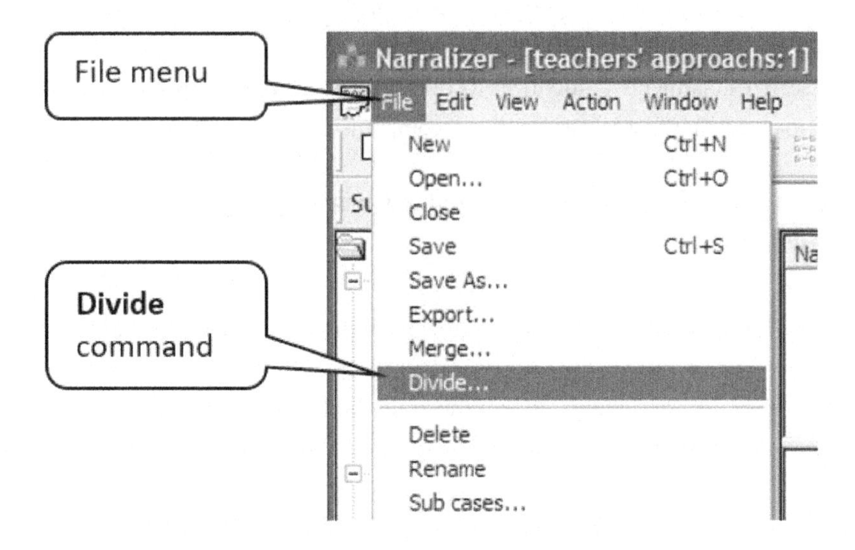

Figure 10E: Divide Command

3. The **Select Sub-Cases** box will open. The left-hand pane contains a list of all sub-cases and sub-case groups. However, this time, **they must not be moved to the right pane**. Click **Use selected** as shown in Figure 10F .

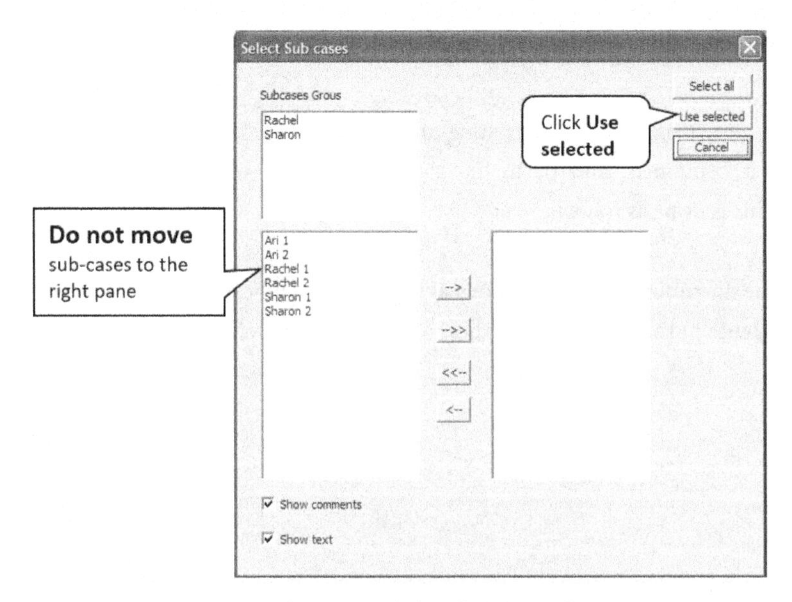

Figure 10F: Select Sub-Cases Box

4. After clicking "Use selected," the **Save As** box is opened. Type the name of the new **Analysis document,** and assign and save it to the appropriate directory.

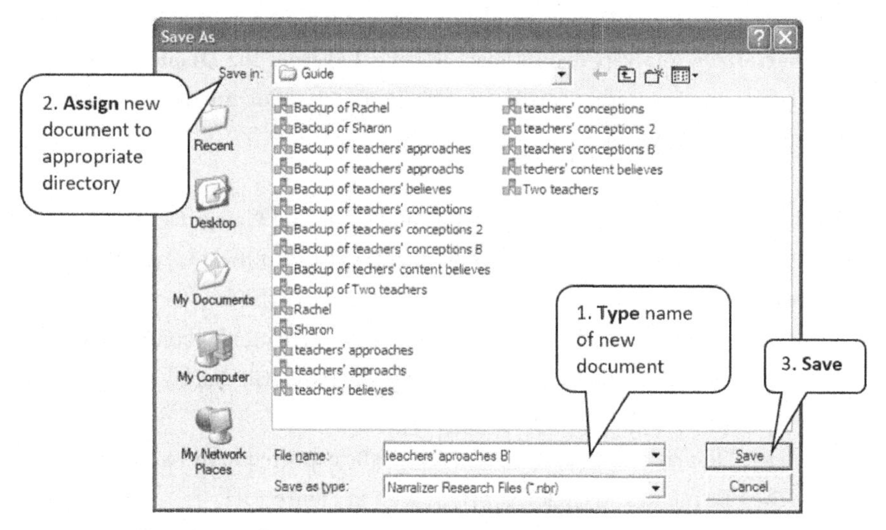

Figure 10G: "Save as" Box for Creating a New Research Document

5. The **Save** command creates the required document without data segments. This document can be used as a Work Document for additional **sub-cases** and can easily be merged later with documents of the same category array (using the **Merge** command, to be described in Chapter 16).

The Display of an Analysis Product in a Word Processed Document

In any qualitative research procedure, using a word processor or **Narralizer,** we use a word processed document to display the data collected and the research findings. Those who prefer to use the word processor for the analysis process are essentially using a word processor throughout the entire research process. This enables the researcher to print an intermediate analysis product, work on it, present it to colleagues, and at the end of the process to show it to the readers-- a simple editing and printing procedure which requires no explanation to the readers.

Export the Analysis Products from Narralizer to a Word Processed Document

As explained earlier, you may display analysis outcomes in the **Display pane**. Using **Narralizer, analysis results** may be **exported** from the **Display pane** to an external destination, in other words, to a word processed document (word.rtf). The procedure is as follows:

1. **Highlight the category** (indication or content) in the **Categories pane** whose data is to be exported as a Word™ document. (This can include all the analysis process by highlighting the main category on the top of the category pane, or part of the analysis process by highlighting the specific indication or content category)
2. Carry out the **Show sub-tree** steps. The **Display pane** will display the data to be exported as a Word™ document.
3. When the **Display** screen contains the data to be exported to the word processor, open the **File** menu and click **Export,** as seen in Figure 10H.

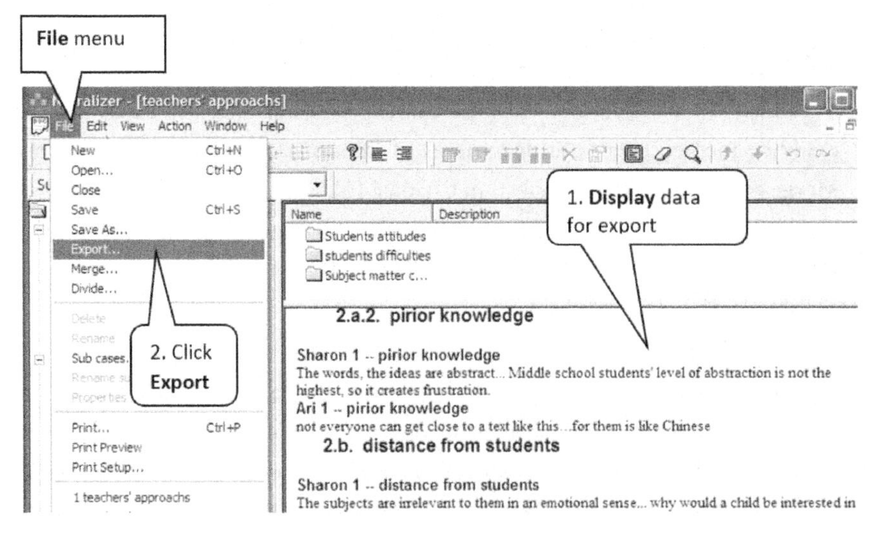

Figure 10H: Export Command

4. When clicking **Export,** the **Save As** box will open. In the upper pane of the **Save As box**, enter the name of the directory where you wish to save the newly-created Word™ document.

The bottom section of the box contains two panes. In **File name**, type the name of the document you wish to save. In the lower pane, **Save as type,** the *rtf suffix (Rich Text Format) appears automatically. **Do not change this suffix.** (See Figure 10-I.)

5. Click **Save** to save the Word™ document as an rtf file. This document can now be used as a word processor work document. All or part of the document can now be printed as required.

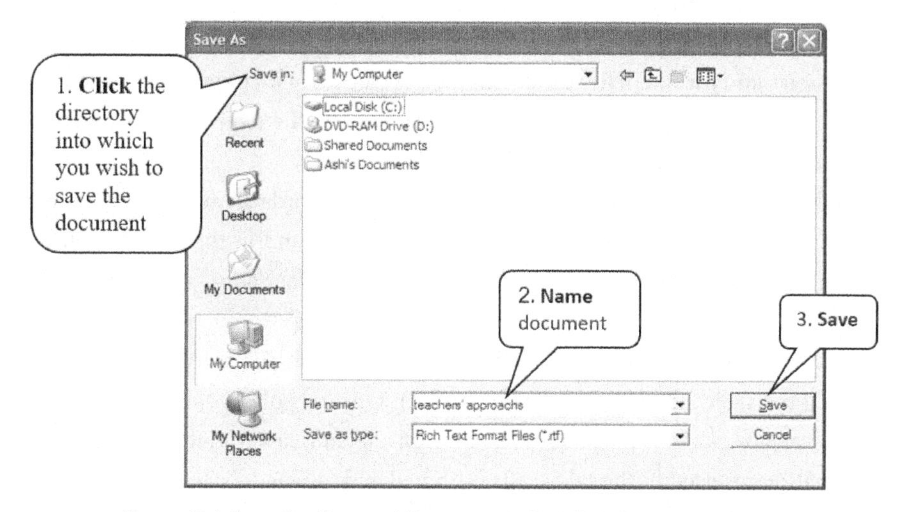

Figure 10-I: Save the Exported Document in Rich Text Format (rtf format)

You now have a Word™ document in rtf format, which can be converted into any word processing document. The newly-created document can be used for all research needs, and its text can be changed and, of course, printed.

Summarizing and Comparing the Analyzed Data

Once the mapping analysis is completed and the data segments are inserted in the appropriate content categories, the researcher can examine the overall picture of the phenomenon under study. Generally, we can find evidence to a rich degree of each of the categories of the data segment on each of the content and indication categories. Researchers can see which of the properties are more or less pronounced. The complete picture of the data, as represented in the mapping analysis array, can motivate researchers to consider new directions for research. Or, the researcher may decide to return and continue to collect data.

One highly recommended process at this point is a comparison between all the sub-cases. As mentioned, a sub-case can be any one of the participants and/or regarding any of the participants at some point of time (first year interview, second year, etc.) and/or in regard to certain research aspects (interview, observation, etc.). It is likely that in each study (one research case) there will be several sub-cases, even if the study focuses on only one participant or a single phenomenon. In order to get the entire picture of the overall study, we can compare between the sub-cases to find the similarities and differences between them, to see the development (or regression), and the like. The researcher has a great deal of material, and sometimes it is difficult to control it all. The danger in this case is the temptation to base the analyzed picture on impressions and not on solid data.

Perhaps researchers may be assisted by a tool which affords them a comprehensive picture of the analyzed data focused on the categories, and in this way be able to compare the sub-cases. A table can be placed as a summarization and comparison of categories and data, as shown in Figure 17J. All researchers, including those who analyze with the **Narralizer**, will utilize the word processing software Word™ for creating the table (through command "table"). One axis of the table will contain all the categories, and the other axis is for all the sub-cases. Make sure that the information inserted in the table cells is limited and targeted, or the table will lose its effectiveness. Note that the cells connected to the indication categories will remain empty. In Figure 17J, the numbers in each of the cells in the table indicate the number of segments in each category for each sub-case and the literal information regarding the degree of prominent expression of each category in each sub-case. Information about the degree of prominence does not coincide with the quantitative information of the number of data segments, but is based on what they were told by the participants.

Sub-cases Categories	Sharon	Rose
Teachers' conception of teaching		
1. Cultural emphasis		
1A. National aspects		
1A1. Heritage	3 very prominent	2 somewhat prominent
1A2. History	5 very prominent	3 very prominent
1B. Religious aspects	0	0
2. Knowledge emphasis		
2A. General knowledge		
2A1. Esthetic	0	1 hardly prominent
2A2. Academic	7 very prominent	5 very prominent
2B. Specific knowledge		
2B1. Text	3 somewhat prominent	2 somewhat prominent
2B2. Language	1 hardly prominent	1 hardly prominent
3. Values emphasis		
3A. Particular values	0	0
3B. Universal values	4 very prominent	3 very prominent
Teachers' conception of teaching		
1. Cultural emphasis		
1A. National aspects		
1A1. Heritage	0	1 very prominent
1A2. History	2 somewhat prominent	3 very prominent
1B. Religious aspects	3 somewhat prominent	1 hardly prominent
2. Knowledge emphasis		
2A. General knowledge		
2A1. Esthetic	2 very prominent	
2A2. Academic	3 very prominent	3 somewhat prominent
2B. Specific knowledge		
2B1. Text	4 very prominent	3 somewhat prominent
2B2. Language	2 somewhat prominent	0
3.Values emphasis		
3A. Particular values	1 somewhat prominent	2 very prominent
3B. Universal values	2 somewhat prominent	1 hardly prominent

PART FOUR

THE CONSTRUCTIVIST-INTERPRETIVE AND CRITICAL-CONSTRUCTIVIST RESEARCH APPROACHES AND METHODOLOGY

INTRODUCTION TO PART FOUR

As mentioned, the Structural and Partial Structural Patterns of Qualitative Research Approaches and Methodology presented in the previous part of the book represents the "qualitative revolution" to a lesser degree than other qualitative Methodologies and approaches (Denzin & Lincoln, 2005b, p. IX). In contrast, it is proper to view the Constructivist-Interpretive and Critical-Constructivist approaches and methodology as representing the "interpretive turn" or the "narrative turn" (Lyons, 2007, p. 611-613) of the recent decades. The idea that the phenomenon under study need not be examined by a predetermined theoretical perspective was a conceptual turning point, if not revolution, in conventional scholarly perception. I have learned from my experience that many research students, and probably no small number of qualitative researchers, have difficulty internalizing the idea that we can maintain a study without a predetermined conceptual framework. In many ways, the constructivist approach is the conceptual opposite of the structural approach and methodology (Fetterman, 1989; Guba & Lincoln, 1998; Schwandt, 2000).

In this part, we present two methodological pattern: The participant-focused methodology (chapter 11) and the participant-critical focused methodology (chapter 12)

As stated, the criteria-focused methodology is based mainly upon analytical research skills, and much less on intuitive research skills, characterized by a process which focuses on external aspects of the phenomenon under study. In contrast, the participant-focused methodologies are based more than any other methodologies on intuitive research skills, while providing relatively limited (but existing and essential) expression to analytical research skills. The participant-focused and participant-critical focused methodology are characterized by a focus on the internal aspects of the phenomenon under study, and less on external aspects. Bruner (1996) described the two approaches as representing two extremes: one is called a "paradigmatic approach" (which focuses on the criteria), and the other is the "narrative approach" (which focuses on the informants) (Craig, 2001).

CHAPTER 11

THE PARTICIPANT-FOCUSED METHODOLOGY (THE CONSTRUCTIVIST-INTERPRETIVE APPROACH)

The participant-focused methodology is identified largely with the constructivist – interpretive philosophy (Guba & Lincoln, 2005). "Interpretive paradigms - associated with intellectual traditions such as phenomenology, symbolic interactionism and ethnomethodology - stress the dynamic, constructed and evolving nature of social reality." (Travers, 2001, p. 8). Knowledge, according the constructivist conception, "is constructed (i.e., built up, brought into being) through our social practices, rather than already in existence, ready to be discovered." (Wiggins & Riley, 2010, p. 138).

The participant-focused methodological pattern relates to ideas that have evolved from the German intellectual traditions of hermeneutics and the Verstehen perception in sociology. The interpretive understanding process involves the ability to identify and recreate the experiences of others in our mind. "At the heart of the dispute was the claim that the human sciences were fundamentally different in nature and purpose from the natural sciences." (Schwandt, 2000, p. 191). According to the supporting Verstehen approach, the goal of scholars of the human being and society should focus on understanding the people and their actions, rather than searching for causal explanations for phenomena, as is common in scientific research (Schwandt, 1998, 2000). "The aim of the inquiry is understanding and reconstruction of the constructions that people (including the inquirer) initially hold." (Guba & Lincoln, 1998, p. 211).

The participant-focused methodology is based "not only upon what people actually do, but what they say they do and what they say caused them to do what they did. It is also concerned with what people say others did and why. And above all, it is concerned with what people say their worlds are like." (Bruner, 1990, p. 16). This methodology assumes that we can describe what people think by listening to what they say (Fetterman, 1989). The researchers who constantly and faithfully listen to the voices of the participants are described by Guba and Lincoln as "passionate participants" (Guba & Lincoln, 1998, p. 215), expressing the willingness of researchers to absorb the words and stories of their subjects. The focus is upon people's speech, based on the assumption that in this way we can empower them (Cortazzi, 1993).

At this point, it is important to emphasize that this is not a methodology, which appeal solely to the intuitive mode of thinking, and focuses exclusively on the internal perspective. As mentioned, qualitative research, throughout the entire scope of its methodologies, is characterized by a simultaneous reference to external and internal perspectives, even though to a different extent in each of the approaches and methodological patterns. It utilizes the natural human investigative skills, and at the same time relates with a critical approach towards these skills. Assigning significance only to the human intuitive investigative skills is a sign of crossing the line between research and non-research, leading towards the direction of journalism and even literature and art, but not research. Figure 11A illustrates this concept:

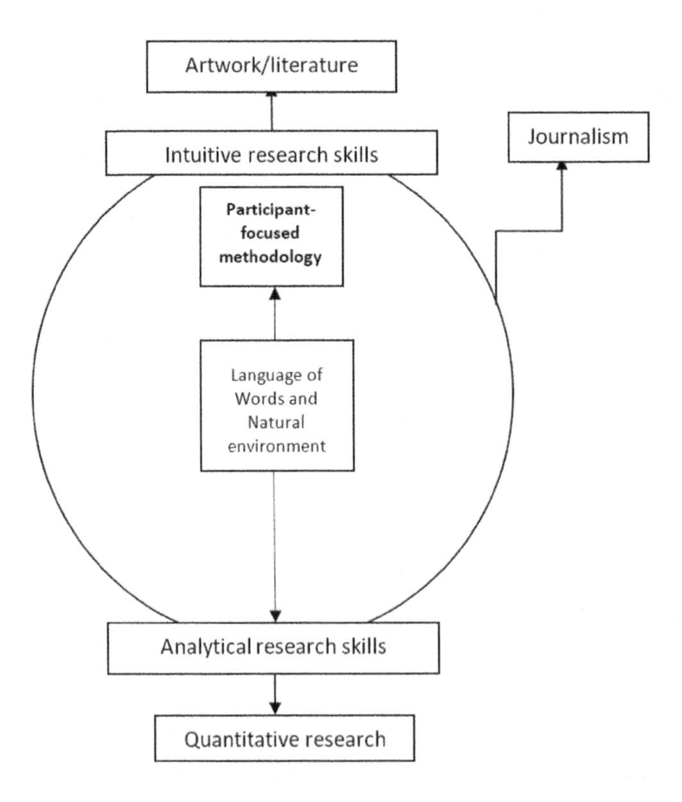

Figure 11 A: The Place of Participant-focused Methodology in the Scope of Qualitative Research

The Basic Constructivist Assumptions

Unlike the criteria-focused methodology, which refers to the phenomenon being studied in an objective manner as reality, regardless of who looks at it, the participant-focused methodology assumes that the reality we attribute to the worlds we inhabit is a constructed reality. Construction of reality is a product of meaning (Bruner, 1996). The reality is therefore constructed by people. "The question of whether the events have actually happened in some real place to some real individual is not so important." (Heikkien, 2002, p. 25). All we acquire is built by us. We organize and interpret our experiences, and actually we design our experience by our interpretation (Bettencourt, 1993). The methodological assumption underlying this pattern is that to be a human being means to be able to represent experiences with language (Seidman, 1991). A fact does not exist unless there is an observer who tells it to him/herself or to others (Lieblich, Zilber and Tuval-Mashiach, 1995).

Reality is perceived as dependent on the vantage point of the observer, so we can't separate the observer from the objects of study and must be objective in our vantage point. The researchers are part of the phenomenon being studied, rather than separate from it. The individual and his/her world is interdependent upon each other (Denzin & Lincoln, 2005a; Guba & Lincoln, 1998, 2005; Lincoln & Guba, 2000; Moilanen, 2002). What seems true and factual is actually constructed inside the minds of people. Construction is relativistic, and therefore this approach is a "view that our representations of the things in the world are socially constructed and can't be seen as simple reflections of how those things actually are" (Sullivan, 2010, p. 23).

The structured reality is best understood as a whole entity whose every element is intertwined with reciprocal connections between them, while the context of the phenomenon under investigation is not separated from the phenomenon itself. The basic assumption is that in order to understand the meaning of any part of the phenomenon being studied, we must understand the significance of the context, and vice-versa. Therefore, a significant interpretation requires a constant transition from the whole to the parts, from the parts to the whole, and back again (Henwood, 1996; Maykut & Morehouse, 1994; Stake, 1995). Each investigated case is perceived as unique, and researchers attempt to be as close as possible to the informants and to the phenomenon under study (Angrosino, 2005; Denzin, 1995; Sciarra, 1999).

The people are working in the world while giving meaning to their surroundings.

The meaning is rooted in the social cultural world where people live and in the interaction they maintain with others in this world. Human nature is not detached from culture (Moss, 1996; Schwandt, 1998, Geertz, 1973). The culture within which we live and work shapes our minds and equips us with the tools with which we build not only our worlds, but also our concepts about ourselves and our abilities. (Bruner, 1996) Reality is best understood in a social-historical context, due to the specific time and context in which the phenomenon occurs. The values of the subjects and the researchers are integrated in the study and not separated from the phenomenon being studied (Merriam, 1998). No one can function in the world and find meaning in his surroundings without the assistance of the symbols of culture systems (Bruner, 1996). Therefore, the participant-focused methodology tries to understand the phenomena being studied through its cultural background (Tedlock, 2005).

As it will be shown in the next chapter, proponents of the participant-critical focused methodology also assume, as do adherents of the participant-focused methodology, that reality is not objective, but a result of construction. But they believe that the participant-focused methodology is conservative. They argue that this methodology is interested only in descriptions of reality, and accepts reality as it is, thus serving the political-economic interests of dominant groups (Carr & Kemmis, 1986; Travers, 2001). Supporters of the participant-focused methodology would answer that being a reality results from construction, and because it can be viewed from different perspectives, we as researchers cannot expect to get a "correct" and "true" picture, but simply take what the specific subjects have ready and able to share with us (Josselson, 2004). Researchers that accept the assumption of the participant-focused methodology would prefer a stance of acceptance and trust toward the participants, even if they are unsure about the ability of participants to present the full picture of their world.

Research as an Interpretative Process

The participant-focused methodology is "willing to describe the life experience" before any explanation of this experience. (Moilanen, 2002; Grbich, 2007 Van Manen, 1990). Researchers want to see and understand the experience of the participants through their eyes, not through those of the investigators or through any other external eyes. This position requires the researcher to be as close as possible to the world of the subjects and to interact with this world (Moss, 1996; Travers, 2001). This stance constructs "symbolic interactionism" with the world of the participants (Holstein & Guberman, 2005). Its meaning is that "the inquirer actively enters the worlds of people being studied in order to see the situation as it is seen by the actor, observing what the actor takes into account, and observing how he interprets what is taken into account." (Schwandt, 1998, p. 233). The ability to maintain a dialogue with the symbolic world of the participants and to experience it from their perspective can be realized only if researchers can enter as deeply as possible to the world of the participants and try to absorb their world (Woods, 1996).

Describing the experience of the participants from their viewpoint is to be confident in what is told, and believe that the story of their experience is indeed typical of what they actually experience. Ricoeur (1970) calls this process the "hermeneutics of faith," as opposed to the "hermeneutics of suspicion" (Josselson, 2004) that characterizes the critical-focused methodology, to be discussed in next chapter.

Describing the participants' world means to understand their world, and in order to do so, the researcher must interpret it (Bruner, 1996; Schwandt, 1998). The product of the participant-focused methodology is never a copy or photo, but always an interpretation (Gudmundsdottir, 1995). Or, as Gadamer (1986) argued, when we interpret the meaning of something, we actually interpret the interpretation. The participants themselves experience the world while finding meaning in it, or in other words, the experience itself is a process of interpretation (Noblit & Hare, 1988) and the researchers encounter the interpretation of the subjects, with the meaning that the subjects have given to their world. In this way, we can say that the researchers create a second-order interpretation, the interpretation of the interpretation of subjects (Van Manen, 1990; Schwandt, 1998). Geertz's view (1973) is that the interpretation and meaning in qualitative research should be similar to that which hermeneutics interpret as a complex text (Schwandt, 1998).

The aim of participant-focused methodology is to interpret and understand the meaning in the context of this occurrence (Moss,1996; Maykut & Morehouse, 1994; Philips & Hardly, 2002). Therefore it is not enough to absorb the words of the participants, which in many cases can be similar or identical to those of the researchers, and thus be a source of errors in understanding. The words must be interpreted to find a meaning from the perspective of the participants themselves (Robert, 2002; Woods, 1996). The significance of participants' actions and their words can be understood accurately only from the point of view of those being taken into account, which can occur only after creating direct contact with the subjects in the social context in which they experience their lives (Seidel & Kelle, 1995).

"What distinguishes human action from the movement of physical objects is that the former is inherently meaningful" (Schwandt, 2000, p. 191). According to Gadamer's hermeneutics conception (1986), there is no original meaning to the phenomenon under study and not even a permanent meaning. Any interpretation not only clarifies the meaning, but also changes it in the process of interpretation. There is no one final correct interpretation (Ofir, 1996), thus the interpretive analysis of the researchers becomes necessary to understand the phenomenon under study (Josselson, 2004). Interpretation is perceived by Gadamer (1986) as a process of dialogue, a meeting between the perspectives of the participants and the researchers, with the resulting picture being the outcome of this meeting (Widdershoven, 1993). Consequently, there may be different interpretations of the same phenomenon under study conducted by different researchers, each having a different research perspective (Guba & Lincoln, 1998; Schwandt, 2000).

Relationships of "I -Thou" Between the Researchers and Informants

One can define the relationship between the researchers and the informants in the participant-focused methodology as relations of "I – Thou" (Buber, 1964), as opposed to the "I – It" relationship characterized in criteria-focused methodology. "I - Thou" relations are characterized by close involvement and a great deal of empathy and partnership between researchers and participants. "I – It" relations, however, are characterized by distance and a willingness to look at the subjects through objective external eyes. The assumption of the participant-focused methodology is that the construction of reality can be adequate only through involvement and interaction of the researchers with the participants (Guba & Lincoln, 1998). The researchers using this methodology are "fully involved in the experience studied" Clandinin & Conneley, 2000, p. 81). However, as shown in Figure 5A above, the researchers need to be able to perform control procedures, take a few steps back, and look at the subjects with a distant research eye, using their analytical skills and not just their intuitive research skills.

Eisner (1985) refers to the proper research ability as "connoisseurship." This concept is taken from the world of wine tasters, characterized by their involvement, expertise and personal experience that allows them to evaluate the quality of the product rather than rely in advance on a list of fixed external criteria. Connoisseurship allows the qualitative researchers to absorb the subtleties in human behavior and the ability to perceive holistic relationships between things. Connoisseurship requires extensive prior experience and a deep familiarity with the subject under study (Schwandt, 1998; Simons, 1996).The participant-focused methodology assumes that the "the meaning of any fact, proposition, or encounter is relative to the perspective or frame of reference in terms of which it is constructed (Bruner, 1996, p. 13). Some people call the "truth" that characterizes this type of research a "narrative truth" (Freeman, 2007, p. 136). Based on this assumption that perceived reality is the result of personal and cultural construction, the notion of multiple perspectives arises. Consequently, one interpretation does not preclude a different interpretation from a different perspective.

The Study's Theoretical Perspective

Contrary to the criteria-focused methodology presented in chapter 6, the participant-focused methodology is not based upon a predetermined theoretical perspective. Here, the assumption is that the theoretical perspective becomes apparent during the study (Creswell, 1998). "In contrast to the more positivist and behavioral empirical sciences, human science does not see theory as something that stands before practice in order to "inform" it. [...] Practice or life always comes first and theory comes later as a result of reflection." (Van Manen, 1990, p.15) Therefore, researchers adhering to this methodology start their investigation by looking at the experience of the participants, with no attempt to force any interpretive theory. While the criteria-focused methodology applies theoretical concepts to the phenomenon under study in a deductive manner, in the participant-focused methodology, the theoretical concepts are raised inductively from the phenomenon under study (Polkinghorne, 1995; Clandinin & Connelly, 2000).

Although researchers following the participant-focused methodology do not arrive at the studied research field with a coherent theoretical approach, they do not reach the field with a blank slate (Pinker, 2002). The constructivist approach not only claims that the perceived phenomena are the result of construction, but also that the observer himself does not ever look in a neutral manner. In other words, researchers never arrive at the site without theoretical insights, but they are always equipped with theoretical lenses, even if they are often unaware of them. Consequently, participant-focused methodology researchers do not begin with the formulation of a theoretical framework, but they are not without reference to a preliminary theoretical perspective. Therefore, before coming to the field of study, a researcher should determine the theoretical perspective through which he looks at reality, and be prepared to change it according to the new insight arising from the data gathered during the study (Gudmundsdottir, 1996; Strauss & Corbin, 1994).

The best way to determine the perspective of the research is by writing a preliminary literature review, but formulating it in such a way to enable changes, additions and omissions to be made throughout the entire research process. "Conducting a literature review is the usual first step in any research project. The reason we do this is to identify what others have found out about the topic that we are interested in, before we start." (Shaw, 2010a, p. 39) The review of the literature should therefore be updated with the progress of the research, as an expression of the updated theoretical perspective. It seems that as part of the clarification of the theoretical perspective, it is appropriate to integrate the work of data collection and analysis by constantly reading additional relevant literature and jotting memos which raise theoretical ideas, even if they are only raw thoughts. Only at the final stage of the research will researchers write the last version of the literature review which appears to be the theoretical lenses by which the readers read and find meaning in the study.

Formulation of Research Questions and Assumptions

As emphasized above, in participant-focused methodology the researchers do not arrive at the study with a coherent and compelling theoretical perspective, nor do they come with a blank slate. However, they have a general research direction and a topic they wish to explore. "The researcher always brings some theoretical preconceptions with him or her. But these are not 'hypotheses' in the ordinary sense. Rather, they should be referred to as (partly implicit) conceptual networks that provide us with some 'lenses' for the perception of the empirical world. [...] These perspectives help researchers select relevant phenomena, and of course researchers with different perspectives will select different phenomena." (Seidel & Kelle, 1995, p. 56). Therefore, researchers begin their study with one or more broad, open questions guiding them in the early stages of the research, but at the same time giving flexibility to respond to the collected data and the unexpected. The initial research questions growing from the initial theoretical perspective are changed, broadened or narrowed throughout the continuation of the research (Strauss & Corbin, 1990). During the study, initial research questions are re-evaluated and revised until they are appropriate to the data being collected and the revised perspective. As the research progresses, a "dialogue" takes place between the data obtained and the theoretical perspective of the researchers, which relies on constantly updated theoretical literature. The relationship between the theoretical perspective and the collected, analyzed data is dynamic and not linear (Stake, 1995; Jorgensen, 1989).

Research assumptions are a concise formulation of the theoretical perspective of the study, and they should constitute the outline to the formulation of research questions. In the participant-focused methodological pattern, based on the constructivist-interpretive research approach, researchers do not enter into the research arena with hypotheses, but with assumptions. The difference between the two is that the assumptions of the study are formulated in a much more open, dynamic manner, accepting change at all stages of the research, while hypotheses bear the nature of prediction, formulated in a focused manner indicating the relationship between dependent and independent variables (as in criteria-focused methodology). In addition, the research assumptions serve as a basis for formulating research questions. Hypotheses, in contrast, are formulated after the research questions, and actually "predict" the research picture (Strauss & Corbin, 1990; Mason, 1996).

Examples of a question formulated at the beginning of the study:

1. What is the perception of Bible teachers regarding the goals of teaching their subject?

This is a relatively open research question of the "what" and "how" descriptive question variety (Holstein & Gubrium, 2005). The purpose of the question is to set the environment in which the researchers seek to get the picture. This kind of question formulation focuses and directs the observation, but also prevents it from extending. However, at the same time it leaves the option to make changes later in the continuing study stages (Mason, 1996). Such a question may be accompanied by the following research assumption:

The professional knowledge of Bible teachers, as well as teachers' knowledge of other subjects, is different from the academic knowledge of Bible experts and is

characterized by a combination of content elements with pedagogical elements, called by Shulman (1986) "pedagogical content knowledge."

Research questions based on the above theoretical assumptions lead researchers to examine the content and pedagogical elements of the teachers' conception of teaching Bible. The process described so far can be viewed in Figure 11B:

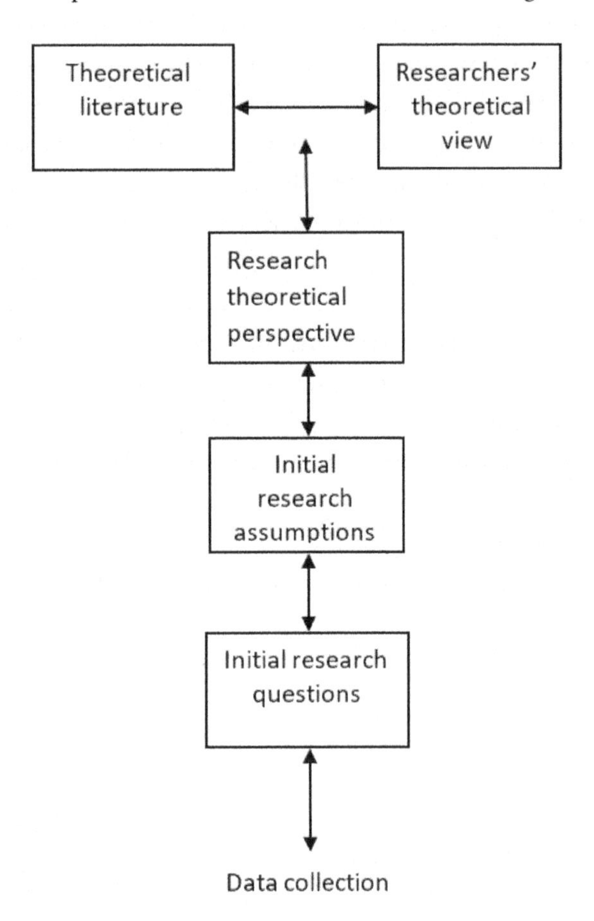

Figure 11B: Outline of the First Steps of the Research Process

It is important to note the two-sided arrows in Figure 11B indicating that the process is not linear (although it has a progress direction phase), but based on a continuing dialogue between the various stages. It is possible at each step to reexamine the previous stage. This picture reflects the fact that the participant-focused methodology provides a central expression to the researchers' intuitive inquiry skills and seeks to be attentive to the criteria arising from the participants' world.

Consequently, as the study progresses and the data is collected from the world of the participants, the picture will become clearer and the research question presented above will be replaced by a more focused one, more attentive to the world of the participants. The following is an example of a focused research question:

1. What are the content ideologies and the pedagogical ideologies that guide the teachers in their work?

This research question is based on a theoretical perspective and research assumptions that were focused during the course of the study while being exposed to the world of specific subjects, data collection and analysis. The revised research question may be accompanied by updating the research assumption, as the following:

"The pedagogical content knowledge (Shulman, 1986) of Bible teachers is largely determined by the teachers' belief systems (the ideology of teachers), both in content and in the pedagogical field. These beliefs are quite fixed and do not change easily in connection with temporary teaching situations."

As the study progresses, the central research question can be split into several sub-questions (Shaw, 2010a). The purpose of the sub-questions is to focus attention to sub-areas. Examples of such questions include:

1 A. What are the teachers' personal beliefs about the Bible, and how are these beliefs reflected in the process of teaching?

1B. What are the teachers' beliefs about the values originating in the Bible, and how are these beliefs reflected in the process of teaching?

1C. What are the teachers' beliefs regarding the capabilities and interests of the students studying the Bible, and how are these beliefs reflected in the teaching process?

First-order and Second-order Questions:

The questions presented above, both the broad questions and sub-questions, are inherently descriptive questions that focus attention upon the description of the phenomenon under study. These are the most basic questions of participant-focused methodological research. In such a methodology, we can distinguish between first-order and second-order questions. The descriptive questions presented above are first-order questions, characterized by the answer we receive directly from the descriptions and explanations of the participants. First-order questions direct the researchers to collect descriptive information as seen through the eyes of the participants, those who experience the phenomenon under investigation. This information may also deal with the description of opinions and positions presented by the participants, as well as the explanation and interpretation that the participants attribute to their descriptions (Shkedi, 2005, 2010).

Unlike first-order questions which are based solely on information collected directly from the respondents, second-order questions draw attention to issues that did not merit a straight and clear answer from the participants. There may be several types of second-order questions, and they differ in their sources. One example is questions based on issues where we observe the participants, but we do not receive the accompanying description from the participants about what we watched. An example of such a question:

"To what extent does the teacher allow the students to express their personal views in class?"

Apparently viewing the class gives us a direct answer to this question, because we can examine the classroom discourse, and whether and how the teacher allows students to express their personal views. However, according to the principles of the participant-focused methodology and the constructivist-interpretive approach, reality is not objective and what we observe can be interpreted in several meanings. What one perceives as being less freedom of expression can be interpreted by others as a show of free expression. Since we are dealing with a methodology that seeks to understand the phenomenon from the perspective of those who experience it, we are not aiming to judge the situation, but expect that the participants will interpret what happened for us. In many cases, this actually takes place during the interview, but in many other cases, there is a lack of direct reference to the specific issue.

But we can learn the meaning that the participants have attributed to the specific situation from their reference to other similar issues in the same research. We treat information obtained by comparing the two situations and define it as second-order information, since we do not receive it directly from the participants. Second-order questions, then, turn the attention of researchers to look at the issues for which we have no straight answers from the participants, but which can be comprehended from other contexts.

Another type of second-order questions are those in which the participants themselves did not relate to the questions directly, nor did they relate to similar situations which would enable a case to case transfer, as described above. It seems that this situation can be studied by cross-reference to other issues which can be linked to existing theories taken from the research literature. An example of such a question:

"To what extent does the personal identity of the participants affect their perceptions about the value source of the Bible?"

In principle, this question may be asked as a first-order question. If the researchers would place this question in the first stage of the study, they would gather appropriate data and may perhaps get a satisfactory answer from the research participants. But the dynamics of the participant-focused methodology create situations in which many research questions are raised in the course of the study, after data has been collected. A question such as the above will be a second-order question, whether the researchers did not determine it before the stage of data collection and did not receive satisfactory responses to that issue. One possible answer to the second-order question like that shown above can be obtained by the reference to data obtained from a combination of two first-order questions. For example, two first-order theory questions:

1A. What are the teachers' personal religious beliefs, and how are these beliefs reflected in the teaching process?
1B. What are the teachers' beliefs about the value origin of the Bible, and how are these beliefs reflected in the teaching process?

The researchers can examine the data of each of these questions, determine the possible resulting picture, and answer the above second-order question.

In many cases, the distinction between first-order and second order questions will be clarified only during the continuation of the research, or even at its close. The question will be turned into a second-order type if during the ongoing process the researchers do not receive the appropriate direct information from the participants. As emphasized above, in the pattern of participant-focused methodology, the final formulation of the questions is clarified only during the study or even at its end.

The Study Design and Population and Field Research Selection

In the previous chapter, it is noted that the criteria-focused methodology is characterized by a linear array of the research sequence, when all the stages of the study follow one another in a predetermined order. However, the design of the participant-focused methodology is based on the absolute flexibility of the research stages' order. In the participant-focused methodology we can find all the elements of the research: determining the theoretical framework, research questions, study assumptions, data collection, data analysis and the summary of findings and conclusions. Probably the order of the components will be similar to studies using criteria-focused methodology, but this is not necessarily inevitable.

However, during the course of the study, the order of research stages is not fixed. A constant dialogue ranges between the elements, and revisions in one stage will create two-way changes in other stages. In the participant-focused methodology, the researcher can move from stage to stage, even before finalizing the previous stages. Within the dynamics of the study, researchers often find themselves returning to earlier stages and altering them in light of insights obtained from more advanced stages of research. The movement back and forth between the elements and the stages is characteristic of the participant-focused methodology (Marshall & Rossman, 1989; Jorgensen, 1989; Lincoln & Guba, 1985).

On the basis of initial theoretical perspective and research questions, researchers can make decisions as to where to collect the data (or what is the research field), how to collect data, and what is the relevant research population (Guba & Lincoln, 1985). As in other methodological patterns, there may be certain difficulties in obtaining the agreement of the participants and especially of institutions and organizations to take part in the research. A degree of suspicion may exist among the potential research informants toward the research, which is often seen as leading them to some kind of examination. However, while the criteria-focused methodology is indeed based on external standards, the participant-focused methodology draws its standards from the informants themselves, and does not demand such criteria and external standards. But the vast majority of the potential research participants has not been previously exposed to this kind of research, and may greet the researchers and research initiators with suspicion. I've learned from my experience that if researchers maintain the principles of this methodology, the participants' suspicion dissipates within a short time and they become open and cooperative.

Compared to any statistical-quantitative research, the qualitative participant-focused methodology is characterized by a limited number of research informants. The random sample so typical to statistical-quantitative research is not relevant here, not only because of the assumption that every case and participant is different and unique, and not just because of its assumption that there is no objectivity in the context of research, but also because of the relatively small number of subjects. Instead of a random sample, we prefer a purposeful sample (Mason, 1996; Stake, 2005).

The preference of a purposeful sample and methodology that seek to view the phenomenon under investigation through the eyes of those who experienced it, where participant cooperation is a significant requirement, influences the criterion of participant selection. "For qualitative fieldwork, we draw a purposive sample, building in variety and acknowledging opportunities for intensive study." (Stake, 2005, p. 451). Purposeful sampling focuses on a deliberate choice of the sample studied which best represents the population (Mason, 1996). The selected study populations are those with a large amount of relevant knowledge about the phenomenon under investigation who may serve as a source of reliable information. They must have a better ability than others to express themselves with high clarity and sensitivity (Fetterman, 1989). These participants should be prepared to devote sufficient time required for the research. My experience has taught that many potential participants are willing to answer all the above requirements, especially when they discover that the researchers of this methodology honestly allow themselves to listen to their story without any predetermined standards.

As with any research methodology, it is necessary to ensure a standard of ethics that is commensurate to stringent rules. (Dushnik and Sabar–Ben Yehoshua 2001; King, 2010) However, it seems that the nature of this methodology, based on the "voices" of participants rather than on external standards, inherently promises a higher level of rigorous, ethical standards. The fact that the research participants can be heard without judgment or any note of criticism, and the fact that the participant-focused methodology encourages the option to present the research materials and analysis for review by the participants, assures a high level of research ethics.

Data Collection Methods

Chapter 1 emphasizes that the central feature of qualitative research is its basis in the intuitive skills of the researchers, which function as the primary research tool (Lincoln & Guba, 1985). This concept stands in complete contrast to the practice of positivistic-quantitative research in which the researcher is completely uninvolved in the research methods and assumes a distant, objective stance. The place of "the human as an instrument" will be expressed in any qualitative methodology, but especially the participant-focused variety. As a result, the data collection tools in this methodological pattern are heavily influenced by internal criteria, namely criteria arising from the direct contact with the participants (although a constant dialogue is carried on with the external aspect of the theoretical perspective, and is not completely negligible). Figure 11E describes the difference between the two methodological research tools:

Figure 11E: Characteristics of the Research Tools in the Participant-focused Methodology

As seen in Figure 11E, the research tool of participant-focused methodology is close to the internal criteria and flexible variety of tools, but maintains a certain distance as an expression of analytical consideration (as explained in Chapter 2).

To ensure the place of the researcher as a research tool, and the dominant role of aspects which originate from the direct encounter with the participants themselves, the researcher becomes an integral part of the investigation as a participant observer, involved observer, or an in-depth interviewer. The assumption is that this is the sole way one can approach the world of the participants and absorb and understand their point of view (Denzin & Lincoln, 2005; Maykut & Morehouse, 1994; Seidel & Kelle, 1995). The researchers obtain the proper close, involved stance through their intuitive research skills, mainly their ability to feel empathy with the research participants (Josselson, 2004; Fontana & Frey, 2005). However, researchers in a participant-focused methodology should be able to separate themselves from the phenomenon under study to be reflective and distant, as emphasized in Chapter 2. The challenge for the researcher is to find a middle ground between involvement and empathy and distance and analytical thinking (Woods, 1996; Maykut & Morehouse, 1994).

What people articulate, and not only what they do and the way they interpret what they are doing, is the main source of data in participant-focused methodology. This means that researchers should not only listen to the subjects, but also learn their language in order to be familiar with their culture and to correctly interpret the words, gestures and actions that are common to the culture (Geertz, 1973; Jorgensen, 1989; Woods, 1996). The data collection process is not mainly a technical process, but a process of interpretation (Clandinin & Connelly, 2000; Guba & Lincoln, 1998). This is a process in which researchers must decide where to give more and less amounts of attention.

Researchers in the participant-focused methodology collect the data through intervention, particularly through interviews and observations, field notes, etc. (Merriam,1985; Yin, 1994). The researchers also collect "naturally occurring data," data derived from research which existed without the intervention of the researchers (Silverman, 2006, p. 56), such as documents, artifacts, etc. To ensure that the researcher is focused on the participant's point of view, it is common to use recording devices such as tape recorders, video cameras and other electronic tools to transcribe the events. This enables researchers to absorb in detail all aspects of the events and the discourse, plus the ability to go back and look and listen to this data even after the occurrence (Shaw, 2010a; Clandinin & Connelly, 2000; Jorgensen, 1989).

In participant-focused methodology, we distinguish between two types of data sources: primary data and secondary data (Shkedi, 2003, 2005). Primary data means all information received directly from the participants through interviews, focus groups, diaries, etc. (This can be data collected by researchers' intervention or naturally-occurring data). Primary data includes information that contains descriptions, explanations, illustrations, attitudes, thoughts and any other descriptive or verbal information provided directly by the participants. Secondary data is any relevant information taken from the research site which is not accompanied by descriptions, explanations and interpretations provided directly by the participants, even if this information deals with the participants themselves. Secondary sources can take the form of observations, documents, artifacts and other relevant research materials. (In this case as well, it can be data collected through the researcher's intervention or naturally occurring data). This type of information is considered secondary because it is not obtained directly from the discourse of the participants and does not include their descriptions and explanations-- thus the data may not be interpreted in adherence to the participants' views. Consequently, this data does not attain the same research validity as data gathered directly from participant discourse.

It appears that the main data collection tool in participant-focused methodology is the interview, since it is a tool intended to collect direct information from the participants. Basically, there is no one type of interview. On one hand are structured interviews, as used mainly in the criteria-focused methodology, and on the other are the open interviews. The interviews appropriate to participant-focused methodology are in-depth interviews characterized by a high degree of openness and attentiveness to the participants, while addressing several issues which are in the general focus of the research (Hugh-Jones, 2010; Flick, 1998; Mason, 1996; Seidman, 1991; Silverman, 2006; Shkedi, 2005). According to similar principles, the researchers can conduct focus groups, a process which is basically a group interview (Flick, 1998; Fontana & Frey, 2000; Kamberelis & Dimitriadis, 2005; Gibson & Riley, 2010). Figure 11F describes the characteristics of the in-depth interview, as compared to other types of interviews.

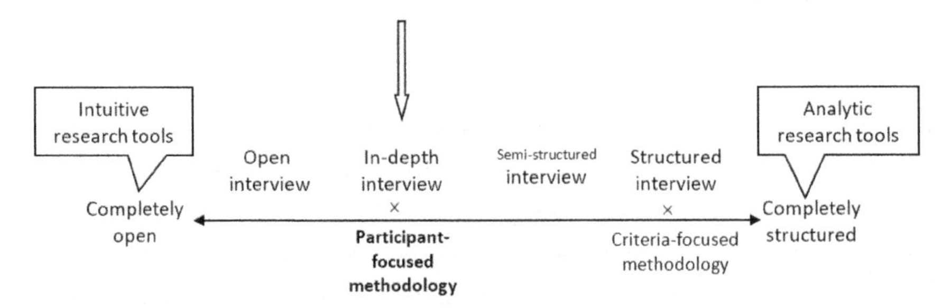

Figure 11F: The Characteristics of In-depth Interviews

The in-depth interview, like other tools commonly used in the participant-focused methodology, gives augmented expression to the researcher's intuitive inquiry skills, as well as a high degree of expression to the analytical skills that ensure the control of the research process. Note that all types of qualitative interviews, the more structured or the open type, express both features. The differences between the methodologies and methods concern the relative significance of each (Shkedi, 2003, 2005). Another primary source can be a participant's diary, but this is certainly not a common source of information because it is based on an initiative of the participants that requires exceptional effort. It should be noted that in some cases, such as when conducting research in the "life story" mode, the researcher will prefer to hold an open interview, i.e. an interview with one broad, open question that invites the interviewees to tell their stories freely and without direction.

Observations are a source of secondary information, an undoubtedly significant source of information in qualitative research. There are different types of observations which differ in the position given to the researchers on a continuum between pure participation to pure observation; from an external position on the investigated phenomena to an involved, participatory position. (Jorgensen, 1989; Marshal & Roseman, 1989 Tedlock, 2005). Since the participant-focused methodology seeks to understand the phenomenon under investigation from the perspective of the participants, the type of observation that characterizes this research pattern will be one with a high level of involvement and even participation. Observations of a "participant observation" or "involved observation" (Shkedi, 2003, 2005) are suited to this type of methodology. Figure 11G describes the characteristics of involved observation and participant observation, as compared to other types of observation:

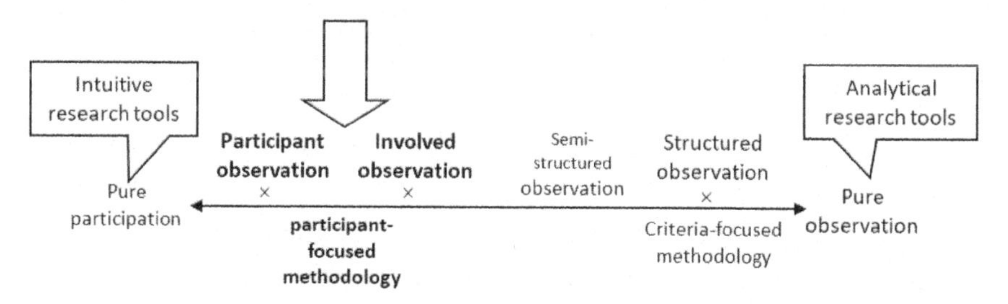

Figure 11G: Characteristics of Involved Observation and Participant Observation

Note that even here, as in the case of the interview, none of the observation types are pure observation or pure participation, but some kind of mixed variety, reflecting the complex nature of qualitative methodology. The choice between involved observation or participant observation, both of which are relevant to the participant-focused methodology, will result from the extent that the researchers stay on the research site. If a relatively long stay is needed to allow researchers to participate and not just to observe, the observation will be of the participant-observation category. Should researchers spend a shorter time on the research site, the method will be involved observation. According to these principles, the researchers will identify and even collect relevant documents and objects.

As mentioned in previous chapters, the data collected in interviews and observations must, of course, be kept in the documentation and analysis. The challenge facing the researchers in participant-focused methodology is to transform the secondary sources into primary sources, and thus increase their validity in the study. We can identify two ways to go about this. One is to increase the duration of the observer's stay in the field being investigated, or in other words, to become a "participant –observer." The assumption is that as the researcher becomes more of a participant in the phenomenon under study and spends more time as a participant-observer, he comes to directly know the phenomenon as perceived by the participants, and the data they possess can indeed be considered as primary sources (Fetterman, 1989; Rachel, 1996; Creswell, 1998; Siverman, 2006). However, most qualitative studies are relatively limited in time, and it is difficult to speak of a real "participant observation." Therefore, another possibility is to conduct "stimulated-recall interviews" following the observations and/or documents and objects collected. In the stimulated-recall interview, the participants "were encouraged to reflect upon what had been observed and to bring their own sense-making processes to the discussions in order to co-construct a 'rich' descriptive picture" (Bishop, 2005, p. 116). As part of the stimulated-recall interview, the participants are referred to video or audio tapes, to a segment of text, documents or objects and asked to describe and explain the recorded material as they perceive it (Shkedi, 2003; 2005). In research that adheres to criteria-focused observations, and in some other qualitative methodologies, the need to use the stimulated-recall interview does not exist, since the predetermined theory may be the anchor through which the participants interpret their observations.

The Research Categories / Themes

As mentioned in previous chapters, the use of categories (or "themes," as some researchers prefer to call them) in qualitative research characterizes all qualitative methodologies. But there is a difference in the status of the categories and in their location in the sequence of the research stages, depending on the characters of each methodology (Polkinghorne, 1995). In criteria-focused methodology, the categories arise directly from the theoretical framework of the study, determined in the preliminary stages of research and maintaining constant definitions throughout the entire study. These categories lead the researchers in defining the research questions, collecting data and analysis, and writing the final report. In contrast, in participant-focused methodology the categories are determined during the study and with the involvement of the researchers in the phenomenon being studied. The categories are not formulated according to external standards, but depending on what emerges from the discourse of the participants. In this process, there is a dialogue between the data and the theoretical perspective, while the analysis process itself has a crucial part in determining the categories.

Categories are used by researchers to sort the data, for distinction and separation within the raw data intended to identify the meaning of the data. In criteria-focused methodology, where the categories are determined before the stages of collecting and data analysis, the process of determining the categories is essentially deductive. In participant-focused methodology, the categories are determined during the process of collecting and data analysis, where the process is essentially inductive (Perakyla, 2005). In the process of collecting data, researchers have essentially made an intuitive data analysis which guides them in the interviews and observations. Therefore, it is likely that already in the stage of data collection, the researchers have very preliminary ideas about the possible categories. These concepts help them to gather data, but do not serve as permanent categories. They may change as this process progresses in the stage of formal data analysis.

The formal process of creating categories in the participant-focused categories is in fact created during the data analysis phase, after the researchers have amassed significant data. As mentioned above, this does not mean that researchers are missing ideas about categories in the early stages. At this point they can also write up their initial ideas as memos, but not structured categories. Creating categories is in fact "the first step in taking an analytic stance toward the data" (Charmaz, 2005, p. 517). While in the earlier stages the researcher also has some degree of analytical reference, the intuitive skills are dominant. At the formal analysis stage, there is a certain balance between the researcher's intuitive and analytical skills.

The process of creating categories is aimed to give meaning to the data by comparing between different data segments to find similarities, differences and connections (Seidel & Kelle, 1995). The categorization process is actually a process of conceptualization; the researchers create a system of concepts that gives meaning to the data (Ryan & Bernard, 2000; Gall et al., 1996). Therefore, the categorization process is linked to the recreating of the theoretical perspective. In participant-focused methodology, the categories are created from the data by examination and analysis, while continuing to clarify the theoretical framework of the study (Araujo, 1995), as illustrated in Figure 11H.

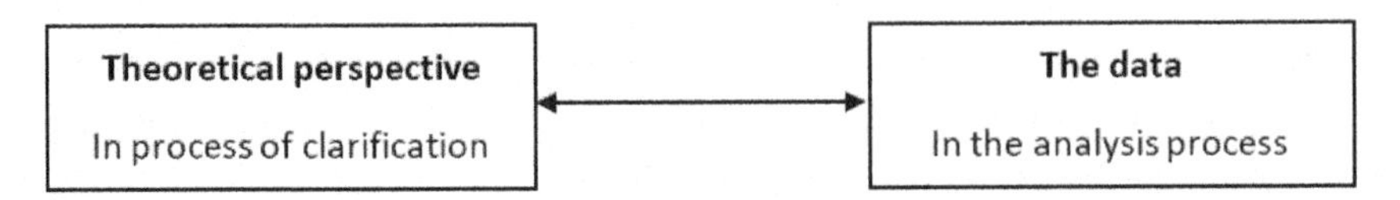

Figure 11H: The Interaction between the Data and the Theoretical Perspective

Categories are created from a permanent "dialogue" between the tentative theoretical perspective, existing in the process of examination, clarification and changing, and the data collected in the field study, as illustrated in Figure 5H. This "dialogue" creates a close encounter between these two elements, each of which becomes clearer by its meeting with another element. In this process, we slowly move from the induction process (with emphasis on the data) to a tension between induction and deduction, when the relative weight of the deduction increases as we advance in the research process (emphasis on theoretical perspective) (Strauss & Corbin, 1990). The theoretical perspective may be changed by the encounter with the data, and the data may take on a new meaning in light of the meeting with the possible theoretical framework. Categories will be determined when a match is created between the theoretical framework and the data, which gives the data a probable meaning. Figure 11-I demonstrates this process:

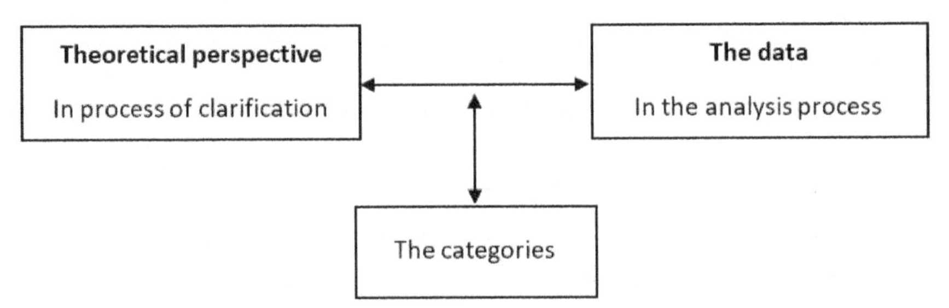

Figure 11-I: The Process of Creating Categories in the Participant-focused Categories

These categories are not detached, but are part of the array that links them together. In this way, we obtain a kind of "category tree," led by the main category ("trunk"), with the upper categories branching out ("main branches"), and sub-categories ("secondary branches"), as demonstrated in Figure 5J. We can identify two dimensions to the categories: a vertical dimension, in which the categories are distinct from each other according to their conceptual level, when the main and upper categories reflect a higher level of conceptualization and the low categories reflect a lower level of conceptualization, which is closer to the data wording. The horizontal dimension indicates all categories belonging to the same "family" (and listed under the same category). In this way, each horizontal level can contain several category families. Figure 11J illustrates a partial list of categories drawn from research on teachers' knowledge.

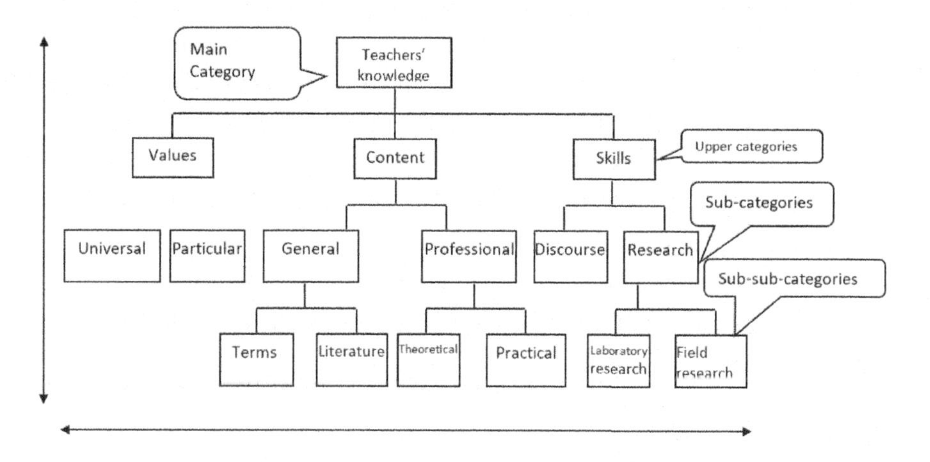

Figure 11J: Category Tree: Vertical and Horizontal Dimensions

As shown in Figure 11J, the category of "Teachers' Knowledge" is the main category. All categories under it, both on the vertical and the horizontal axis, reflect the characteristics of this category. The main category is broader in its nature, reflecting the higher conceptual level. The categories placed in the lower parts of the tree are characterized by a lower conceptual level and closer to the participants' language formulation. The line under the main category contains three categories, "skills," "content," and "values," whose horizontal arrangement indicates their equality, i.e., the three belong to three separate but equal "families." Conceptually, these categories reflect a lower level of conceptualization than the main category above them.

In the same way, we define the relationship between the upper categories and the sub-categories, and between the sub-categories and the sub-sub-categories.

Data Analysis

In participant-focused methodology, as the name implies, the categories are "grounded" in the data collected from the lives and contexts of the participants" (Gordon-Finlayson, 2010, p. 154). The category creation is carried out through the process of analysis and as an integral part of it. Therefore, in this methodology, researchers were not approached at the analysis stage with predetermined categories, but let the data "speak." The first phase of data analysis in the participant-focused methodology is the division of the data into small segments, according to subject, without regard to the relationship between the different segments. At this point, each data segment is like an initial category, but is still not a real category. By dividing the data into initial categories (data segments), we prepare the foundation for the entire category array in the next analysis phase. At the end of the initial analysis phase, we obtain a collection of data segments, where each segment name reflects its contents. Figure 11K illustrates the distribution into data segments under the main category. It is understood that the actual number of segments is several times larger. As can be seen in Figure 11K below, the array of the data segments reflects the connection of all data segments to the main category of the study, with no attempt to indicate further connections.

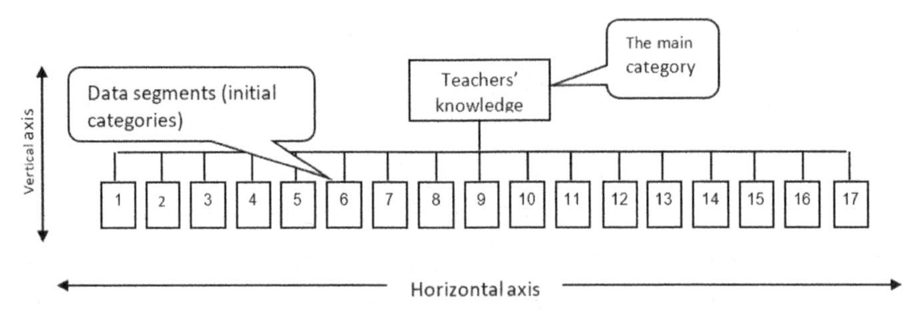

Figure 11K: Division into Data Segments (Initial Categories)

In the second stage of the analysis, the mapping stage, researchers construct a complete array of categories. This category array is created by finding thematic connections between the various data segments. This connection reflects a situation in which a number of segments deal with the same subject. The fact that they deal with the same subject (or with a close thematic connection) justifies joining them together in one upper category under the main category. In this way, we conduct several upper categories ("families"). The location of each upper category expresses its relationship to other categories, and they are positioned according to their relationship on the horizontal and vertical axes (Gordon-Finlayson, 2010; Charmaz, 2005). Each upper category can be re-divided into sub-categories, and then into sub-sub-categories. At the end of the process, all the data segments (primary categories) will be organized under categories. (It is quite possible that some data segments will be placed in more than one category.) We thus create a "category tree" with the summit featuring the upper categories, with a higher level of conceptualization. At the base of the tree are the sub-categories, with a lower level of conceptualization. The data segments are located in the lower categories. Figure 11L displays the mapping analysis array sorting the data segments (primary categories) to the lowest sub-categories. The process of positioning the data segments into the appropriate sub-categories overlaps simultaneously in the process of creating categories.

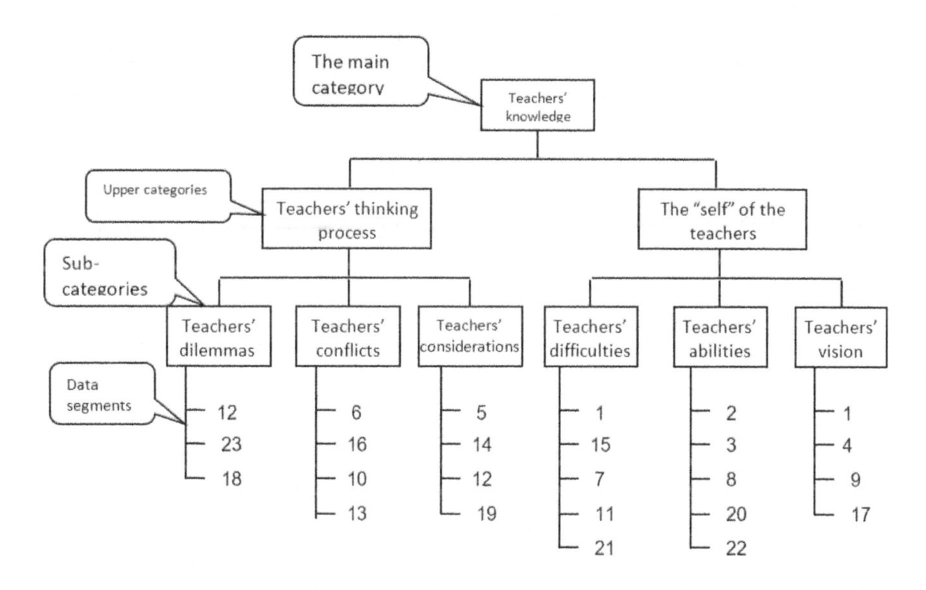

Figure 11L: The Mapping Category Tree

The mapping analysis stage, as its name expresses, maps all the data segments into categories. This is the basis for the next analysis stage, the focused stage, where researchers determine the focus of the study. In other words, which category best reflects the general phenomenon under investigation, or which categories can be the main and upper categories and which categories and segments will be under them. In this sense, the mapping stage may offer several options for directions to the research (a number of options for selecting the main and upper categories), and researchers must decide what they deem the most significant and richest. In the course of this step of choosing the appropriate categories, we obtain a new category "tree" with a new main category that reflects the theme of the study. Figure 11M illustrates the focused tree which developed from the mapping tree (see Figure 11L). It is important to note that the main category of the focused analysis is necessarily one of the categories of the analysis maps, and can definitely be a new category emerging from the insights gained from the process of analysis and based on the mapping analysis phase.

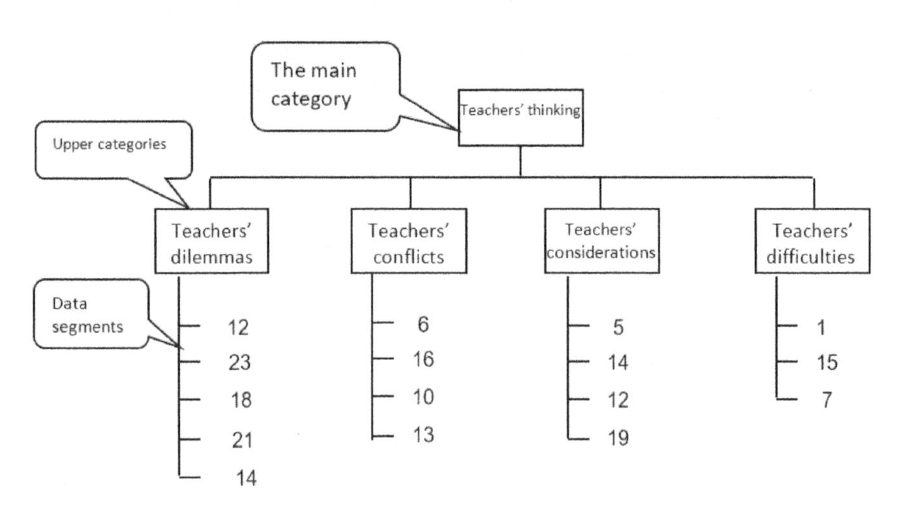

Figure 11M: The Focused Category Tree

In many cases, the researcher using the participant-focused methodology is satisfied with a descriptive picture of the phenomenon under investigation. In that case, we can certainly see the end of the focused analysis phase as the end of the entire analysis process. The categorical-descriptive picture can be the basis for the next stage of the research, the organization of the research products for displaying the description of the phenomenon under investigation.

However, in many other cases, the researcher continues to go through the analysis to another stage, that of theoretical analysis. The theory in this context is a conceptual explanation for the phenomenon under study. This is not a predictive theory and not characterized by a conceptual density, as is customary in positivist theoretical concepts. Rather, this theory reflects the conceptual characteristic of the phenomenon under study that can apply to immediate and specific situations only. The phrase "theoretical explanation" is suitable for the nature of this theory (Strauss & Corbin, 1990; Flick, 1998; Dey, 1999; Charmaz, 2005; 2006). It is based directly on the descriptions of the subjects. The theoretical picture is created in a process of translating the descriptive picture and the descriptive categories into theoretical concepts that provide a theoretical explanation for the phenomenon under investigation (Araujo, 1995). Figure 11N illustrates the translation process:

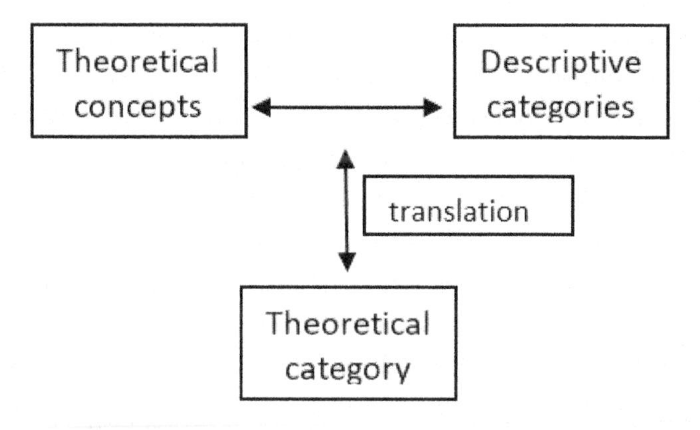

Figure 11N: Translation of Descriptive Categories into Theoretical Categories

At the theoretical analysis stage, the researcher compares the descriptive category array accepted in the previous analysis stages with theoretical concepts taken from literature in order to translate into a theoretical category array (Chamaz, 2005; 2006). The picture that emerges in this process is a theory stemming from the data which is a "grounded theory," (Glaser & Strauss, 1967; Gordon-Finlayson, 2010), one deriving from the description and picture of the participants' discourse. This creates a categorical array whose higher levels are the categories formulated with theoretical concepts and lower levels are more descriptive categories. For a more detailed explanation, see Chapter 9.

Sometimes the researcher is not satisfied with conducting a theoretical explanation derived directly from the data by way of translating the descriptive categories, but wishes to provide a theoretical explanation by way of second-order analysis (Shkedi, 2004). The theoretical analysis obtained by way of translating descriptive categories into theoretical categories points to a direct relationship between the theoretical explanation and the descriptive picture. However, in many cases we cannot create a direct link, even though researchers notice that they can suggest a theoretical explanation based on several separate descriptive pieces taken from the descriptive category array and to employ a theoretical explanation taken from theoretical literature. (For extensive reference to this option, see the section "First and Second- Order Questions" above.)

In practice, the second-order theoretical analysis presents a main category that expresses a theoretical link between two or more groups of descriptive categories. We call this process a second-order theoretical analysis, since the theoretical explanation is not implied directly from the participants' discourse, but taken from theoretical literature and considered suitable for the descriptions received. There is a substantial difference between the process of second-order theory, and the theories that guide the criteria-focused methodological process. While the process described in this chapter deals with the theory that emerges after the process of collecting and analyzing data, in the process of criteria-focused methodology, the theories are identified before the process of collecting and analyzing data and guide the research from its first to last step. In criteria-focused methodology, we do not create grounded theory, but come to the research with a predetermined theory (Shkedi, 2003, 2005). For a more detailed explanation of the second-order analysis, see Chapter 9.

Final Research Report

The research report in participant-focused methodology usually includes all the acceptable components of the research report, as presented in the previous chapter. In many cases, the order of the components in the final report will be "standard," which means it starts with a literature review, and continues with a presentation of the research questions, methodology, research context, findings, etc. However, there is no required order, and sometimes researchers may change the order to highlight the message that they wish to deliver. The nature of the participant-focused methodology may also point to the title of each report component. For example, instead of the common research name "Findings," this methodology may prefer to call the section "Description of the Cases" or "Analyzing the Cases," etc. (Crowley, 2010).

It seems that in participant-focused methodology, the case-report component might be the most challenging. Since this methodology is in a linear order but with interlinked components, the researchers continue the work of analysis and insights in the stage of the case-writing, which may perhaps even cause them to return to the formal stages of the analysis. Therefore, the writing phase should not be seen only as a communication process, but an integral part of the study. "Writing is inquiry [...] We write to know. We write to learn. We write to discover." (Ely, 2007, p. 570)

There is not one type of case description. As with any methodological pattern of qualitative research, the research genre preferred by researchers largely determines the nature of the description. If the research genre is a "case study," and the study is limited to one case, the case report will be the most appropriate form of description (Creswell, 1998). If the research genre is a "multiple-case study," the description of many cases would be the appropriate description (Stake, 1995). If the research genre is "ethnography," the cultural depth description will be the preferred form (Silberman, 2006). If the research genre is "narrative research," the description will show the story of a single or several subjects. If the research genre is "grounded theory," (Strauss & Corbin, 1994) the researcher chooses a theoretical description as the preferred form of summary, and so on for other research genres.

However, within each of these genres, the diverse options will be also varied. We can equally emphasize all the issues under study, or highlight several issues to some degree. We can combine the theoretical component with the descriptive component, or separate them. We can combine bibliographical references in the description or leave them to the literature review. The description can be in style of focused-categories, namely based on a description of the number of categories as the backbone of the description, or written descriptions which focus on a plot, i.e., the dimension of time would constitute the main axis of the description. We can, of course, combine several of the above options. Research dealing with a number of cases or number of subjects can be viewed in a parallel and equal form in all the different cases, or we can select a number of representative sub-cases and display them intensively while presenting the other sub-cases in a more comprehensive way (Shkedi, 2003; 2005). Chapter 20 suggests a detailed explanation of how to organize the analysis outcome of the study as an appropriate source for writing the final report.

Research Quality Criteria

As mentioned in the previous chapter, the criteria-focused methodology refers to "objective" as a concept that defines the quality of the research (Kirk & Miller, 1986). In contrast, the participant-focused methodology, which consists of the constructivist-interpretive approach and seeks to view the phenomenon under investigation through the eyes of those who experience it, sees the concepts of truth and factual as necessarily subjective. The constructivist approach assumes that any conception of knowledge and the evaluation of knowledge occurs within a conceptual system in which the world is described and explained, and therefore cannot be a general objective description. Each description is a necessary perspective by its nature (Schwandt, 2000; Lincoln & Guba, 2000; Smith & Hodkinson, 2005).

The accepted concepts of quantitative literature examining the quality of such research are validity, reliability and generalization. Dealing with the unique nature of the participant-focused methodology and the constructivist-interpretive approach, several researchers offer alternative concepts as a norm for research quality, such as credibility, transferability, dependability, and others. All these terms are supposed to grant the research the quality label of "trustworthy" (Guba & Lincoln, 1998; 2005; Arksey & Knight, 1999). Other researchers prefer to use the standard terms of validity, reliability and generalization, but to accompany them by emphasizing the unique nature of qualitative research (Stake, 1995; 2005; Greenwood & Levin, 2005 Shkedi, 2003; 2005)

The concept of validity relating to the question of to what extent the findings are consistent with the theoretical perspective of the study and interpreted accordingly, can be applicable to the participant-focused methodology. By considering the data, the research process and the research final report, the researchers, colleagues and readers can examine the relationship (or lack thereof) between the theoretical perspective (shown in a literature review and research assumptions) and the findings and their interpretation. Since we are concerned with qualitative research that shows its process and products in the language of words and not in the language of mathematics, researchers not only need to ensure congruence between the theoretical perspectives of the research process, but to expose them to a constant test, both to themselves, their colleagues, judges, and the readers of the study.

Researchers need to show the chain of evidence (interviews, observations, documents, etc.) of all stages of research to examine if they are maintaining the association between the data and its analysis with the theoretical perspective presented the final report (Mason, 1996; Huberman & Miles, 1994). "[...] an outsider should be able to look at your audit trail and see exactly how you got from the raw data to the claims made in the analysis which led to the conclusions drawn." (Shaw, 2010B, p.183) However, the external person who reads the study, its process and findings, must consider it through the theoretical perspectives formulated by the researchers rather than through any external "objectivity" in theoretical perspectives. While in criteria-focused methodology it is possible to use external judges not involved in the study to examine the research "objectively" and determine the degree of validity, this process is impossible in the participant-focused methodology. This methodology, based on the assumption that every research process is a result of dialogue between the participants and the researchers and the picture that emerges is an expression of a unique encounter, cannot be dependent on "objective judges" from an external perspective (Merrik, 1999). Nevertheless, it is certainly desirable, if not necessary, for the research process and the product with all its components to be exposed to the consideration of peers, those who are willing to devote time and to view the phenomenon through a unique theoretical perspective in which the researchers looked at the phenomenon. This is the appropriate way to examine the validity of the study.

Reliability is the ability to repeat the same research process of the study and to achieve the same results, even if the research is carried out by other researchers (Yin, 1994). It is difficult, or even impossible, to achieve the classical concept of reliability in any qualitative methodologies and in the participant-focused methodology in particular (Schofield, 1989; Merrick, 1999). The way in which researchers seek reliability is similar to the way they claim validity. The researchers document their research process and present a chain of database evidence, and present a detailed description of the findings in the final report (Pidgeon, 1996; Mason, 1996). In this way, the researchers themselves, colleagues and external readers will be able to examine the reliability of the study from a stated theoretical perspective of the researchers (Peshkin, 2000; Dey, 1994).

As in validity and reliability, even the concept of "generalization" (external validity) seeks a unique meaning when dealing with participant-focused methodology. Qualitative researchers suggest alternative concepts to the term "generalization" that emphasize the uniqueness of qualitative research. Apparently, there is a contradiction between the concept of generalization and the sense of this methodology which is based on the constructivist interpretive approach, stressing the uniqueness and specialization of each phenomenon under investigation. Of the three types of qualitative research generalizations, the case-to-case generalization is perhaps the most relevant to this pattern of methodology (Firestone, 1993). Since the participant-focused methodology is characterized by rich, extensive descriptions of the phenomenon under investigation, it is possible to compare a new live case to the specific case studied. The process of case-to-case generalization can occur in a kind of a dialogue between the investigated case and new live cases, with this dialogue taking place between the readers of the research report and the description of the research report (Schofield, 1989). Two other types of generalization, analytic theory (generalization through theory) and generalization to the population will be more problematic in the case of participant-focused methodology. Analytic generalization is possible only if the final research product will not be merely a descriptive picture, but also present a theoretical picture. Generalization to population will be possible only if the study deals with a relatively high number of subjects and investigated cases (Arksey & Knight, 1999; Marshal & Roseman, 1989; Dey, 1993; Firestone, 1993).

Triangulation of sources and methods and varied types of data collection using different methods in different combinations (Fontana & Frey, 2005) may enrich the quality of research and strengthen the argument for the qualitative validity, reliability and generalization (Shkedi 2003, 2005).

CHAPTER 12

THE CRITICAL-CONSTRUCTIVIST FOCUSED METHODOLOGY (THE CONSTRUCTIVIST-CRITICAL APPROACH)

The critical-constructivist focused methodology (or in short the critical-focused methodology) is identified largely with the constructivist–critical research approach (Guba & Lincoln, 2005). This methodology expresses a research mode comprised of several paradigms based on different philosophies, which we shall refer to by the overall title "the constructivist-critical approach." The source of these critical philosophical approaches comes from the academic school called the "Frankfurt School" of Germany in the 1920's. (Creswell, 1998; Kincheloe & McLaren, 2005). These paradigms reflects different philosophical schools which vary from one another in their philosophical sources and assumptions, known by such names as critical theory, post-modernism, neo-Marxism, feminism, post-colonialism, queer theory, post-structuralism, deconstructivist theories, and others (Denzin & Lincoln, 2000; Guba & Lincoln, 1998).

Each of the critical schools has a different set of arguments, and often there are profound disputes between them. But beyond the disagreements and debates, it appears that they have a common research goal that could be called in one word "emancipation." Emancipation as the goal of the study can be seen as an emancipation of the people from the shackles of conventional knowledge and discourse in society, and equally as an emancipation of the human from political and social shackles (Kamberelis & Dimitriadis, 2005). Several paradigms place greater emphasis upon political and social emancipation, while others stress the freedom of human discourse, and most emphasize both. The methodology specified in this chapter is therefore related to quite a number of different research paradigms that appear in literature under different names.

The methodological patterns of qualitative research described in this book are presented in a sequence in which one side indicates a penchant for intuitive research skills, internal research perspectives, and the participants' points of view, while the other side gives expression to analytical research skills, an external research perspective, and the theoretical points of view (Fetterman, 1989; Gall, Borg & Gall, 1996). As such, the participant-focused methodology is placed at the end of one side of the continuum, while criteria-focused methodology is located at the opposite pole. The critical-focused methodology, like the participant-focused methodology, is based on constructivist research assumptions, which consider the phenomena under study to be a result of human construction. The participant-focused methodology learns the meaning of such phenomena by listening to the discourse of the participants, and describing and understanding them from a close proximity. The critical-focused methodology also heeds the participants' discourse and sees it as a central information source, but considers it as external critical criteria that assign meaning to the participants' discourse that is not necessarily identical to what the participants themselves understand and explain. The critical-focused methodology comes to the study field with assumptions about the social, cultural and political forces that affect the discourse of the participants, without their necessarily being aware. According to the logic of these approaches, the researchers' advantage is that they, unlike the participants, are aware of these forces and can look at the discourse while assuming the existence of these forces. From this, as we see in Figure 12A, the methodology can be placed near the pole of intuitive research skills and internal aspects, but with a shift towards analytical skills and an external perspective which characterize the critical theories.

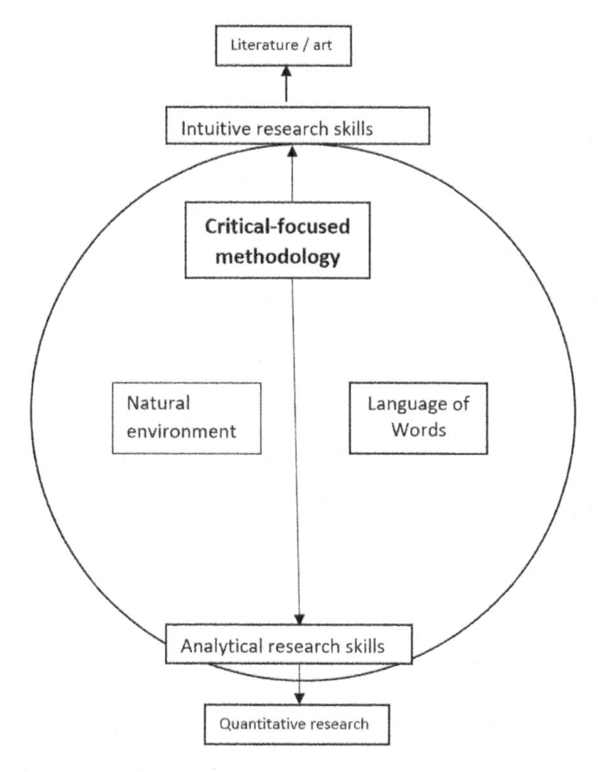

Figure 12A: The Location of the Critical-focused Methodology in the "Space" of Qualitative Research

It is worth remembering that Figure 12A places the critical-focused methodology in the realm of the methodological aspects, not in the context of the research paradigm. Thus, this does not connote that the critical-focused methodology is characterized by a lesser degree of constructivism in comparison to the participant-focused methodology.

The Critical Approach versus Traditional Approaches

The critical-focused methodology includes studies that express approaches considered by the critical theoreticians to be an alternative to the traditional approaches. These approaches are alternatives to the positivistic and post-positivistic research approaches on one hand, and alternatives to the constructivist-interpretive approaches the other (Phillips & Hardly, 2002). The philosophy behind the critical-focused methodology originated from the criticism of the dominant positivist approach and its great influence on the ideologies of the twentieth century. The positivist approach that had yielded success in the fields of natural sciences led researchers to adopt it in the social sciences. This philosophy is regarded by the critical theorists as being dangerous to human society, by its tendency to interchange the processes of reasoning into techniques. Science becomes an ideology that guides social activities (Carr & Kemmis, 1986). On the other hand, they also criticized the interpretative-constructivist approach as being a conservative approach only interested in describing reality and accepting reality as it is (Carr & Kemmis, 1986). According to this view, the interpretive-constructivist approaches are "only being concerned with the surface feature of society, and for either serving the interests of economically dominant groups or not engaging with political or moral issues." (Travers, 2001, p.115). Critical approaches expressed in the critical-focused methodology try "to tear away the masks and illusions of consciousness" (Josselson, 2004, p. 13th) that the interpretive approaches offer.

Common to all types of research employing the critical-focused methodology is the assumption that the knowledge is not objective. Habermas, one of the philosophers associated with the critical approach, rules out the possibility that knowledge is a result of pure intellectual work. He argues (1971) that knowledge is created on the basis of needs and interests. Habermas believes that it is helpful to sort knowledge into three different types, according to the interests guiding its creativity (Carr & Kemmis, 1986; Kemmis & McTaggart, 2005): technical knowledge, characterizing the positivist and post-positivistic approaches (expressed by the criteria-focused methodology and partial criteria-focused methodology); practical knowledge characterizing the interpretive-constructivist approaches (expressed by participant-focused methodology); and emancipation knowledge, characterizing the critical approaches (expressed by critical-focused methodology).

Technical knowledge reflects the desire of the human being to be in control of the world, and is expressed in the accumulation of scientific knowledge. This is the knowledge of the human scientific and technological realms. In Habermas' opinion, this knowledge reflects a necessary interest, but it cannot be the only legitimate knowledge, certainly not as far as in society and human life. Practical knowledge reflects the desire of mankind to understand and clarify the human world, to understand the conditions necessary for action in the world, and to maintain communication and meaningful dialogue between people. Thus, practical knowledge creates interpretative, contextual and dynamic knowledge which can guide people in everyday life. However, this knowledge, with all its importance, does not consider the possibility that knowledge itself "may be systematically distorted by the prevailing social, cultural or political conditions" (Carr & Kemmis, 1986, p. 134).

Emancipatory knowledge reflects the assumption that knowledge is not fixed, but is a product of historical, cultural and social conditions. "Knowledge is seen as historically and culturally specific." (Sullivan, 2010, p. 30) These contexts can be changed over time and affect the construction of knowledge (Guba & Lincoln, 1998). Culture imposes the workings of the mind, and the decisions that people take reflect the kind of cultural consensus or dominant view of the elite in society. This situation exists even if the people are not aware of the deep sources of their considerations and the forces operating upon them. The interpretive-constructivist approaches see indeed that people work in that context, but believe that they can actually be aware of the context. The critical approaches, in contrast, see the context as a power which is beyond the consciousness of the people. Sometimes it is expressed in differentiating between the micro level, which refers to an individual person, and the macro level, which refers to society as a whole. The assumption is that social-cultural forces (macro level) are activated in people (micro level), but it is not perceived in many cases by the individual person (Travers, 2001). The purpose of the critical-focused methodology is to understand the macro level, the context in which people are operating, and at the same time to understand the micro level, the perceptions of the specific people (Moss, 1996; Phillips & Hardly, 2002).

Epistemologically, there is a clear difference between the approaches that characterize the critical-focused methodology and other methodologies, as discussed in the previous chapters (Kincheloe & McLaren, 2005). According to the conception that guides the criteria-focused methodology, the researcher is mainly a service tool of the theoretical perspective and functions primarily as a distant objective observer, and the investigated knowledge does not depend upon the researcher. According to the assumptions that guide the participant-focused methodology, the researcher is involved in the phenomenon being studied and wishes to see it through the eyes of those who experience it, and the investigated knowledge is a result of interaction between the researchers, the phenomena, and the people being studied (Carr & Kemmis, 1986; Guba & Lincoln, 2005). According to critical-focused methodology, the researchers and the phenomena under study are also in an interactive relationship. Their values, as in the case of participant-focused methodology, are involved in the research (Guba & Lincoln, 1998, 2005). But critical-focused methodology assumes that the researcher "has a superior insight into the nature of human happiness and well-being, and how to achieve this, than most ordinary members of society." (Travers, 2001, p.114). These insights are the result of critical observation, and not of superior human qualities. While the participant-focused methodology gives decisive weight to the participants, the critical-focused methodology tries to look beyond what is said by the participants and give great credence to the social, historical and political factors surrounding them, which are seen as background elements. (Carr & Kemmis, 1986; Philips & Hardly, 2002; Travers, 2001).

Between Theory and Action

The critical-focused methodology is based on approaches that offer a different relationship between theory and action than are implied by the positivist and interpretive-constructivist approaches. According to the positivist approaches, the theory guides the action and suggests a direction to human activities. According to the interpretive-constructivist approaches, the theories do not offer a direction, but portray characteristics and contextual conditions based on past investigated situations as a basis for future action. It is expected that past experience will be taken into account by the practitioners, and they will make the decisions on how to act (Schon, 1983, 1987). The critical-emancipation approach requires more than a characterization of the contextual situations; it requires granting the practitioners those methods by which they can distinguish the powers that distort and limit the expression of liberated people (Carr & Kemmis, 1986).

Understanding the concept "ideology" and its place in the minds and actions of people can help us understand the expression of the critical approaches. Ideology is perceived by the critical approaches as a representation of the social structure. This representation acts upon individuals and affects their minds and actions without their being aware of its origin. According to this view, people delude themselves into thinking that the sources of power are caused by agents such as God, market forces, natural sources and the like. Ideology is like a film of reality, and as such overturns and distorts the reality (Idan, 1990). On this basis, ideology is seen as a mechanism of control over the people. However, according to Zvi Lam (2000), people are unable to supervise ideological thinking, but the ideology is their supervisor. This is demonstrated by the fact that only a few adherents of ideologies abandon them, and only then in very extreme situations. "Researchers operating with an awareness of this hegemonic ideology understand that dominant ideological practices and discourse shape our vision of reality." (Kincheloe & McLaren, 2005, p. 310) Based on this concept of ideology, the critical-focused methodology seeks to clarify the ideological origins, and to release people from their unconscious grasp of the hegemonic ideology.

The crucial difference between the research objective of the critical-focused methodology and other methodologies is expressed in the famous statement of Karl Marx: "The philosophers have only interpreted the world in various ways. The point, however, is to change it." The critical approach argues that explaining the world (the purpose of the positivist approach) or understanding it (the purpose of interpretive-constructivist approaches) are the only components in the process, but not the total purpose of the research process. The study is not a process that focuses on the world, but a process that occurs in the world (Carr & Kemmis, 1986). Thus, the research work of scholars engaged in critical-focused methodology will strive to emancipate people and society. It "should articulate a politics of hope. It should criticize how things are and imagine how they could be different."(Denzin, 2000, p. 916)

The aim of the critical-focused methodology "is the critique and transformation of the social, political, cultural, economic, ethnic, and gender structures that constrain and exploit humankind" (Guba & Lincoln, 1998, p. 211). The way to emancipate people from false ideas and ideologies is to make "them aware of the social institutions that have caused them to think in this way." (Travers, 2001, p.115). The implications of this methodology exceed theoretical criticism, and direct toward critical action which will be reflected directly in changing the social activities (Carr & Kemmis, 1986).

Knowledge as Power

A central concept in understanding the critical approaches is the concept of "power." "Critical research attempts to expose the forces that prevent individuals and groups from shaping the decisions that crucially affect their lives." (Kincheloe & McLaren, 2005, p. 308) "Power," according to the Marxist theory, means imposing one group's ideological desire upon that of a different group. This is evident in the oppressed groups embracing the world views of the oppressed groups, even though these views are incompatible with their interests. According to the Marxist view, this power is exposed and blunt, and usually reflected by a move from above to below, from the oppressor to the oppressed (Eagleton, 1991). Foucault (1980) presents a more complex picture of power, where power is an unperceivable network that penetrates even unto our most private utterances. Power is not necessarily something visible that is imposed upon others, but a network or web of common relationships in society. Power relations are perhaps the most hidden phenomena in the social body (Foucault, 1980).

Individuals belonging to Western society have internalized the social norms, so that they usually do not experience them as being perpetrated by institutions. These practices are actually seen as natural, and it is hard to imagine how our life would have been without them (Mills, 2003). According to Foucault, knowledge and power are closely related. Power will not be activated without knowledge, and it is impossible that knowledge does not bring about power (Bar-On & Sheinberg-Taz, 2010). Those in a position of authority who are perceived as "experts" will produce the knowledge and determine what the truth is. The disciplines which are determined by the academic experts will define what can be considered as acceptable knowledge within the defined content area and according to rigid methodologies that the experts created. According to Foucault, universities and schools are the institutions designed to preserve this kind of "legitimate" knowledge (Mills, 2003).

The concept "power" prompted another major concept: "hegemony." According to Foucault, hegemony is a social situation in which those who are subjected to others adopt their values and ideology and accept them as their own. This leads to accepting their place in the hierarchy as natural or as something that works in their favor. Foucault argues that knowledge should not be considered as something objective and detached, but seen as an entity that always operates in the interests of certain groups. Foucault brings a variety of examples from various fields, including one from the field of psychotherapy. He claims that this medicine has produced much relief from suffering, but at the same time has placed the medication for madness into the hands of psychiatrists and health care professionals, and consigned a greater label of disgrace to mental illness. Some confusion exists in the distinction between sanity and insanity, and between normal and abnormal, and each case where seemingly unusual behavior exists is perceived as a case of mental illness (Foucault, 1986).

The critical researcher's task is to explore and uncover the hidden powers of the hegemonic ideology. "Each agrees that the dominant perspective's pretense of neutrality serves to divert criticism of the dominant ideology. By giving us ground to stand on outside the dominant approach, the critical approach enabled us to examine critically the technical production perspective, to identify its blind spots, and to understand its political and social implications." (Posner, 1998, p. 96). Karl Marx, for example, argued that the bourgeoisie of the 19th century not only enslaved the proletariat, but also controlled the institutions where ideas are voiced, such as schools and the mass media. "The objective of the critical theorist is to challenge this 'dominant ideology' by suggesting that things can be different, but also by explaining why it is that they come to hold false beliefs." (Travers, 2001, p.114). One of the characteristics of the critical-focused methodology is the desire to "give voice" to marginalized or oppressed groups, to allow them to voice their stories and thus protest against hegemonic groups that protect the existing social order which all take for granted.

To refine the methodological difference between the critical-focused methodology and other methodological patterns, we can apply a definition from the hermeneutics of Scripture (Josselson, 2004). Ricoeur (1970) distinguishes between the hermeneutics of suspicion, which he attributes to Marx, Freud and Nietzsche, and the hermeneutics of restoration of a meaning. The hermeneutic of restoration of a meaning is characterized by the willingness to listen and absorb as much as possible of the author's message, based on the cultural background in which it operates and works. This is the hermeneutics of faith, and appears to be consistent with the participant-focused methodology. The hermeneutics of suspicion demonstrated by Cohen (1999) uses Freud's book *Moses and Monotheism.* Here Freud compares the story of Moses and other ancient mythical stories with a lack of confidence in the way the text presents itself as a tale of a unique history, and assumes that the Biblical author is unaware of the psychodynamic forces that affect his writing of the Bible story. Freud, as a researcher, sees its role to help the reader uncover the hidden psychodynamic reality in the story. It seems that the hermeneutics of suspicion is consistent with the critical-focused methodology, which seeks to release the stories from the meaning that the saying or writing gives.

Various Critical Approaches

As mentioned, under the heading "critical-focused methodology" are various critical theoretical approaches, and this section aims to present them, or at least the most prominent among them. But before introducing the different approaches, we shall focus upon the post-modernism approach, which is associated with the critical approaches but not identical to all of their components. Post-modernism is a common term among many scholars and philosophers to respond against and even reject the modern period and a relation to modern thinking and values. Post-modern thinking seeks to challenge the modern assumptions about the fundamental universal basis of the human society according to the modernist view. According to postmodern conceptions, "facts" do not exist, and all knowledge is relative to the context in which it was created (Josselson, 2006). Post-modernism tends to be a relativistic concept, non-rational and nihilistic of human reality, emphasizing the principle of multiple voices, diversity and contradictions of human reality, and rejecting the existence of meta-narratives. The post-modernists reject the idea that a human is an autonomous subject which structures the social order and changes it. Post-modernists argue that the modern reality of an individual who can and wants to take control in the world has been replaced by the post-modern conception in which the human, his desires, horizons and experience are only text created by the language in which he lives. Beyond philosophy and research, post-modernism is expressed in many other areas such as literature, art, film, architecture and more.

The similarity between the post-modern ideas and critical-emancipatory methodology is quite striking. Some of the streams associated with the critical-focused methodology can be considered, and perhaps even consider themselves, to be post-modern, although there is no identical likeness between the two. While the critical-focused methodology places emancipation as the goal of the research, the post-modern approach is content with understanding reality, and does not necessarily include the ideas of change and emancipation. There are also expressions of the post-modernist approach in other qualitative methodologies, such as the participant-focused, self-focused, and structural-focused methodologies. But even in these cases, there is no identical similarity between the post-modern approach and any of the qualitative methodologies.

Following is a brief overview of some of the characteristics of several critical approaches. It should be noted that the intellectuals identified with the various critical approaches are philosophers, and the translation of their concepts and ideas into research methodology is carried out by researchers who were affected by the philosophic concepts.

Neo-Marxism originated from the classical Marxism and its growth from the Frankfurt School, which sought to re-examine the ideas of classical Marxism. Essentially, neo-Marxism accepts the social-economic analysis of classical Marxism. On the other hand, neo-Marxism has reservations about the idea of determinism that accompanies classical Marxism, and rejects the classical Marxist predictions about the future. The hegemonic concept described above, which has been adopted by all schools of criticism, is rooted in the neo-Marxist tradition. According to neo-Marxist notions, the concept of hegemony relates to the capitalist social class who hold the status and attitudes that enable them not only to continue to get rich, but also to take over the minds of the masses and penetrate within them the belief that the current social-economic order is correct, appropriate and possible (Shlalsky & Arieli, 2001; Greenwood & Levin, 2005; Josselson, 2004).

The **critical hermeneutic** is a hermeneutics conception, theory, philosophy or art that deals with interpretation, understanding and giving meaning to texts. At first, "hermeneutic" dealt only with written texts, especially sacred texts, but in recent decades the term "text" has expanded to include human behavior, art and movies (Dargish & Sabar-Ben Yehushua, 2001; Levy, 1986). The range of hermeneutic tendencies spans the conservative or objective hermeneutics at one pole, and critical hermeneutics at the opposite pole, with various other hermeneutic approaches in-between. (Gallagher, 1992; Widdershoven, 1993). Critical hermeneutics is largely associated with the writings of Habermas (1971), which purport a "hermeneutic of suspicion" that assumes that the text expresses, consciously or not, the interests of the power holders (Josselson, 2004). The role of the hermeneutician is to emancipate the mind from cultural and political constraints that serve the dominant class. It should be noted that other methodological approaches give expression to hermeneutic conceptions streaming from other trends (e.g., the criteria-focused methodology gives expression to conservative hermeneutics, and the participant-focused methodologies give expression to dialogic hermeneutics).

Critical-feminist research seeks to emancipate women from the "false consciousness" that establishes the male as the dominant group. This false consciousness results from women themselves conveying hidden messages from generation to generation that perpetuate the non-egalitarian position of women in society. The critical feminist argument considers women as an exploited group that accepts the situation as natural and normal as a result of men's dominance of the media and education. Critical-feminist research believes that it has the tools to understand the situation of women in society (Grbich, 2007). The role of critical-feminist inquiry is "[to show women that] they could lead fulfilling lives if society was organized differently, and exposing the institutions and social processes that have caused them to accept the economic dominance of men" (Travers, 2001, p. 134). Foucault, for example, shows in his critical work how social pressures shape women's bodies and sexuality, and then unconsciously force them towards certain behaviors (Mills, 2003). Critical-feminist researchers seek to expedite the passage of the scientific community and the feminist community from the positivist paradigm to other modes of research, and express the experiences of women from their own perspective (Zellermayer & Peri, 2002).

Post-colonialism offers a different concept of colonialism from that which was accepted from the 19th till the mid-20th century. (Kincheloe & McLaren, 2005). Classic Marxist analysis views colonialism as a very high level of social-economic exploitation, expressed in the European Empire's takeover of countries in Asia, Africa and South America. Post-colonialism believes that there is no reason to consider colonialism as a power exploitation of the indigenous population by taking over their country (this pretty much disappeared in the second half of the 20th century), but it should be seen as a complex, subtle process that also includes the production of knowledge and information. The Western world, being a hegemonic world of knowledge and information, imposed information systems on the reality of the controlled countries, and labeled them as primitive compared to Western culture, perceived as normative and progressive (Mills, 2003). The post-colonialism theory "helps us see the worldwide oppression against the 'other'" (Ladson-Billings & Donnor, 2005, p. 285). Edward. W. Said (1978) indicates that there is no necessary connection between the theory of the Western world about the East and the reality of the East. Western researchers have created myths about the East (for example, the "noble savage" of Rousseau or the studies of Margaret Mead), constituting a kind of power underlying the studies (Idan, 1990). The critical-focused methodology therefore seeks to emancipate the observation of the non-Western world from the distorted perceptions of the West.

Queer theories refer to a group of people perceived by hegemonic Western thinking as falling within the margins of society. "Queer" is originally a derogatory nickname connoting the unusual, strange or weird, associated to people with a sexual orientation contrary to what is perceived by Western culture as the norm. It is a common name for homosexuals, lesbians, bi-sexuals, trans-gender people, and more, and has been adopted into the critical theories by those seeking to change society's common sexual conceptions (Plummer, 2005). Sexuality, according to Foucault in his *History of Sexuality* (1996), functions as a means of power which includes the means to monitor a distinction between what is acceptable and what is unacceptable. The accepted discourse about sexuality dictates that whenever the sexuality issue is brought up, the issue of sin arises alongside. The fact that sex is connected with sin helps to suppress sex, because a person is forbidden to sin. That is, by suppressing or prohibiting a particular discourse on the subject, we control it. According to this view, research that frees the debate on gender into modern accepted discourse concepts can emancipate groups who are depressed because of their sexual orientation.

Post-structuralism grew out of the French structuralism of the mid-20th century, but attacked its scientific assumptions. The researchers associated with this concept are mainly Foucault, Derrida, Barthes and Lacan. Unlike structuralism, which believes that "any system is made up of a set of oppositional categories embedded in language," (Denzin & Lincoln, 2005a, p. 27) post-structuralism purports that there are no such stable categories. According to the post-structuralist, there is no single, absolute truth based on the assumption that the human is an autonomous, rational, stable entity with an unlimited capacity for action. Unlike the structuralist approach, which seeks truths free from circumstantial contexts, post-structuralism emphasizes the contextual factors and the role of power in social relations. Post-structuralism links language, subjectivity, social structure and power. The important issue is language. Language is not a reflection of social relationships, but creates meaning and social reality. Different languages and different types of discourse give different meaning (Richardson & St. Pierre, 2005). Unlike other critical approaches presented above, post-structuralism does not stress a right and just social order, because it assumes that the ideological meta-narratives that offer solutions to problems of humanity and society have collapsed. The critical-focused methodology which is based on a post-structuralist approach yields petite stories and specific types of discourse that reflect a complex, multifaceted reality (Shelasky & Alpert, 2007). Although all the qualitative methodologies deal with discourse analysis of the participants, it seems that those who adhere to the post-structuralist approach will tend more than others to be precise about elements of the text, and turn to a structure textual analysis (Peters & Burbules, 2004). This is reflected, among other ways, in adopting the deconstruction method, as shown in Chapter 10.

The Study's Theoretical Perspective

Critical-focused methodology is related, as noted, to various philosophical approaches that share an orientation of criticism and emancipation. From a methodological viewpoint, there is much in common among the scholars adopting the various critical approaches. As in other qualitative research methodology, the study begins with formulating a theoretical perspective. Although these approaches perceive reality as constructivist and created by the people, the proportion of the external theoretical perspective in the critical-focused methodology is higher than in participant-focused methodology, which is also characterized by a constructivist perception. However, critical-focused methodology assumes that critical researchers have greater insight about nature and human existence and their formation than the research participants. Therefore, unlike the participant-focused methodology where the researchers see their role as helping the participants to tell their stories, critical-focused methodology researchers attempt to look beyond what is said by the participants and give great weight to the social, cultural, historical and political factors in understanding of the phenomena under study (Carr & Kemmis, 1986; Phillips & Hardly, 2002; Travers, 2001).

Accordingly, the critical-focused methodology assigns a certain degree of significance to the external aspects that researchers believe are reflected in the phenomenon under investigation, even if unknown to the participants. This is clearly shown in Figure 7A above. Thus, the critical-focused methodology researchers "recruit" critical-theoretical concepts to look directly into the phenomenon under investigation. The a-priory place of theories is at the macro-level, relating to society and the forces operating within and above it, while the micro-level is more attentive to what emerges from the participants' discourse. Clearly there is a permanent dialogue between the micro and macro levels, with the assumption that social-cultural forces (macro-level) influence the people's thoughts and activities (micro-level) (Travers, 2001).

The theoretical perspective of the critical-focused methodology seeks to present the participants and readers with ways to identify the forces which distort and limit the free expression of the people (Carr & Kemmis, 1986). Prior to their entry into the study, researchers adhering to the principles of critical-focused methodology must clarify which ideological forces affect the people in creating their perception of reality, and seek to release people from these unconscious forces throughout the research process. The surrounding social, historical and political forces are understood not only as background factors but as an activated power in day-to-day human life. Thus, the practical implication is that before entering the research field, researchers will have formulated a written preliminary literature review (Shaw, 2010a). However, researchers of critical-focused methodology, like their counterparts in participant-focused methodology, will conduct a dialogue with the participants and alter and adapt their theories in light of the participants' discourse. Thus, the literature review, which reflects the theoretical perspective of the study, will be completed only in the final stages of the study. Figure 12B illustrates the process of creating a theoretical perspective of the critical-focused methodology.

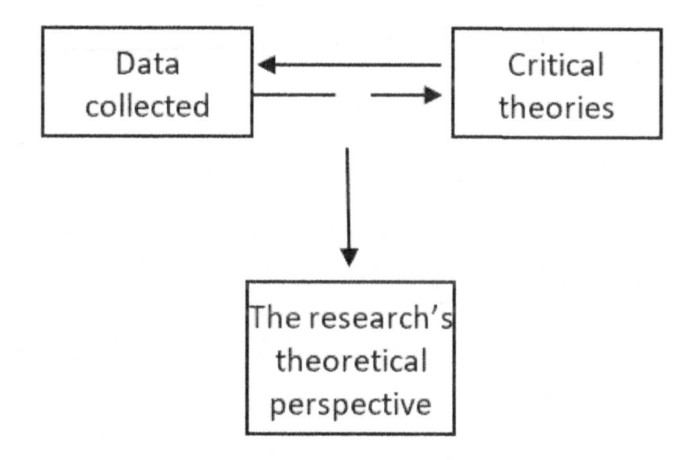

Figure 12B: The Process of Construction of the Theoretical Perspective in Critical-focused Methodology

It is important to note that Figure 12B clearly indicates the significant position of the critical theory in the process of creating the research's theoretical perspective, but there is also prominent importance attributed to the unique internal view arising from the empirical data. Combining these two or a dialogue between them creates a dynamic process of constructing a theoretical perspective, while making changes during the study in light of data collected.

Formulation of Research Questions and Assumptions

Research questions in critical-focused methodology emerge from the unique theoretical perspective of this methodology, namely, the questions that give expression to a cultural-political-social context as well as those concerning the specific subjects. As a result, the research questions will involve the wider circle of the social forces, but at the same time they will touch upon the specific situations of the participants. While the researchers in the participant-focused methodology prefer to formulate research assumptions and not research hypotheses, emphasizing the dynamism and flexibility of their approaches, researchers in the critical-focused methodology will raise much more solid assumptions and might not even be deterred from imposing hypotheses (Kincheloe & McLaren, 2005). For example, researchers examining classroom discourse will not hesitate to create hypotheses about possible classroom discourse, originating from their critical-theoretical perspective. If their theoretical perceptions view discourse between people as being tied to the social-political hegemonic forces that determine the nature of the discourse and give people the illusion of freedom of opinion, they would hypothesize that this would be found in the classroom. However, unlike the researchers of criteria-focused methodology, which is almost entirely bound to hypotheses, there will be more attention paid to the participants, and researchers will find a willingness to test the hypotheses in light of specific research situations (Kincheloe & McLaren, 2005).

The distinction between macro-questions and micro-questions can help in shaping the research questions. The following research question demonstrates a macro-question:

1. **What is the influence of the political, social and moral public debate on selected content in the Bible studies curriculum?**

Within this question lies the assumption that Bible teachers are influenced by the hegemonic concepts expressed in the compulsory central curriculum, even without being aware of this effect. To understand the perceptions of the teachers, researchers should reveal these forces. The researchers will seek answers to these questions in external texts (like curriculum material), but also, of course, in the discourse of the participants themselves.

Another example of the question emphasizes the macro:

2. **Based on the curriculum, what are the goals that society has set for the school and how does the school manage to achieve those goals?**

The micro-questions, dealing specifically with the participants, can be formulated in a similar manner to questions offered in the participant-focused methodology, but they take into account the critical theoretical assumption and the research picture reflected through the two macro-questions presented above. For example:

3. **What are the personal beliefs of teachers about the Bible, how have these beliefs evolved, and how are they reflected in the teaching process?**

4. **What are the teachers' beliefs regarding the place of students in the teaching process, how are these beliefs consistent with the school ideology, and how are these beliefs reflected in the teaching process?**

During the data collection process, an apparent contradiction may arise between the picture emerging from the macro-questions and the micro-questions. For example, teachers may talk about the dialogue they develop with their students, a situation that contradicts the picture resulting from the macro-questions. As part of the critical-focused methodology, researchers seek to present this contradiction before the study participants, lead them to talk about it and/or inform the readers of this contradiction in the researchers' explanations. Similarly, researchers might examine their critical theoretical assumptions and ask themselves if their pre-assumptions are indeed reflected correctly in the emerging research picture.

The Study Design and Population Selection and/or Field Research

Unlike the participant-focused methodology, which bases research on information obtained directly from the population studied, the critical-focused methodology assumes that the relevant information is beyond the information that the participants can provide. Researchers also need macro-information. The parable of the fish and the water is a good example: the fish depends on water and cannot live outside it, nevertheless, "The fish is always the last being to discover water, but as an observer, we can see it clearly." (Josselson, 2004, p.14). Participant-focused methodology researchers, even though they "see the water," will "explore only the fish." They will attempt to find whether the fish is "aware" of the water and how, and will display the picture reflected through the "eyes of the fish." In contrast, critical-focused methodology researchers will turn to two sources of information--that extracted from the "water" and that obtained from the participants (macro and micro information).

It seems that in critical-focused methodology, there is great importance attached to selecting the research site and its population. The choice should be the purposeful sample, consisting of the best representative participants of the population which the study focuses upon, where and who can teach us about the phenomenon under study (Mason, 1996; Stake, 2005). In the case of this methodology, it seems advisable to choose those participants who are willing to conduct a dialogue with the researchers and open up to new insights regarding the forces affecting their minds, without feeling threatened by these insights. As in all qualitative research, it is essential to ensure an adequate level of ethics and rules (Dushnik & Sabar - Ben-Yehoshua, 2001; King, 2010). The fact that researchers come into the research with a sense that they have a particular advantage with the study participants, having knowledge about the social forces that affect the study participants without their awareness, should increase the sensitivity to carefully maintain the rules of ethics.

The research array is not according to linear and fixed-order components. The research components do not necessarily follow a predetermined order. The study design is much more flexible. The distinction between the macro-aspect, which relates to the context of the study, and the micro-aspect, referring to the participants themselves, does not determine the linear order, although the context is perceived not only as a background effect, but also as a dominant factor (Travers, 2001). It is possible to move from stage to stage, from the macro-to-micro aspect, before finishing the previous stage. Often, researchers find themselves returning to earlier stages and altering them in light of insights obtained from more advanced stages of the research. (Jorgensen, 1989; Lincoln & Guba, 1985; Marshall & Rossman, 1989). Figure 12C illustrates the connections between the components of the research presented so far. Note that the arrows in the direction from left to right are long continuing arrows, which reflect the centrality of the critical theories and their impact on all other components. Arrows in the opposite direction (from right to left) are truncated, which means that there are significant two-way effects of each of the components.

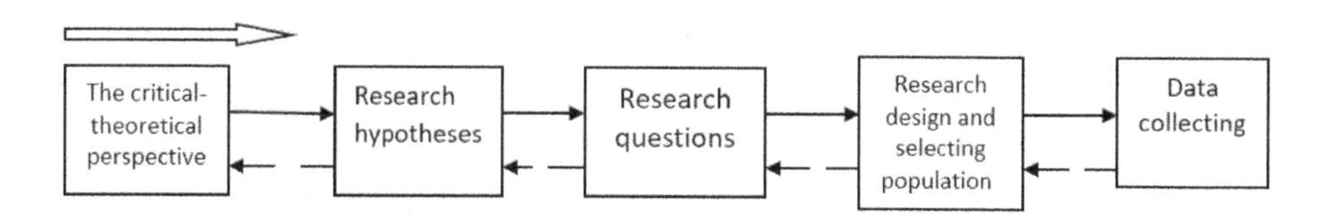

Figure 12C: The Connections between the Components in Critical-focused Methodology

Data Collection Methods

In critical-focused methodology, the data collected should clarify the macro-picture (social-cultural context aspects) of the phenomenon under investigation, and the micro-picture (referring to the participants and their world). It seems that in addition to data collected by way of intervention such as interviews, observations and field comments, significant substance will be given to the "natural data" (Silverman, 2006), data that exists in itself and not as a result of initiatives by researchers,- such as documents, artifacts, articles and the like. It is likely that researchers of this methodology will place substantial credence upon "media information" such as reported by the mass media of radio, television, newspapers and the like. At the same time, there will also be a significant presence of "Web data" - data from various Internet websites, including online forums on relevant topics (Gibson & Riley, 2010). The significant use of such data arises from the need to display an in-depth research context, which is a significant component in understanding the political-cultural-social forces that shape the phenomenon under investigation. From this standpoint, we could place the character of data collection tools between the external and the internal criteria, with some tendency toward internal criteria, as is seen in Figure 12D.

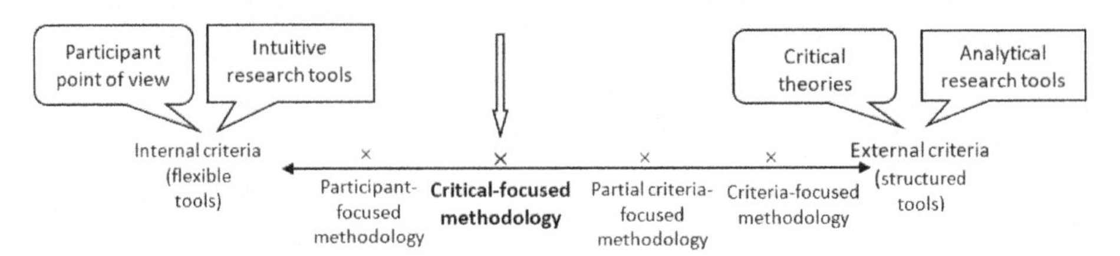

Figure 12D: Characteristics of the Research Tools in Critical-focused Methodology

The nature of the initiated data collection in the critical-focused methodology poses the researchers as "participating observers" or "involved observers," but unlike the observers in interpretive-ethnography, the observers here are not content with becoming part of the investigated society but try to influence the shaping of reality and consciousness of the participants. The researcher seeks to establish a dialogue with the participants in order to bring about a change in their consciousness towards the awareness of social and political forces that act on reality, and thus "emancipate" them (Carr & Kemmis, 1986; Travers, 2001). Accordingly, the researchers will find the degree of balance between the participants' perspective and the critical point of view with which they come to the study.

It appears that in critical-focused methodology, there is a significant, though not exclusive, importance attached to observations and data products such as documents, artifacts and the like. Researchers adhering to the critical theory approach the observation with their critical conceptions, paying particular attention to the situation observed and the specific participants. Sometimes there is no correspondence between the meaning that the researchers attach to the phenomenon under study and the meaning that the participants give. While in participant-focused methodology, researchers use the simulated-recalled interview (Shkedi, 2005) in order to acquire the participants' description and explanations of what happened, in critical-focused methodology, researchers tend to understand what happened largely on the basis of the critical theories that guide them, which also explain what has disappeared from the participants' awareness.

The interviews are conducted in the nature of a dialogue between researchers and respondents. These are not structured interviews using a series of questions prepared in advance, but interviews characterized as a conversation. During the interviews, the interviewers try not only to listen carefully to the stories of the participants, but also to open the eyes of participants to the external forces that influence them, and assist the participants to become free from the forces that bind their minds and to see reality in a different light. These researchers "are sometimes advised to take the opportunity to tell their own stories to respondents" (Silverman, 2006, p. 118). It seems that the interviews can be characterized between in-depth interviews and semi-structured interviews. See Figure 12E.

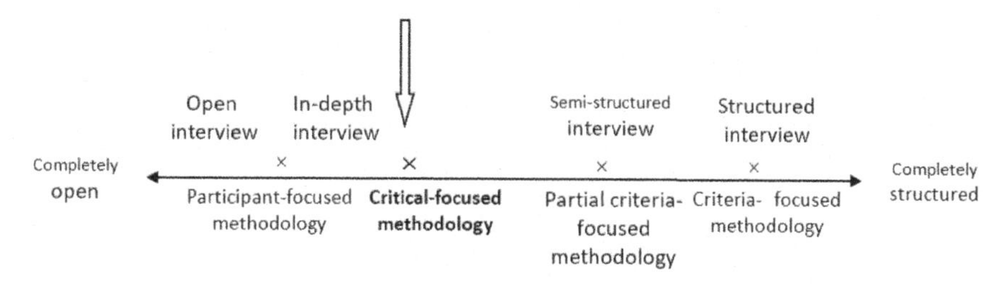

Figure 12E: Characteristics of the Interviews in Critical-focused Methodology

As emphasized, in order to ensure the quality of the study and refrain from relying upon intuitive impressions only, it is important to record all the data, preserve and transcribe it in a way that enables researchers to examine and analyze the data as necessary and at any time, and place the database at the consideration of colleagues and the academic community.

Research Categories/Themes

It is important to understand the creation process of the research categories, the time of their creation and the way to use them according to the unique characteristics of this methodology. Researchers enter into the study with initial research insights translated into very general categories with which they view and interpret the phenomenon under investigation. There are only a few pre-defined upper categories, and all other categories will be created later in the research process. Moreover, in adherence to the critical-constructivist concept, researchers will examine the upper categories and their ultimate relevance to the phenomenon under study, and where necessary, make changes during the data collection process. Figure 12F illustrates the category array in research dealing with teaching Bible studies. The main category is about the role of the ideologies of Bible teachers, and the upper categories relate to the components of this category. The categories arise largely from the research questions, following the critical approach.

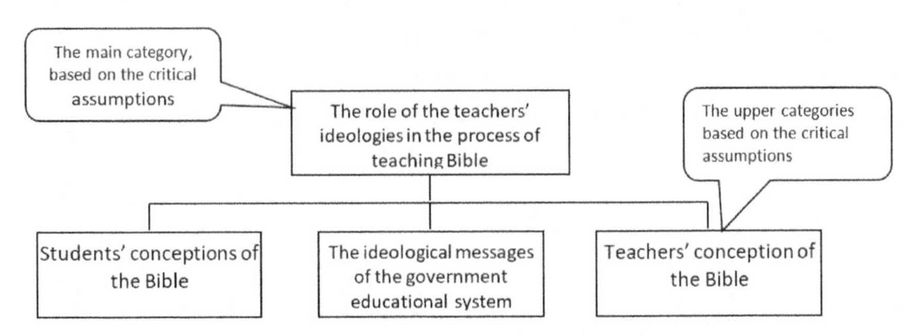

Figure 12F: An Example of Categories that Guide the Research in Critical-focused Methodology

Researchers in the critical-focused methodology, although "equipped" with the critical-theoretical perspective, are different from their colleagues using the criteria-focused methodology and do not consider the theories as a constant, steady, uniform guide, and certainly not predictive. Thus, these researchers will enter into the study field only with the upper categories, which are used merely as general lenses of the phenomenon under study and will be constantly tested. The full category array will be created while collecting data, and especially in the analysis process, which is an essentially inductive process (Perakyla, 2005).

Figure 12G illustrates the category tree. At the top of the tree, below the main category, there are three upper categories (themes) illustrated in Figure 12F which express the critical-theoretical approach of the researchers towards the phenomenon under investigation. These categories guide the process of data collection. As a result of the data collection process and with the addition of the analysis process described in the next section, a more complex network of categories will be created. Figure 12G illustrates four levels of categories: the main category and the upper categories mentioned above, and two additional levels, the sub-categories and sub-sub-categories. (Actually, there may be a larger number of levels.) The two upper levels of categories express the critical theory (macro view), and as we "go down" through the levels, the expression obtained from the participants themselves (micro view). The category tree displays the relationships between the components in a vertical axis and in the horizontal dimension, illustrating the relationship between the macro and micro views, and creating a complete picture.

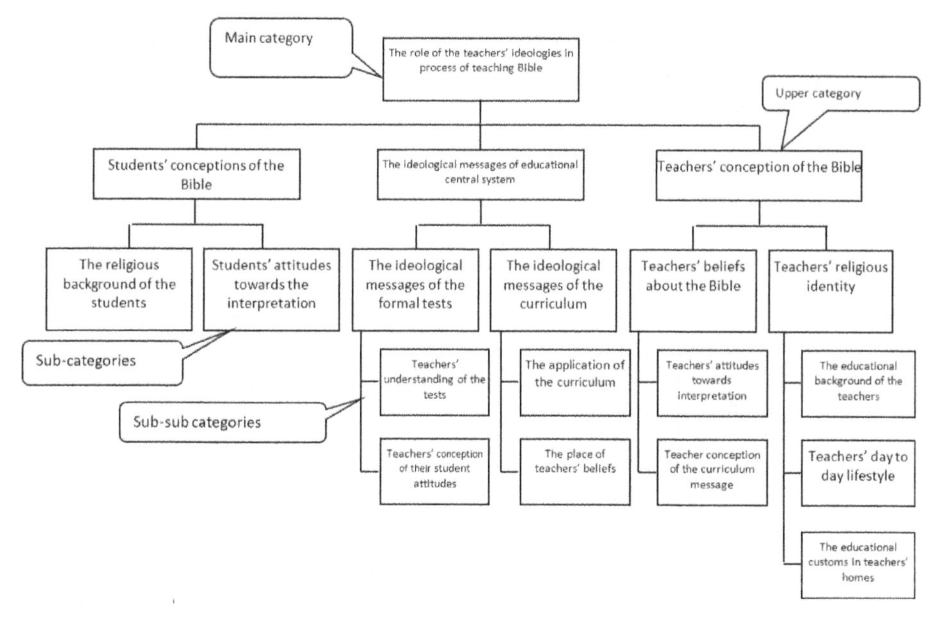

Figure 12G: The Category Tree in Critical-focused Methodology

Data Analysis

The data analysis process in critical-focused methodology may be similar to the process in participant-focused methodology, thus a complex process which involves several stages. As stated in the previous section, in the critical-focused methodology, researchers do not come to the analysis process with a complete category array, but with a few upper categories reflecting the critical-theoretical perspective of the phenomenon under investigation. The researchers then analyze the data using several upper categories determined before data collection but tested and found to be appropriate (or changed and honed if necessary) during data collection. Researchers do not set a detailed and complete category array prior to analysis (such as in criteria-focused methodology), and one of the objectives of the analysis phase is to create categories of data. The first step in data analysis, the "initial analysis stage," deals with the division of texts (interviews, observations, etc.) into initial categories (data segments) which are not yet real categories, and the association of each initial category to an upper category, including delineating names to express the content of the data segment. The resulting image creates a situation where under each of the upper categories are a large number of initial categories (data segments). Figure 12H illustrates this process. Note that some initial categories may be associated with two upper categories or more.

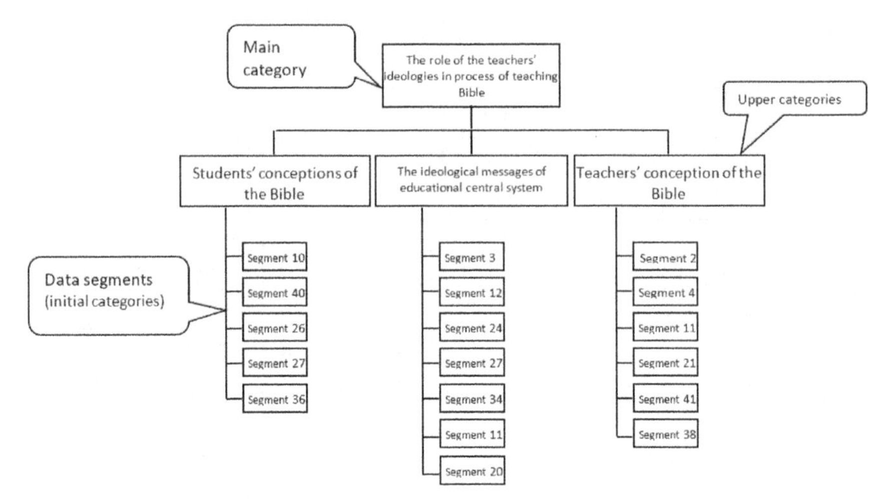

Figure 12H: Initial Analysis Stage in Critical-focused Methodology

Already in the initial analysis stage, researchers might conclude that on the basis of the theoretical upper categories, there are several data segments that are irrelevant to the issue under study or that their relevance is only marginal, and will conduct an initial selection during the analysis.

In the next stage of the analysis, the mapping stage, the researcher creates the complete array of categories by finding thematic connections between the various initial categories. These relationships indicate that all the text segments dealing with the same subject (theme) are associated with the same category. Each category will reflect the shared content of the text segments belonging to the category, while also addressing the upper category to which they are associated. Figure 12-I illustrates the mapping analysis tree. The tree is based on the three upper theoretical categories (in actuality, there may be fewer or more upper categories). This mapping tree in Figure 12-I includes four levels of categories: the main category, upper categories, sub-categories and sub-sub categories (in actuality, it is likely that the division of some categories contain a large number of levels). At the end of the process, when all the data segments are organized in categories, a complete mapping tree is obtained. At the top of the tree are categories with higher conceptual levels, clearly reflecting the critical-theoretical perspective. At the bottom of the tree are the categories with lower conceptual levels. The data segments are located in the final lowest categories.

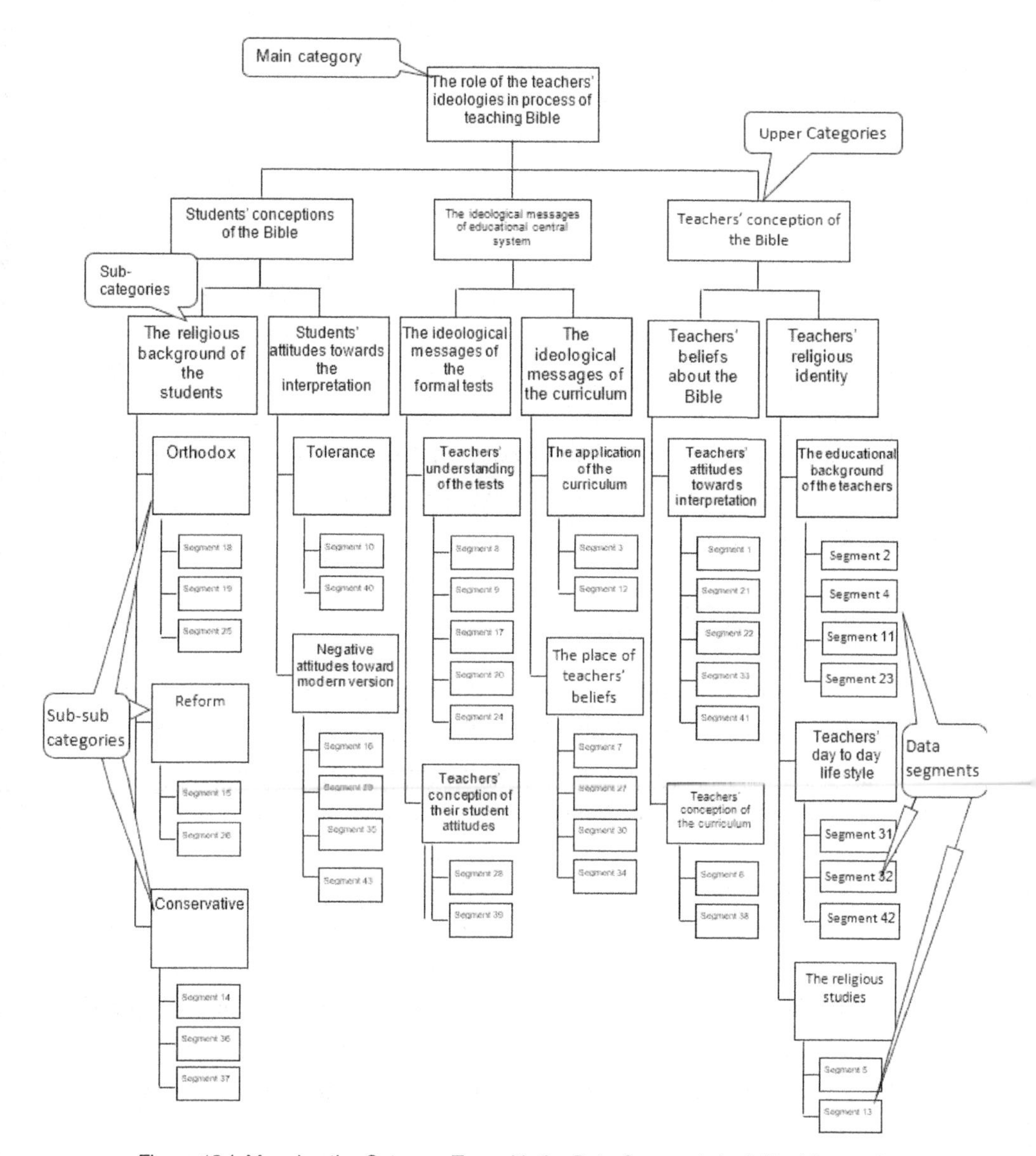

Figure 12-I: Mapping the Category Tree with the Data Segments in Critical-focused Methodology

As mentioned above, in critical-focused methodology we start the analysis process with at least several upper-level critical-theoretical categories expressing the focus of the research.

Therefore, the third stage of data analysis, the focused stage, is not necessary in the critical-focused methodology. However, sometimes the scope of the categories and data segments attached to them is too broad, thus the focused stage of analysis is very helpful. In the next stage of analysis, the theoretical stage, the theoretical categories are examined once more in light of the critical-theoretical insight gained, with any resulting changes being made to the array of categories. In other words, the critical-theoretical categories formulated at the beginning of the study are not necessarily characterized by stability, and the researchers are not obliged to maintain the constancy of these categories. It is entirely possible that during the process of analysis (and perhaps even in the earlier stage of data collection), in light of data collected from the participants, the researcher hones the categories and perhaps changes, adds or even drops categories, as well as adding or dropping data segments accordingly. This process, carried out throughout the research process, reaches its peak in the theoretical analysis phase.

Research Report

As with any research methodology presented so far, all the common components of the final report are relevant here: the literature review, research questions, research assumptions, methodology, presentation of the population and the field studied, findings, discussion and conclusions. This order of elements does not have to be carefully maintained if the researchers conclude that a different order can better communicate the message to be conveyed. Unique to the research report in critical-focused methodology is the highlighting of two related elements: the critical-theory and the research description (Merriam, 1985, 1998). The research report seeks to describe the phenomenon under investigation while stressing the significant role of cultural-political-social forces, shaping the specific phenomenon being studied and the consciousness of the participants. In order to construct the specific nature of this report, there are many options available for the writing.

Researchers can focus their description around a few cases and describe each one in depth, and accompany each description with theoretical emphases, before, in, or after the description; they can be linked to the categories and describe the phenomenon, emphasizing its elements and theoretical meaning, and find various kinds of combinations of in-depth descriptions with theoretical descriptions focused on the categories.

Throughout the writing, researchers repeatedly test the logic of the categories and incorporate any necessary changes (thus they actually continue the process of analysis). They make a selection, emphasizing certain sub-categories over others, choosing the exact quotes that they want to display and converting other quotes into their own language, which may be integrated with theoretical-academic language. The appropriate description of the critical-focused methodology is a description that uses the language of the participants with the critical-theoretical language, and combines the description of the phenomenon with the theoretical explanation.

Figure 12J brings an example, one of many possibilities, of a description that combines the critical theory with the description of the phenomenon. This is a study of the issue of teacher education, and the critical-theoretical argument that the past experiences of the teachers and student-teachers from their days as schoolchildren shape their ideological perception about education and about teaching. These ideological conceptions are deeply ingrained in their perceptions as teachers, without being aware of their sources, and almost without the possibility to change them. Given this, the critical-theoretical approach argues that the impact of teacher education is nothing compared with the memory of these early experiences and the ideology that developed in their mind (Lortie, 1975). In light of these critical-theoretical assumptions, the researchers interviewed student-teachers about two central issues: their experiences as schoolchildren and their current vision about the ideal school. Figure 12J brings an excerpt from the Final Report.

Orna, like any high school graduate, experienced a few thousand hours observing teachers' work during her school years. For twelve years, almost every day of the week, hour after hour, Orna found herself, like other students, watching her teachers. This is not necessarily a deliberate viewing and not even a conscious process, but without realizing it, Orna and other students comprehend, for better or worse, the teaching practices of their teachers. Not everyone "catches" the same thing - what appears to one as interesting and attractive looks dull and is rejected by others, and still others don't even notice it. But all have experienced and absorbed, and these experiences have been imprinted in their minds and teaching patterns. Now, when they are student teachers and confronted with the task of teaching other kids, these teaching patterns emerge even without their being aware of it. These patterns prevail over any information, theories and ideas that Orna and her colleagues are exposed to in the teacher's college. It seems that everything is painted in the colors of past experiences, almost without enabling it to be changed (Lortie, 1975). Orna wants to become a math teacher. She came to teacher-training with a high motivation to change the world of mathematics education. As she spoke about her memories of school, she summarized her view by saying:

"I had wonderful teachers ... I learned with very good teachers who knew the curriculum and knew how to teach it. Mathematics was for them the language that they knew the best. It was fun to learn from them. Then I realized that I really wanted to study mathematics and become a math teacher, and I love it. Math is the place to express myself."

Therefore it is not surprising at all that when speaking about her vision of teaching, we see that Orna tries to restore her experiences as a student:

"I think a teacher must learn all the time, to come to the class with up-to-date knowledge ... I feel I owe the students a high level, which I feel that I can bring them... I believe that if I know the material and know how to teach the material, students will be with me and love mathematics. "

It seems that Orna adopted the teaching pattern of her teachers even without being conscious of it. She has good memories, so they are a role model for her.

Figure 7J: Description Section in a Final Report in Critical-focused Methodology

Research Quality Standards

A research study in the critical-focused methodology reflects a direct correlation between the critical-theoretical concept and all research stages that come after it and/or are related to it. In spite of that, the status of theories in the critical-focused methodology is not similar to that of criteria-focused methodology or in partial criteria-focused methodology. The "reality" perceived by these two is objective, and therefore the theories guiding the study are connected linearly and consistently throughout the long process. In contrast, the critical-focused methodology perceives reality as a process of human construction steered by social forces. (Grbich, 2007). Thus, it is only a general theoretical system, and the categories are reformatted only at their highest levels, and can develop and change as the study continues and the data picture becomes -clear. (Lincoln & Guba, 2000; Schwandt, 2000; Smith & Hodkinson, 2005). As defined in previous sections, this methodology is equipped with two points of view, an external perspective, the critical theories from outside which relate to the macro-level, and the internal perspective, relating to the story of the participants at the micro-level. These two perspectives determine the nature of the study and the standards for examining the quality of the research.

The term "validity," which denotes the extent to which research findings are consistent with the theoretical perspective of the study and interpreted correctly within it, is applicable in critical-focused methodology. The researchers as well as their readers can examine the connection (or the lack of connection) between the critical-theoretical perspective and the emerging research picture. Researchers should show the chain of evidence (interviews, observations, documents, etc.) for each stage of the study, to examine the desired balance between theory and the research description (Huberman & Miles, 1994; Mason, 1996). While in criteria-focused methodology the researchers can be helped by external judges to determine the research validity, in critical-focused methodology this review is problematic. This methodology is based on the assumption that reality is constructive and the result of a unique and dynamic construction, and that perception depends on a critical-theoretical perspective. Thus, it is impossible that impartial judges examine the study from an external-objective perspective (Merrick, 1999). Those who read the study who are not involved with the critical theory perspective of the researchers could not get a picture similar to that reached by the researchers themselves. It is essential that all research components be exposed to the peer review of those involved with the critical-theoretical assumptions and be willing to devote the time to see the phenomenon from its unique perspective and consider the validity of the study (Shaw, 2010a).

Reliability means the possibility of repeating the same process of research and getting the same results, even if the research is conducted by other researchers (Yin, 1994). Reliability, in its "classic" sense, is difficult to achieve in qualitative methodologies in general and particularly with critical-focused methodology (Merrick, 1999; Schofield, 1989). The way in which researchers claim reliability is similar to the way they claim validity. The researchers should document their research and present the database, revealing the processes of analysis and presenting a detailed description in the final report (Mason, 1996; Pidgeon, 1996). In this way, the researchers themselves, and external colleagues and readers, can examine the degree of the reliability (Dey, 1993; Peshkin, 1993).

As in the case of validity and reliability, even the concept of "generalization" has a unique meaning in critical-focused methodology. Of the three types of qualitative research generalizations, the analytic generalization (generalization through theory) may be most suitable for this research methodology. Comparing with similar critical research approaches may strengthen the external validity (generalization) of the study. Generalization from case to case is also very relevant to this methodology (Firestone, 1993). By comparison between the investigated case and other live cases in other situations that we seek to apply to our findings, we can generalize our research findings. This process is like a kind of dialogue between the investigated case and other live cases (Schofield, 1989). Obviously those who want to examine the generalization must adopt the critical-theoretical approach characteristic of the study, and see this in light of the phenomenon to which they seek to compare the findings. Generalization to population will be more problematic in the case of critical-focused methodology, unless the researchers indeed studied a large population (Arksey & Knight, 1999; Dey, 1993; Firestone, 1993; Marshall & Rossman, 1989).

A researcher's attention to triangulation, that is, the use of a variety of information sources and methods of examination of the study, (Denzin & Lincoln, 2000; Fontana & Frey, 2005; Stake, 2000) would strengthen the argument for validity, reliability and generalization. This requires the researchers not only to increase the sources of information and methods of data collection and analysis, but also to maintain the database and document all the data analysis processes to enable transparency for the re-examination all stages of the research by the researchers themselves, their colleagues, and those interested in examining the study.

CHAPTER 13

THE FIRST STEP OF ANALYZING IN CONSTRUCTIVIST BASED METHODOLOGICAL APPROACHES: THE INITIAL ANALYSIS

This analysis track starts after the phase of reading and arranging the data (Chapter 5). These methodologies focused on the analytical process without predetermined categories, and the categories are determined during the analysis process itself. The analysis process begins with the initial analysis phase described in this chapter.

Categories, as noted, contain two dimensions: one referring to the content of the textual segments, and the other referring to the relationship between the different segments and the different generated categories. Thus the process of generating categories involves two steps:

A. Distribution of data into segments, each of which is a unit of meaning
B. Organizing the segments in an analytical order that reflects the relationship between the segments (units of meaning) so that similar units will be included in that same category, which is positioned according to its relationship with other categories. Thus, the initial stage of analysis focuses on dividing the data into units of meaning, the first of the two steps mentioned above.

The initial analysis is characterized by offering ample expression to the intuitive researchers' skills. This will be reflected in what sometimes resembles "brainstorming" rather than an analytical controlled process. Strauss & Corbin (1990, 1994) call the first phase of the analysis "open coding." However, this step is not completely open, but subject to the same principles of all other analysis stages. This step is not free from the conceptual perspective of the researcher (i.e., the conceptual assumptions of the researcher). The analysis process is a result of an "ongoing discussion" between the data and the conceptual perspective of the researcher, respective to the unique principles of every methodological pattern.

Identifying the Main Research Category

The initial analysis begins by determining the main category. In this case, (contrary to the situation in which we start the research with predetermined category arrays, as explained in Chapters 6-8) we set up a main category with a temporary identification. The name of this category may change as the research advances and the entire research picture becomes clarified. This temporary category will reflect the focus of the study as perceived by the researcher at this stage of analysis. What can this main category be? A research study starts from the researcher's interest in a particular issue. Researchers design the study according to a research problem. Determining the focus of the investigation in the study's early stages establishes the conceptual boundaries of the study (Lincoln & Guba, 1985; Stake, 1995). Thus the temporary main category should reflect the topic of the research and its focus. The fact that the name of the category may change eliminates the need for precise and clear wording from the start. Most likely at this stage of the study, the researcher will use the name of a research project

Determining the Main Research Category Using the Narralizer

This operation is explained in detailed in Chapter 8, and will be repeated here in brief. To determine the main category, open the **Narralizer** to automatically obtain a **Narralizer research document**. Upon opening the **Narralizer**, the temporary name "**Narralizer 1**" will appear at the top of the categories pane. Determining the main category name (which may also be the name of the document analysis) is made by the command **Rename** in a process identical to that explained in Chapter 8. After determining the main category name, place this analysis document in the appropriate document directory.

Dividing the Data into Units of Meaning

In the initial analysis phase, we divide the raw data (interviews, observations, documents, etc.) into units of meaning (data segments) by an in-depth reading of the data. Each unit will be defined by several words that express the contents of the unit. In other words, each unit of meaning will be given a name that reflects its theme/subject. Researchers must endeavor to look for all possible themes in the data. Strauss & Corbin (1990) liken the initial phase to a starting a puzzle, where each data unit (segment) is like a single puzzle piece that is not integrated with other units. Integration of the separated units will be carried out only in the analysis stages that follow. It can be said that in the initial stage, all the segments "belong" to the same main category, as shown in Figure 13A.

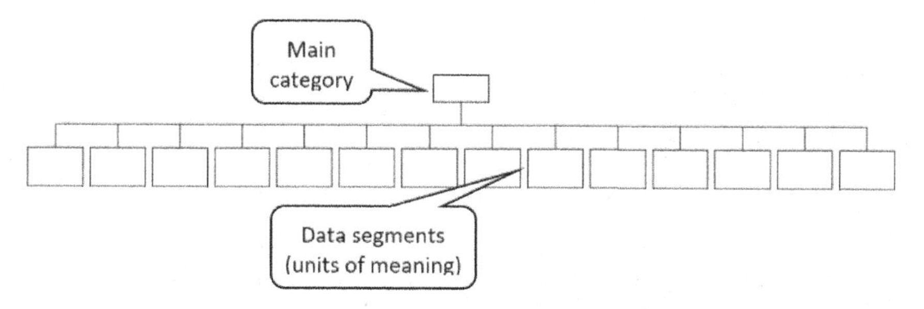

Figure 13A: Dividing the Data to Units of Meaning

Despite the fact that analysis is basically an act of interpretation, we can say that the initial division of units of meaning is seemingly devoid of interpretation. However, any interpretation of any given unit of meaning is provisional at this stage (Pidgeon & Henwood, 1996). Since the analysis process in the initial stage does not address the relationship between the segments, we cannot relate here to the different segments as completed categories. At most, we can call them initial categories, but not completed categories. During the initial stage, the data must be carefully examined - word for word, line by line and sentence by sentence - but without losing the whole picture and context of the entire data documents (Charmaz, 2000, 2005). Researchers ask questions about each of the segments, such as "What it is?" "What does it represent?" and try to identify segments (a single sentence or a few sentences) that relate to the same theme.

Labeling the Units of Meaning

When identifying units of meaning, the researcher gives names that reflect each unit's character and content. In the initial stage, the researcher seeks names taken from the world of the participants, terms that may appear in the relevant segments. Terms taken directly from the participants' discourse may bind the analyzed segments to the specific context of the data, and prevent the researcher's avoidance of study discourse (Pidgeon & Henwood, 1996).

Figure 13B illustrates the process of assigning names by analyzing an interview. The interview took place with a student teacher who teaches a high school Bible class. On the right, segments are marked and the names assigned.

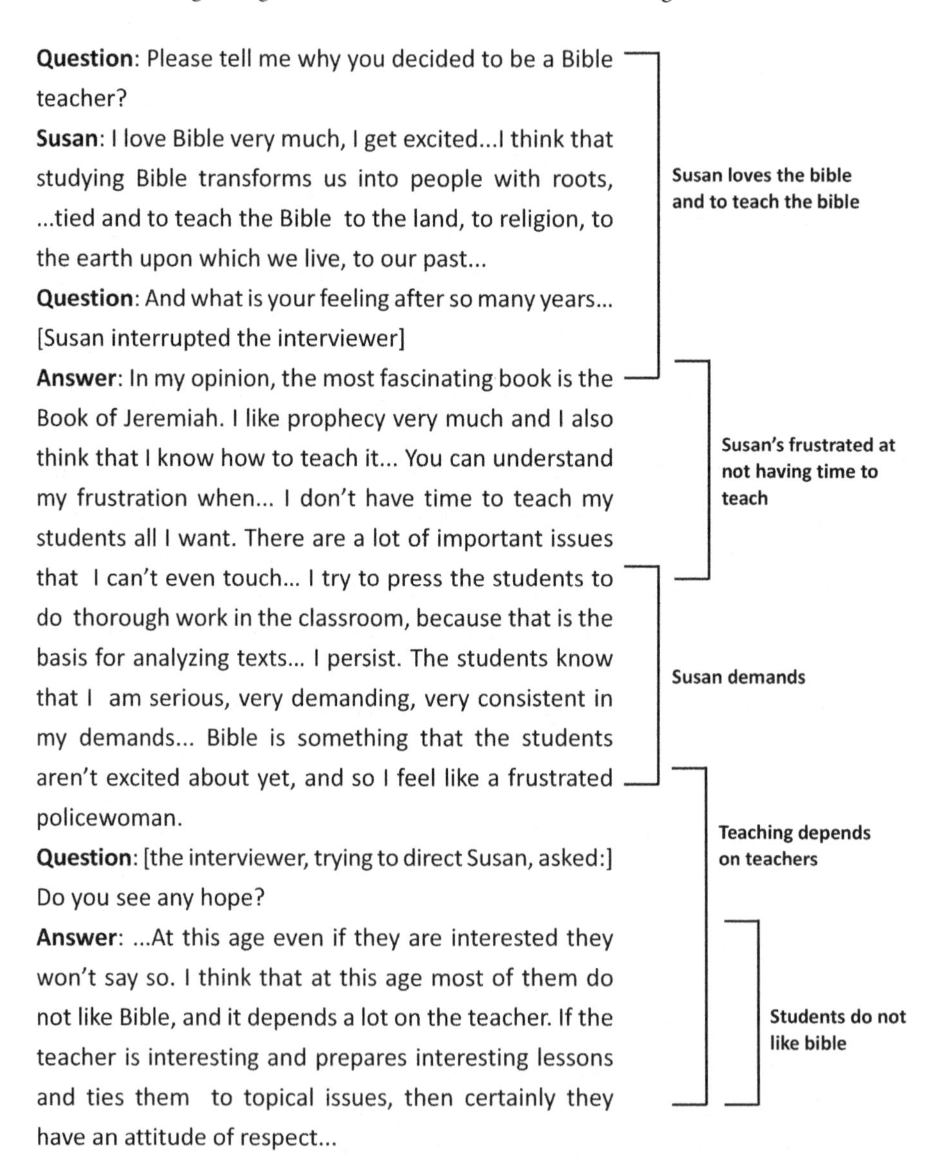

Question: Please tell me why you decided to be a Bible teacher?
Susan: I love Bible very much, I get excited...I think that studying Bible transforms us into people with roots, ...tied and to teach the Bible to the land, to religion, to the earth upon which we live, to our past...
Question: And what is your feeling after so many years... [Susan interrupted the interviewer]
Answer: In my opinion, the most fascinating book is the Book of Jeremiah. I like prophecy very much and I also think that I know how to teach it... You can understand my frustration when... I don't have time to teach my students all I want. There are a lot of important issues that I can't even touch... I try to press the students to do thorough work in the classroom, because that is the basis for analyzing texts... I persist. The students know that I am serious, very demanding, very consistent in my demands... Bible is something that the students aren't excited about yet, and so I feel like a frustrated policewoman.
Question: [the interviewer, trying to direct Susan, asked:] Do you see any hope?
Answer: ...At this age even if they are interested they won't say so. I think that at this age most of them do not like Bible, and it depends a lot on the teacher. If the teacher is interesting and prepares interesting lessons and ties them to topical issues, then certainly they have an attitude of respect...

Figure 13B: Dividing the Raw Data into Units of Meaning

As mentioned above, when the data is divided into units of meaning, it may possibly be detached from its original context and be identified by the terms of the researchers themselves (Araujo, 1995; Dey, 1999). Therefore, when sorting the materials into segments and giving them names, we must safeguard that the original meaning and contexts are preserved. One of the characteristics of the initial analysis is that although we divide the data into units of meaning and seemingly remove them from the context in which they were spoken (interview), written (in logs or documents) or occurred (observation), the original sequence is maintained, as seen in Figure 13B. In this way, we can see not only the division into segments and the names of the segments, but also the original location of each segment in the source document.

It is clear that choosing a segment name is not the result of a proven formula, but rather reflects the understanding of the researchers. Thus, the segment beginning with the words "Bible is something that the students aren't excited about ..." may be called "students do not like Bible" to use a direct quote from the teacher (as shown in Figure 13B), but we can also attach a more general phrase "students' attitudes toward Bible." However, this segment should not be called "students' behavior problems." Not only is this type of name a judgment (as the interviewee herself did not define this behavior as disorderly), but it is a conceptualization that is irrelevant to the immediate context. Be certain in the initial analysis to avoid giving names with an overly-broad conceptualization.

If we analyze an interview consisting of questions and answers, the names of the segments should not be based on the interview questions. Sometimes one answer may engage two issues in an integrated manner, so the segment would require at least two different names. Sometimes an answer to one question can relate to two different topics that appear distinct from each other, thus we would need to divide the answer into two segments or more. Moreover, often a respondent's answer does not focus precisely on the question being asked, but extends or ignores it. Thus we recommend reading the data carefully, trying to understand what the subjects intended to say, and base the division of the segments and names, first and foremost, on the words of the participants, regardless of the essential question asked or the title given to the document being analyzed.

Words can be different in terms of cultural significance and in terms of the participant or the external observer. Thus, sometimes the researcher may give the data an interpretation which sounds reasonable, but which is actually wrong. Therefore, researchers must try to see the data from the culture of the participants, rather than forcing ways of understanding that are appropriate to other cultures, where the same words or concepts can have different meanings (Woods, 1996). When we give names to the data segments, we should select names taken from the world of participants, and which approximate the language of the participants. As such, we help to maintain the original context, which is necessary at least in the initial analysis.

Dividing the Data into Segments and Naming with the Narralizer

The process of dividing the data into segments using the **Narralizer** takes place simultaneously in at least two documents: the source document (interviews, observations, and so on) and the **Narralizer** research document. Following is the analysis process:

1. After preparing the research document and giving it a name (as explained above), open the source document as well. In the source document, **highlight** the first data segment (unit of meaning) and press **Copy**.
2. Turning to the analysis document, **highlight** the name of the **main category** in the **Categories pane** (in Figure 14C, the category is "teachers conceptions") Click **Add text segment** by placing the cursor on the category and right-clicking the mouse. Alternatively, use the **Action** menu, or click the **Add text segment** icon, as shown in Figure 13C.

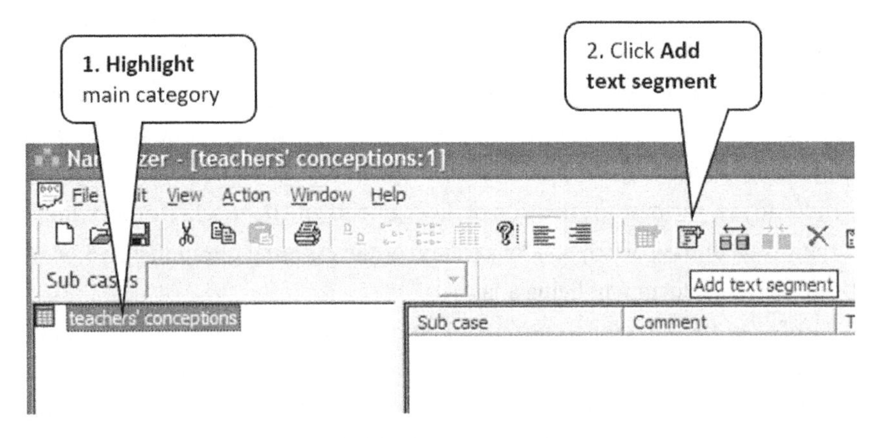

Figure 13C: Add Text Segment Command

3. After clicking the **Add text segment command,** the **New Text Segment for Category [...]** dialog box is opened. When the dialog box opens, the segment that was marked and copied is displayed automatically (see Figure 14D).

   Bear in mind that the last segment highlighted when the **Copy** function was used will always appear automatically. If we forget to **Copy** the new relevant segment in the source document, the dialog box will show the last segment highlighted using **Copy** (or **Cut**).

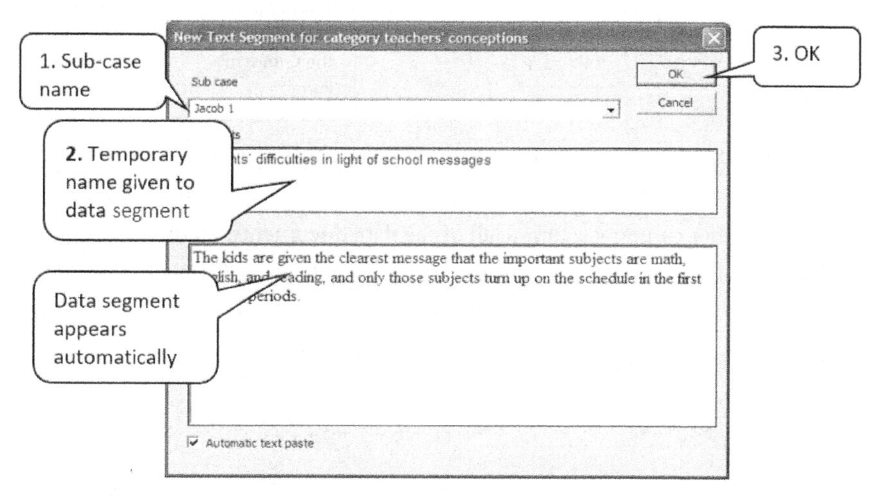

Figure 13D: New Text Segment for Category [...] Dialog Box

4. Write the Sub-Case name or use the right arrow of the Sub-Case screen to **highlight** the name of the **Sub-Case** (in Figure 14D: "Jacob 1"). **Type the name** to give to the segment in the **Comments pane**. This is a temporary name, and may be changed later when we create categories.

   Bear in mind that we can type more than one name in the Comments pane if a segment deals with more than one subject, and add comments that can help us identify and analyze the segment.

5. Click **OK** to transfer the data in the box into the **Narralizer** document database. The icon of the data segment will appear on the screen along with the name of the sub-case to which the segment was assigned. The first words of the segment

appear under the Text column, allowing the researchers to check that they have indeed inserted the right segment.

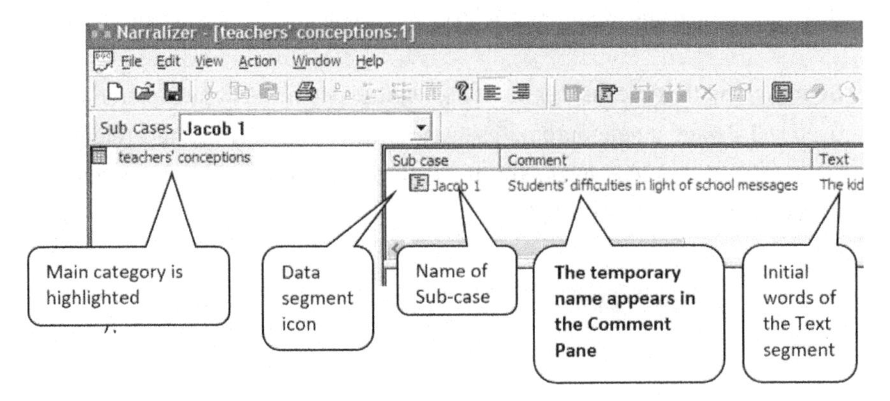

Figure 13E: The Identification of the Data Segment as Appears in the Content Pane

6. The researcher can now examine all of the data documents, identifying segment after segment by undertaking the actions described above. As mentioned, at this stage all of the data is shown under the same category, namely the main category. When more data segments are added, the **Content pane** looks like Figure 13F.

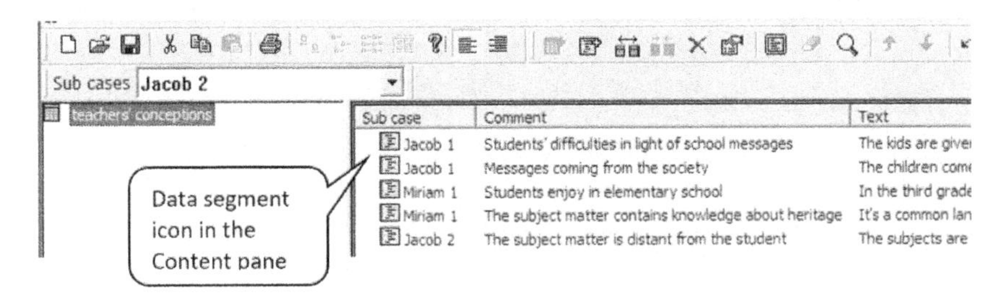

Figure 13F: More Data Segments Added to the Research Document

Length of Data Segments

The initial analysis is the basis for all stages of the data analysis in all methodological patterns that do not enter into the study with predetermined external categories, but create the categories during the analysis process. Careful work in this phase will increase the efficiency of the analysis in the next stages. Sometimes segment division in the initial phase focuses the researcher upon a narrow perspective which acts to block other analysis options that may appear in subsequent analysis stages. Therefore, when researchers decide on the length of the segments, they must examine the alternatives open to them and the "cost" involved with every alternative.

Although researchers read line by line and sentence by sentence and may find meaning in any short phrase, it is not efficient in analysis-based content to divide the data into segments of a line or a sentence (not to mention a few words) except when special circumstances justify it. The reason is that a short segment taken out of its total context could be misunderstood. A data segment is likely to be several lines which are focused on at least one subject. We cannot establish rules for the length of every unit, and obviously segments will differ in length. In structural-based analysis, as compared to content-based analysis, it is worthwhile to occasionally create short, meaningful units composed of one sentence, parts of sentences, and sometimes even one word.

As mentioned above, the proposed method of analysis in this book, with or without the **Narralizer**, allows freedom of movement throughout all stages of initial and subsequent analysis, so that a decision in the initial stage does not block options in the subsequent analysis steps. Given this, the most flexible approach is probably to divide the data into "medium" size segments, reflecting initial data segments that are not too narrow, but also not too broad. However, none of these approaches holds the monopoly on efficiency, and whether to start with broad, medium or small division segments must be decided due to practical considerations rather than principle.

An Initial Analysis Based on Several Upper Categories

The initial analysis is characterized in the absence of categorization and levels of categories, so that all pieces of data "belong" to a main, comprehensive category. Here, researchers strive to avoid allowing their initial assumptions or any theoretical assumptions to affect the identification of the data into a non-reversible perspective. However, some researchers actually intend for their theoretical assumptions to part in the analysis process, side-by-side with the participants' perspective, by a way of openness for the data to "speak" directly. This process may characterize researchers in critical-focused and often self-focused methodologies, as well as with those researchers that place themselves between participant-focused and partial criteria-focused methodologies. An example of such a category tree is illustrated in Figure 13G.

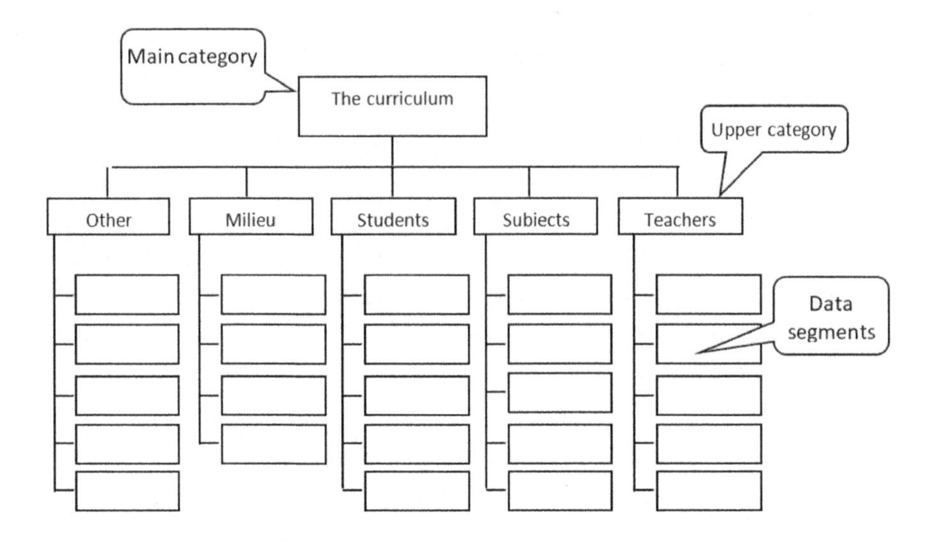

Figure 13G: Initial Analysis with Upper Categories

The example in the Figure 13G is taken from a study in which the researchers entered with a general theoretical perspective, striving for it to be reflected in the initial analysis. In this example, the theoretical perspective deals with a curriculum offering four basic theoretical components: teachers, content, students and the milieu (Schwab, 1969, 1973). The researchers sought to sort out the units of meaning (data segments) of the initial analysis into these four components, each of which would become an upper category. In this way, the researchers let the data "speak" and at the same time highlighted the general theoretical concept. The fifth upper category, the "other," is defined as temporary, and intended to house all the units of meaning not yet attached to one of the four upper categories.

As mentioned, the process of dividing the data document into segments is very similar to the one illustrated above, where all the segments are entered under one main category. As explained, the data will be divided into units of meaning and will be associated with each of the five upper categories. There are cases in which the same segment will be inserted in parallel to at least two upper categories, but under each upper category the same segment will be assigned a different name.

Using Narralizer for Dividing Segments into Several Upper Categories

1. Open a new document in **Narralizer** and record the main category by the **Rename** command, as explained above.

2. To split the main category into several upper categories, click on the **Split Category** command (as shown in Chapter 8). The dialog box is now open for you to insert the names of each of the upper categories, one after another. The example in Figure 13H represents dividing the main categories into three upper categories.

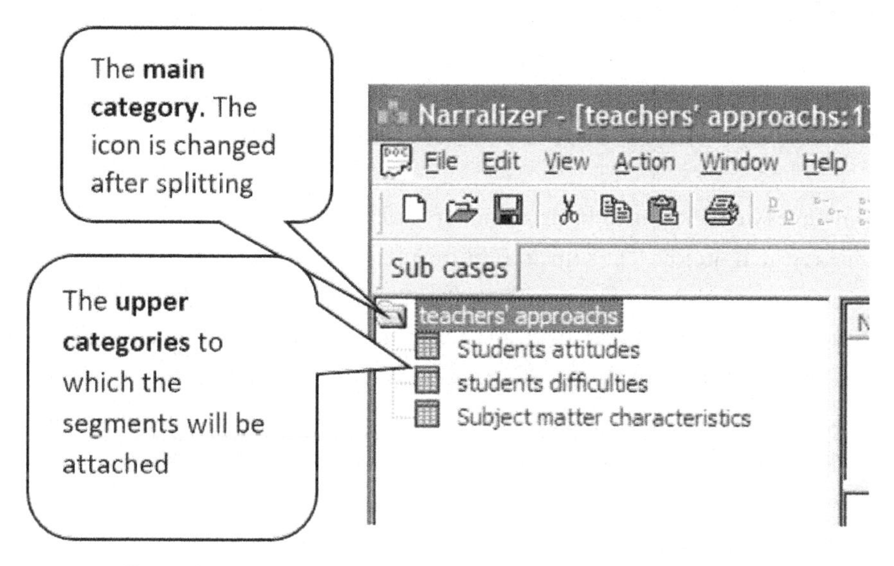

Figure 13H: The Categories in the Initial Process of Dividing Segments into Several Upper Categories

3. Now **insert and give a name** to each segment and attach it to the appropriate upper categories, as explained above and illustrated in Figures 13C-D. As noted, some segments may be inserted to two or more upper categories. Figure 13-1 illustrates the highlighting of one of the upper categories during the division process.

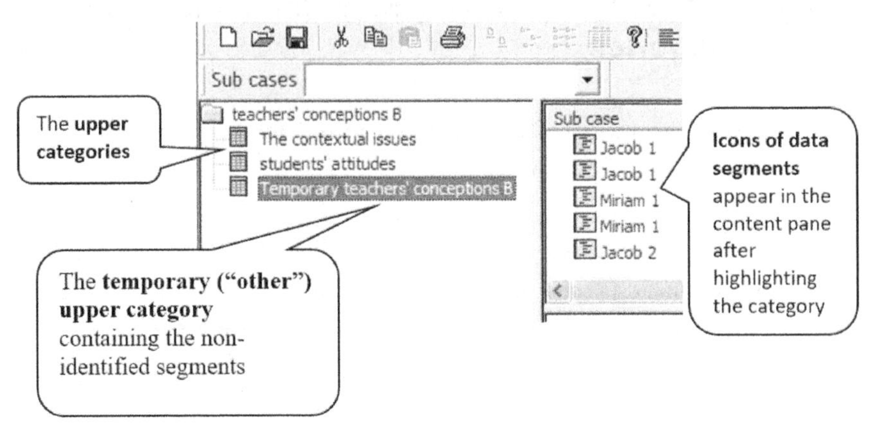

Figure 13-I: The Upper Categories with Attached Data Segments

An Initial Analysis of Several Sub-Cases

Sometimes we want to investigate a number of sub-cases (and/or participants) in the same research project. The principles of the initial analysis of one case and several cases are identical. At the initial analysis, the researchers divide the data into units of meaning in each separate case, with an open mind and according to equal standards. It is important to avoid forcing the segments' division from the first cases upon the cases to follow. However, since all cases deal with a similar phenomenon, it is likely that the researcher harbors a common perception that influences the nature of the segments' division in all sub-cases, even without his or her awareness or prior intention. Thus the names of the various segments are often similar (though not necessarily identical). As long as the researcher is constantly attentive and sensitive to differences between the various sub-cases, rather than treating analysis as a simple technical task, the dual study will succeed. If researchers conduct the initial analysis with several upper categories, it is important to always relate to all upper categories with all sub-cases.

Bear in mind to ascertain that identification marks appear in each segment of all of the sub-cases (for example, "Ilana C1"). Without the identification marks, the analysis process will lose meaning.

Setting the Sub-Case Name of the Data Segment Using Narralizer

Narralizer helps to very carefully maintain the identification marks of each data segment throughout the study. When clicking on the command Add Text Segment, a dialog box is opened (see Figure 14J). In the upper pane, the **Sub-case pane,** insert the identification marks of the segment. If the name of the sub-case has been set earlier in the document analysis, click on the arrow triangle on the side of the pane, and a list of all the sub-cases appears. Choose the appropriate sub-case and click on it (in Figure 13J, "Sharon 1"). If the sub-case is not listed, type it in the Sub-case pane.

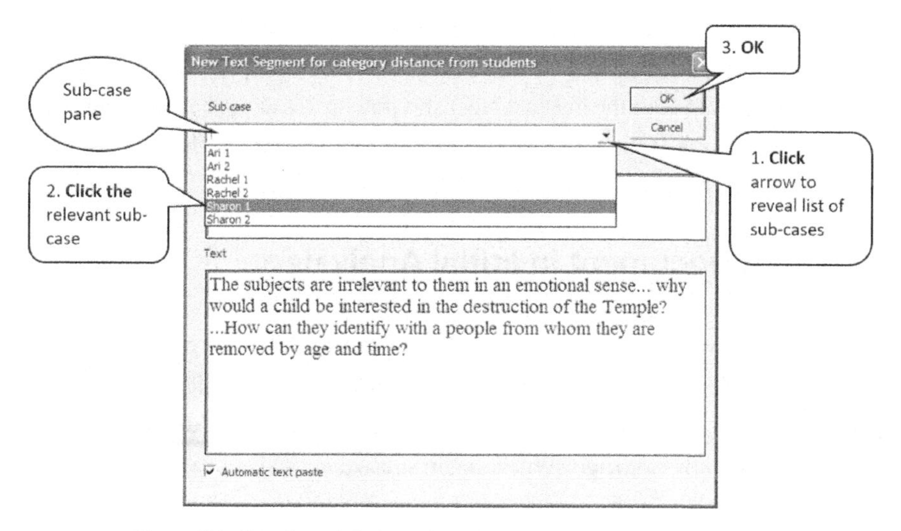

Figure 13J: Choosing a Sub-Case Name from the List for the Data Segment

End of the Initial Analysis Process

Generally the initial analysis phase ends when the researcher finishes dividing data into segments of all relevant sub-cases. Sometimes researchers can decide to limit the initial analysis to a particular part of the sub-cases (for example, to analyze 20 to 25 sub-cases out of 50 or even less). The main benefit of reducing the scope of the initial phase is to save time (and sometimes money). As mentioned earlier, the main purpose of the initial analysis is to highlight the direction of the analysis stages, i.e., the initial analysis is an interim analysis phase and not a final stage (except in certain cases of exploratory research which may see the initial analysis as final step, as will be discussed below). Therefore, if the researchers find that they can point to a clear direction of the analysis based on the initial analysis of a limited number of sub-cases, they can move to the next stages which focus on constructing the category array from the initial data segments, and then to analyze all sub-cases (including those sub-cases which were not part of the initial phase "sample") into the constructed category array (see Chapter 15). If during the subsequent stages of analysis the researchers find themes which were not listed in the initial stage, they will add them to the list of categories (an issue which will be explained below).

There are also instances where researchers are in the midst of a lengthy study which already began with an initial analysis. In this circumstance, the initial analysis was carried out in some certain cases during the early stages of the study. The category array was constructed on the basis of the initial analysis process (a process that will be explained in the next chapter) and now serves as the basis for analyzing the other sub-cases. If the researchers find additional themes in the new raw data, they will enrich the category array by adding new categories.

Saving a Document in Initial Analysis

The initial analysis is the first step that opens the analysis process. The analysis process is carried out at this stage, even if it will be changed later. Researchers should keep the initial phase documents for reference and to ensure that they will not stray from the data's original meaning in subsequent stages of analysis (Pidgeon & Henwood, 1996). In this way, it will be possible to ensure the transparency of the process as a condition to claim qualitative validity and reliability. Moreover, as we have emphasized in previous sections, in the initial analysis phase, all data segments are stored in their original order (of the interviews, observations, etc.). In the advanced stages of the analysis, the segment will be detached from its authentic order (even if we use numbers to enable tracking of sequences). Maintaining the initial analysis documents will allow us to examine the original places of each of the data segments.

Saving the Initial Analysis Document in Narralizer

Narralizer allows us to save the initial analysis document (and documents in later stages) to be easily accessed and read. Every few minutes the **Narralizer** performs an automatic "save," and automatically maintains a backup document. Because every step of the analysis is based on the findings of the preceding phase, in the transition from the initial analysis phase to the subsequent analysis stages we must continue to work on the same picture of analysis located in the initial analysis document. But we also want to keep the initial analysis document as is. Thus, perform the **Rename** or **Save** commands to give a new name to the document. In this way, a new document is created while keeping and saving the initial analysis. Continued analysis will be carried out in the new document.

From Initial Analysis to Report Description: The Case of a Pilot Study

The initial analysis is intended to serve as the first step in creating categories and to give researchers an orientation towards the entire research process. Therefore, there is no expectation to prepare a report description at the end of this phase. Nevertheless, there is at least one situation in which researchers may base a study on the initial analysis stage and go straight to writing a report description, where researchers are conducting a pilot study intended to result in a descriptive report.

Each research project is based on prior knowledge. This knowledge is grounded in experience and personal knowledge and/or relevant professional academic literature. Sometimes there is a quite big gap between what we already know and what we want to know. This gap may adversely affect the ability of the researchers to focus on their research. A pilot study is a means of bridging the gap by clarifying the focus of the study and highlighting the research issues to be addressed. This exploratory research instills greater confidence within the researchers to prepare their research proposal and research design, and may direct them to new areas of relevant literature.

According to Kirk and Miller (1986), we can distinguish four stages in qualitative research: planning, data collection, analysis and reporting. This distinction places the planning phase as the opening stage of each study. Researchers often lack the knowledge and operational data to design the study, and especially to determine what can or should be the focus of the study. The value of the pilot study as a means to clarify the focus of the research and its questions is very significant when we are required to submit a proposal to an academic committee or to obtain a grant, for such a proposal must be accompanied by clear definitions.

There are several ways to perform a pilot study. If the proposed research is designed to focus on a single subject or a small number of subjects, we can choose one of the issues or one participant, conduct one interview, and if necessary, make an observation. If the research is designed for a larger number of participants, the most effective way to perform exploratory (pilot) research is to work with three to five participants (about 10% of expected respondents). The data collection phase of the pilot study should be similar to the process of research data collection as a whole, though more limited in scope. Researchers can use the pilot study to examine not only the nature of the data gathered, but also the amount and extent of information that may be a received from a single interview or observation. This information may form the basis of the decision of how many interviews and observations and/ or how many participants are required for the research in order to guarantee its contribution and maintain the effectiveness of the research process.

While the data collection process in the pilot study should be very similar to what is planned for the study as a whole, it seems that exploring research analysis processes can be simplified. The purpose of the pilot study allows the researchers to limit themselves to execute only the initial analysis process. Through the process of initial analysis, we can get an idea of how to target the entire study and clarify the focus of the planned research. Upon completion of the initial analysis process, we can describe the research picture obtained. This does not mean to display a rich description as characterizes qualitative research, but a description that points to different research directions, including authentic data segments if necessary. The initial analysis data used in the pilot study can also be utilized later in the main study's analysis research process.

CHAPTER 14

THE SECOND STEP OF ANALYZING IN CONSTRUCTIVIST BASED METHODOLOGICAL APPROACHES: FROM INITIAL TO MAPPING ANALYSIS

The constructivist based methodological approaches that base their categories on the collected data start the analysis process by dividing the data to units of meaning in the initial stage, as described in the previous chapter (Chapter 13) continue to the phase of creating the categories in the mapping stage.

From Initial Analysis to Mapping the Category Array

As emphasized, the process of creating categories is comprised of two components, dividing the data to segments and identifying significant links between the units to create a category array. In the process described in this chapter, all the units of meaning are compared to each other, and the relationship between them and other units of meaning is examined Here, researchers consider the data segments and sort them into categories, forming connections between the constructed categories on the horizontal axis, and the categories and their division to sub-categories on the vertical axis.

In the initial analysis, all the data segments are located at the same level, regardless of their potential relationships. The researcher's attention is focused on investigating the contents of the data (or its structure), paying less, if any attention to relations between the segments, categories, and between categories and their sub-categories. In the mapping analysis phase, however, the segments are associated to categories, and categories are distinguished from each other and divided into different levels (Flick, 1998; Pidgeon & Henwood, 1996).

Practically, the researcher looks at the names of the units of meaning (data segments). For the sake of efficiency, the researcher can concentrate all the names of the segments in running order on a word processor document and/or paper. This list includes the names of all segments from all the documents to be analyzed together, without reference to any specific document (unless the researcher intends to make a different analysis of specific documents which would be analyzed separately). At the first step, the researcher can notice that some segment's names are very close to others and appear as belonging to the same "family." In this way, the researcher will sort all the segments of the same "families," where each family is actually a category (or "theme") that includes several segments of data. Figure 14A shows a running order of segments' names for the purpose of constructing categories.

1. Class characteristics
2. Student characteristics
3. A student faints in the classroom
4. I did not know she was sick
5. One of the students takes care to the student who fainted
6. The relationship between the students
7. Students talk about social experiences
8. A class that includes students with difficulties
9. Nobody informed me that I am teaching a class with difficulties
10. I did not know how to treat and what to demand from the class
11. I was not prepared to teach such a class
12. I consult with a psychologist
13. The connection with the psychologist
14. Difficulties in connecting with the psychologist
15. It was an advantage that I didn't know the difficulties of the class
16. Methods of teaching such a class

Figure 14A: List of Segments' Names as a Basis for Creating Categories

The list of segments' names illustrated in Figure 15A is, of course, very limited. In every research project there will be a longer list, but this limited list is sufficient to demonstrate the process. This list of segment names is the basis for creating categories. Because the segment's name reflects the segment's content, we can observe the list of names and try to organize them by "families" (categories), namely, to combine bits dealing with the same subject. At any time we can return to the initial analysis document itself and examine the data segment, and not simply refer to the segment's name only. Figure 14B illustrates categories that can be created from the list of segments displayed above. Next to each category is the order number of the segments taken from Figure 14A.

A. Who are the students? (segments 1, 2, 6, 7)
B. The class traits (segments 8, 9, 10, 11, 15, 16)
C. Getting help in the school (segments 13, 15, 16)
D. Students' relationships (segments 2, 5, 6)
E. Students' difficulties (segments 8, 16)
F. Dealing with an exceptional event in the classroom (segments 3, 4, 5)

Figure 14B: The Categories Created from the Segments

It is important to note that some of the segments (units of meaning) are located in more than one category. This is because the pieces of data are quite rich, and often include simultaneous references to several topics. It is also possible that some segments cannot be added to the other segments, and which therefore will be defined as a category by themselves.

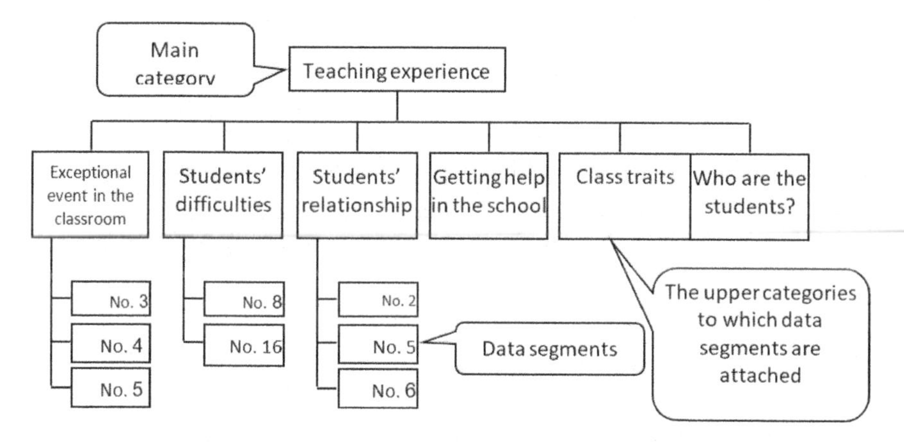

Figure 14C: The Category Tree Created from the Data Segments

In order to insert the segments into the category array created, the researchers listed the category tree in vertical order, as illustrated in Figure 14D.

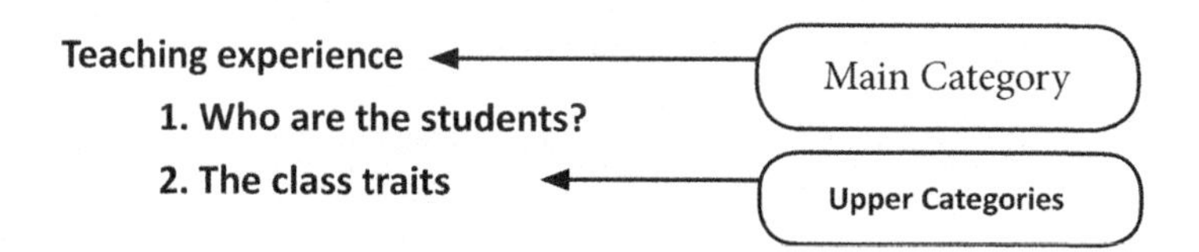

3. Getting help in the school

4. Students' relationships ← Upper Categories

5. Students' difficulties

6. Dealing with an exceptional event in the classroom

Figure 14D: The Vertical Category Array

Researchers then scrutinize the data documents. Although they have already determined which category is associated with each data segment, they can go back and examine each piece of data and, if necessary, change the categorical affiliation. Using "Copy & Paste," they place all data segments under the appropriate categories, while ensuring that each segment includes its identification marks. Figure 14E illustrates several segments under their corresponding categories.

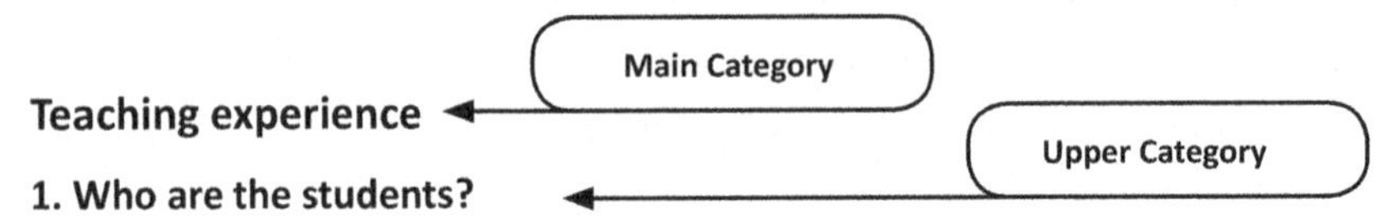

Teaching experience

1. Who are the students?

Very nice children. Fifteen. They are in the first year of high school. They didn't know each other before they came to school. (Jerry, A1)

2. The class traits

This is a smaller class. What I heard from other teachers is that they all come from family and social environments with a lot of difficulties. They have high potential[...] There are a lot of students with serious learning difficulties, too. Some of them are also very smart but with a short fuse [...] (Lea A. 1)

3. Getting help in the school

4. Students' relationships

[...] The class received her very warmly. It is one thing. Or a student with difficulties and a strong student learned together before an examination. [...] They enjoy talking to each other: "What happens at the party tonight?" (Susan, A. 1)

5. Students' difficulties

6. Dealing with an exceptional event in the classroom

Figure 14E: Inserting the Segments into the Category Array

Note that the segments are not added to the main category, which is an indication category, i.e., a category that does not "hold" pieces of data, but shows only the data characteristics. The main category is "holding" the upper categories which in the above example are content categories, that is, categories that can "hold" data segments.

In the process of attaching the data segments to categories, researchers are not tied to the names they gave in the initial analysis stage and they may consider attaching new meanings to certain data segments. Moreover, during this stage they can also change the name of the categories, add categories, attach segments to better-suited categories, and so on.

Toward Creating Upper Categories Using the *Narralizer*

Before moving data segments to the upper category located in the Categories pane, consider the list of segments as a whole, focusing on their names.

1. Place the cursor on the main category in the **Categories pane** and **highlight** it. (If the initial analysis was done by using some upper categories, place the cursor on the upper categories, one after another). All the content segment marks will be put on the content pane, as illustrated in Figure 14E. We can now see and read all the segment names by using the scroll bar.

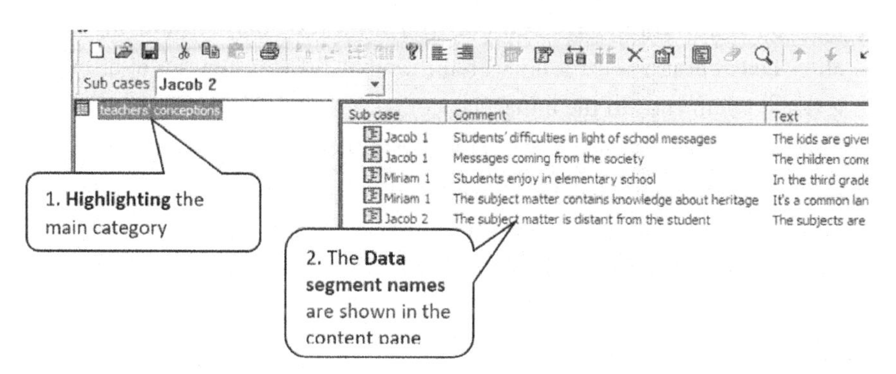

Figure 14F: List of the Segments in the Content Pane

2. In most of the research projects, the list of the initial data segments is very long. It is difficult to work with them through the content pane even by using the scroll bar. **Narralizer** allows us to see this list on the display pane and even to print it.

A. **Highlight** the main category and click the **Show Sub Tree** command through the menu that appears when clicking the right mouse button, as shown in Figure 14F, or by the Action menu. You may also run the command by pressing the appropriate symbol on the Command bar.

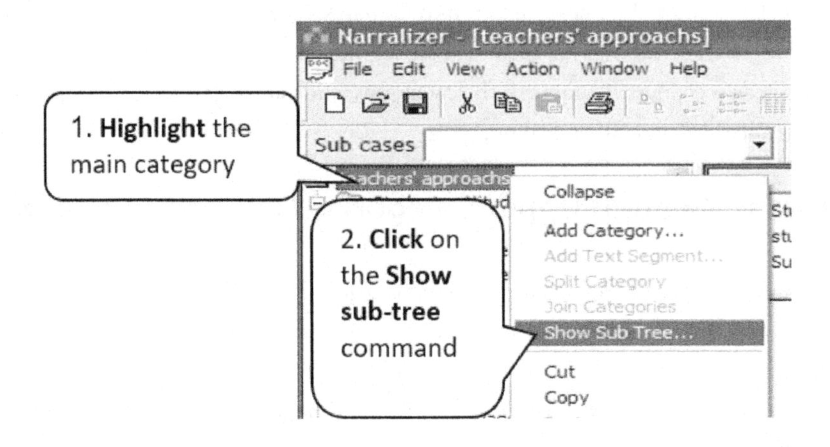

Figure 14G: Show Sub-Tree Command

B. Following the **Show Sub-Tree** command, the **Select sub-case dialog box** is opened, as shown in Figure 14H. At the bottom of the box are three different notations: **Show comments, Text only** and **Show text**. **Show comments** and **Show text** appear automatically with the symbol "V," which means to display text and comments. In the initial stage of analysis, the segment names were inserted into the comments pane, so we want them to appear on the display pane. However, to avoid having the full text of the segments appear, **remove the checkmark symbol "V" from the "Show text" sign** by clicking on it. Then, click **Select all,** and the dialog box is closed. The display pane will display all the segments' names belonging to that category.

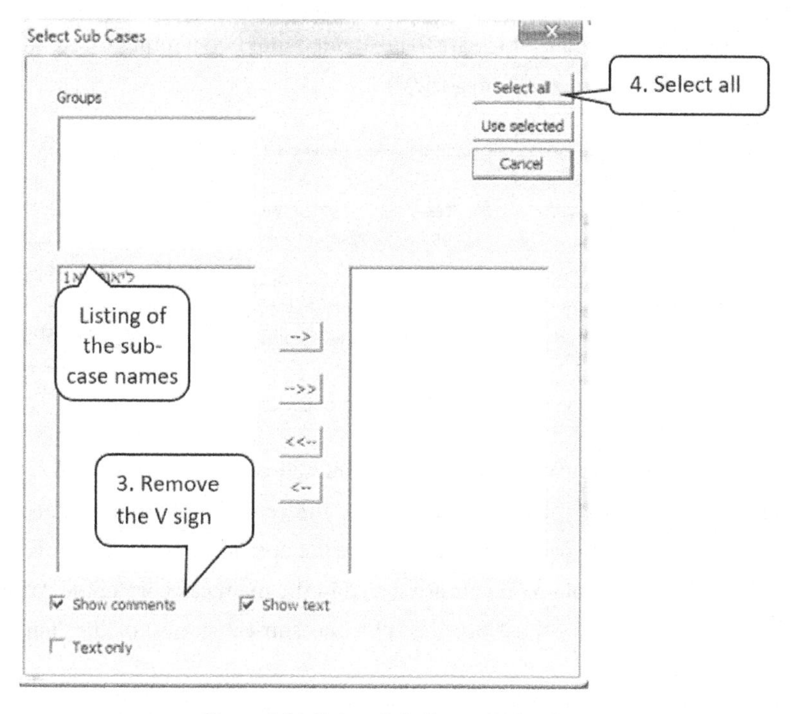

Figure 14H: Select Sub-Cases Dialog Box

Creating Upper Categories Using the Narralizer

In order to create the upper categories using the **Narralizer**, open the initial analysis document, but change the name by using the **Rename** command or **Save As** to a new name that indicates the mapping analysis. In this way, the initial analysis document will be saved, and we actually added a new document with a new name. You may use the same name as was given to the initial analysis document, with a new identifying mark to clarify that this is a different analysis document (for example, if the name of the document in the initial analysis was "Teacher conception," we may call the new document "teacher conception mapping," "teacher conception-B," or add a date or any other identification). In the new document, the mapping category array will be created.

In the process of creating category arrays, the segments are divided into families. We do this by identifying a common theme shared by several different segments. Each family represents a category to which segments may be assigned.

Figure 14-I illustrates an example of a small section of the list of data segment's names in the **Content pane** which are to be divided into two families (or as they are called in the analysis process, "categories").

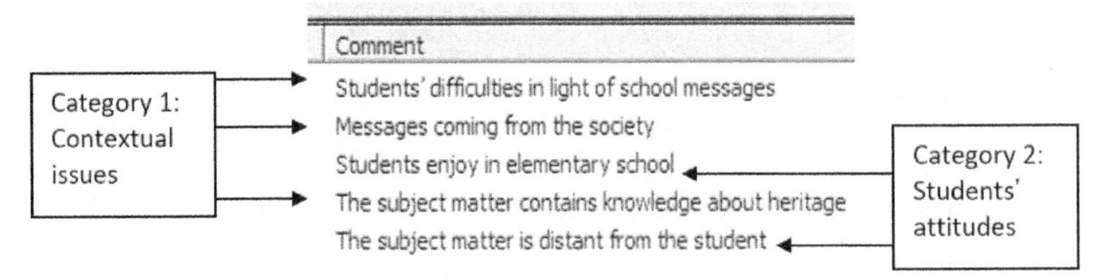

Figure 14-I: Consideration of the Option of Dividing Data Segments

As mentioned, the example in Figure 14-I contained very few segments. A real list of segments to be sorted can contain dozens or hundreds of names. Thus it becomes easier to consider all the data segments by using the operation explained above. (See Figure 14F-G) All the data segments are stored in the main category of the Analysis Document. This category must now be split into sub-categories, or the "families" discussed above.

1. **Highlight** the **main category**. Now split the main category using the **Split Category** command (see Chapter 12). This command opens two boxes in which to type the names of the two sub-categories. Similarly, you may add as many sub-categories as desired by using the **Add category**. Figure 14-j demonstrates the main category, the split two upper categories, and also the temporary category (see explanation below). In the same way, you may create more upper categories.

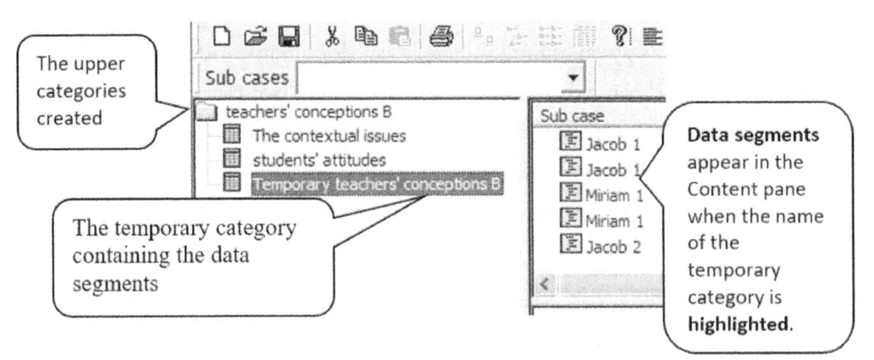

Figure 14-J: The Process of Splitting the Main Category

Note that in addition to the two sub-categories created, a new category was also created (in the example above: "Temporary teachers' conceptions-B"). This category stores all of the data segments. It is a temporary category, which will be deleted after the researcher has moved the data segments into the appropriate sub-categories, an operation that will be explained below.

2. Once the Category array has been created in the **Categories pane**, you may then place the data segments in the appropriate category:

When the cursor is pointed at the **Temporary category,** the data segments and their temporary names will be displayed in the **Content pane**. Now take each segment and place it in the appropriate category. There are two ways to do this: by dragging and by Cut & Paste.

To drag the data segment to the appropriate category, **highlight the data segment** in the **Content pane** (in Figure 14K: "Students' difficulties in light of school messages") and **drag** it using the cursor into the desired upper-category (in Figure 14K: "The contextual issues").

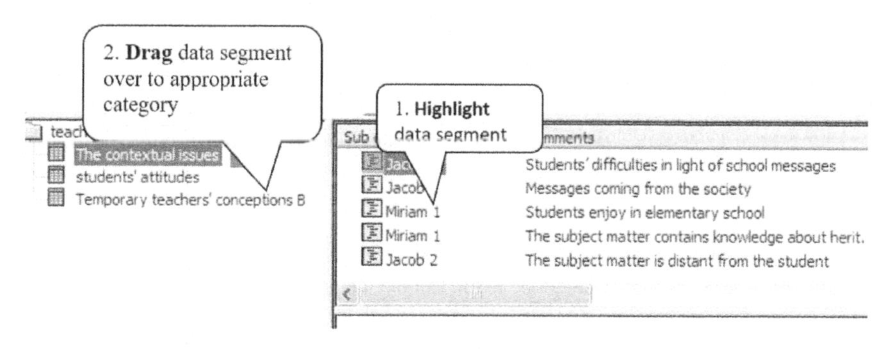

Figure 14K: Drag Data Segment to the Targeted Category

When **highlighting the category** to which the data segment was assigned, the **Content pane** will display the segment belonging to that category, as shown in Figure 14L:

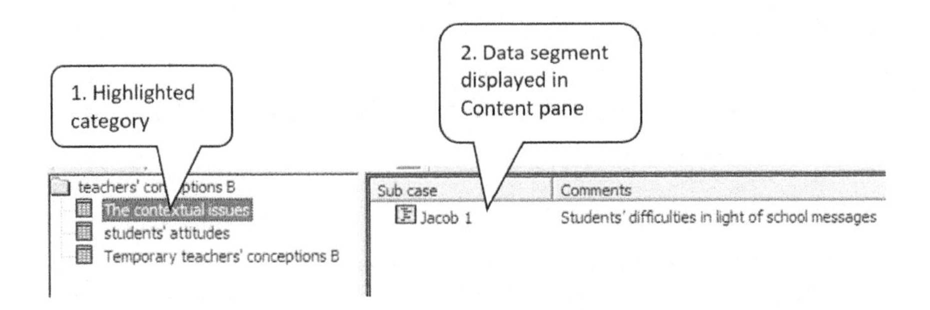

Figure 14L: The Data Segment Displayed in the Content Pane after It is Dragged There

Thus, we can transfer the data segments to their appropriate categories until the temporary category is empty and can be deleted by placing the cursor on it and pressing **Delete**.

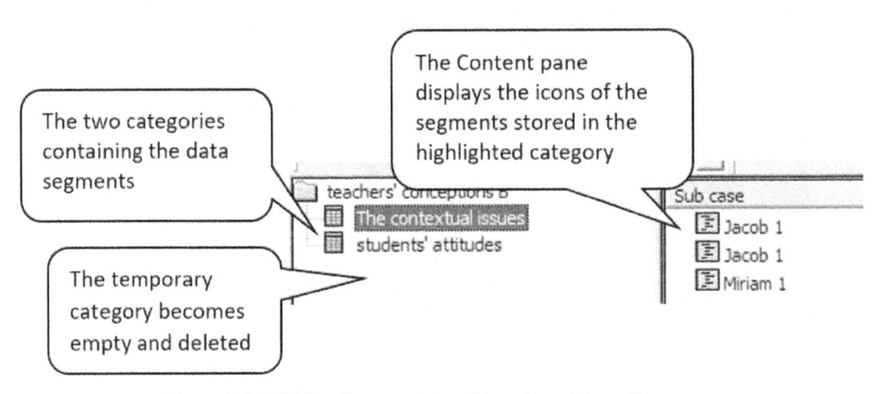

Figure 14M: All the Segments are Transferred from Temporary Categories to the Upper Categories

Bear in mind: Researchers sometimes find it hard to decide where to place certain data segments. This decision can certainly be postponed and the data segments left in the temporary category without being deleted.

This procedure can also be carried out using the **Cut & Paste** function. It is better to use **Cut & Paste** rather than dragging if you wish to assign a segment to more than one category. Data segments can also be transferred using the Copy function, though this is not advisable. Although the Copy function allows us to place the data segment in the target category, it also leaves it in the original category (in our case, the temporary category). Removing the segment from the temporary category gives you control of the segments that you already removed and guarantees that you will not transfer the same segment several times.

The Completion of the Category Array

Once all data segments have been placed in the upper categories, a situation is created in which each upper category includes a considerable number of pieces of data. The researchers may re-examine the segments in each of the upper categories to consider the possibility of splitting the upper categories into several sub-categories (each one category to at least two sub-categories). The researcher will insert the new sub-category into the word processor analysis document under the upper category. The data segments are moved from the upper category to the appropriate sub-categories by using Copy and Paste. Upon completion of the split to sub-categories, the splitting process can continue, and the sub-categories split to sub-sub-categories. This process will continue to the next level until the mapping analysis process is completed.

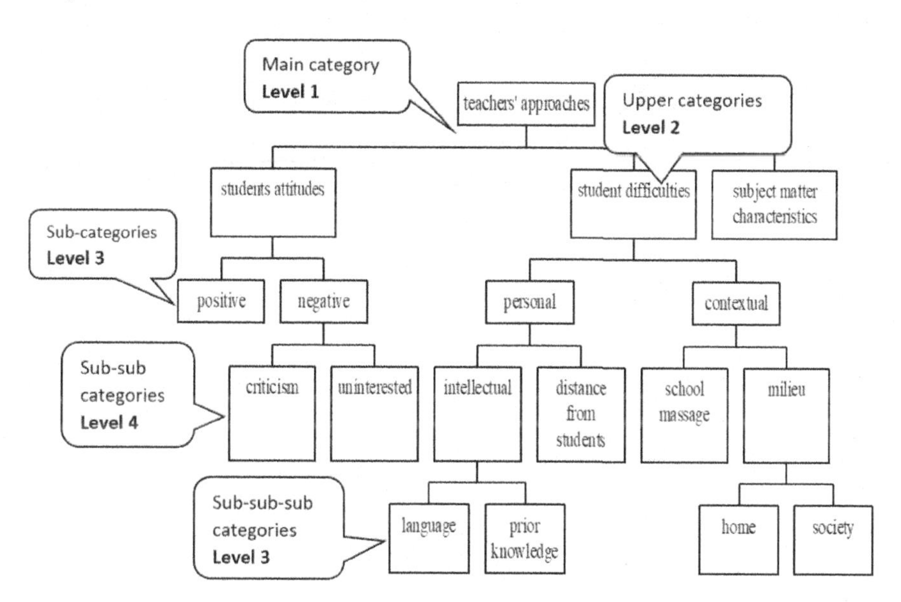

Figure 14N: The Mapping Category Array Tree

The Completion of the Category Array Using the *Narralizer*

To examine the possibility of dividing content categories at any category level, they must be viewed with their content segments. Perform this by **highlighting the category**: Main, upper, sub-categories, and so on, in the categories pane and activate the command **Show Sub-Tree** (as already explained), as illustrated in Figure 14-O.

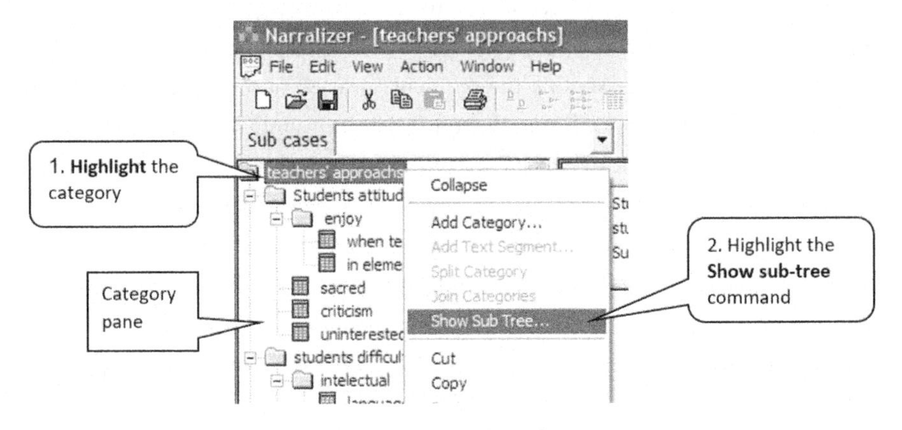

Figure 14-O: Show Sub-Tree Command

The display pane will exhibit all data segments belonging to the same category, as shown in Figure 14N. We can move the scroll in the right side of the pane and read all the relevant segments, or even print it.

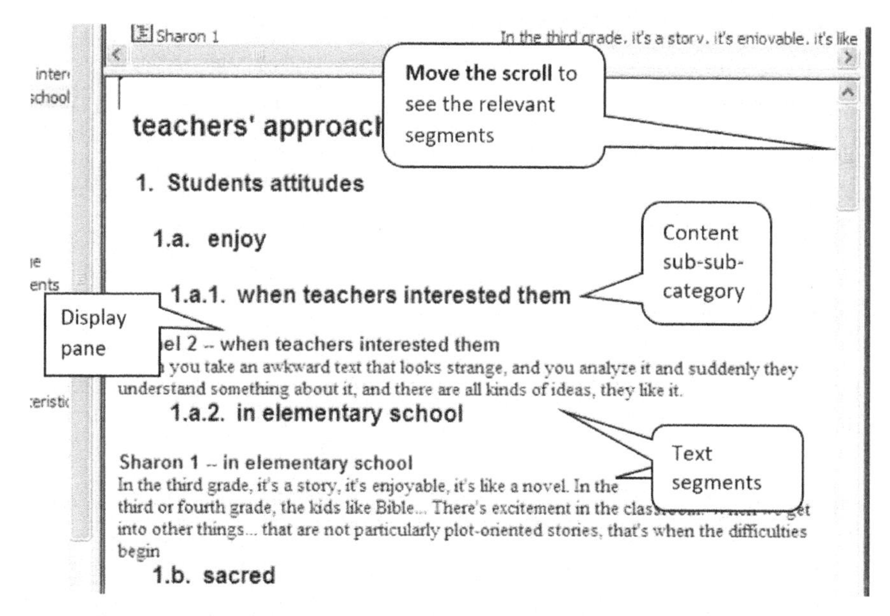

Figure 14P: The Category and Data Segments as Shown in the Display Pane

Read the segments carefully and determine if it is necessary to split them using **Split category** and **Add category**, and then move the segments to the new categories as explained above.

Changing the Data Segments' Names and their Location in the Categories

At this stage (and at all other stages), researchers may decide to change the names of certain data segments, if they feel that the names given to these segments during the initial analysis do not match the complete picture of the mapping analysis array. Any title given to the data segments (and categories) is completely reversible. In the same way, at any time during the analysis at lower levels, you may step back and consider whether the placement of specific segments under this category or another is the appropriate place. If you come to the realization that the segment does not fit the category where it was previously placed, simply change the location of the segment and transfer it to a different category by using Cut and Paste.

Changing the Data Segments' Names and their Location in the Categories Using Narralizer

With **Narralizer**, we can change the names given to the data segments or add a new name by using the **Properties** command, operated as follows:

1. Highlight the data segment icon in the **Content pane**.
2. Click the **Properties** command in the Edit menu or right click the mouse, as shown in Figure 14-Q.

Another alternative is to double-click on the data segment icon in the Content pane

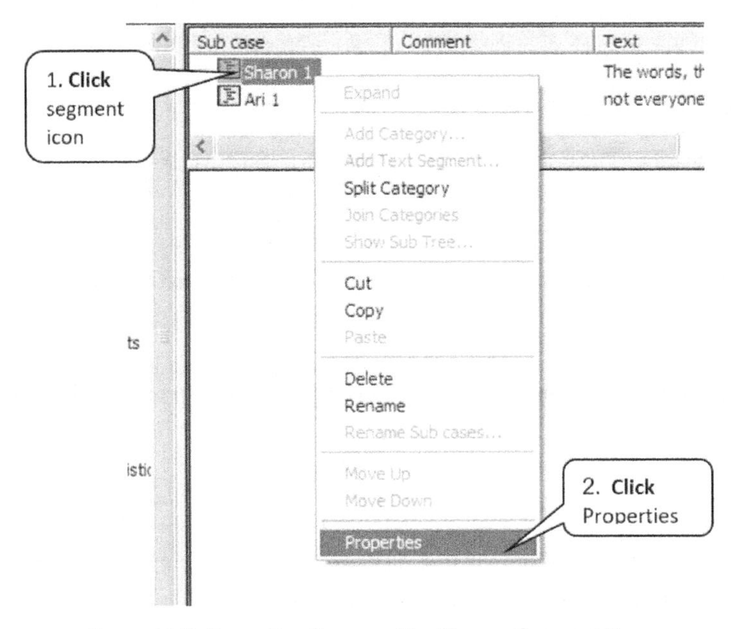

Figure 14-Q: Properties Command for Change Segment Name

3. The **Text Segment Properties** box will open (see Figure 14P), allowing us to change the segment name in the **Comments pane**, delete or add names.

   In the same way, notes may be written in the Comments pane, revising the text. There is also the option to change the **sub-case name** from the sub-case list, if necessary.

4. Clicking **OK** will close the box, saving all your changes.

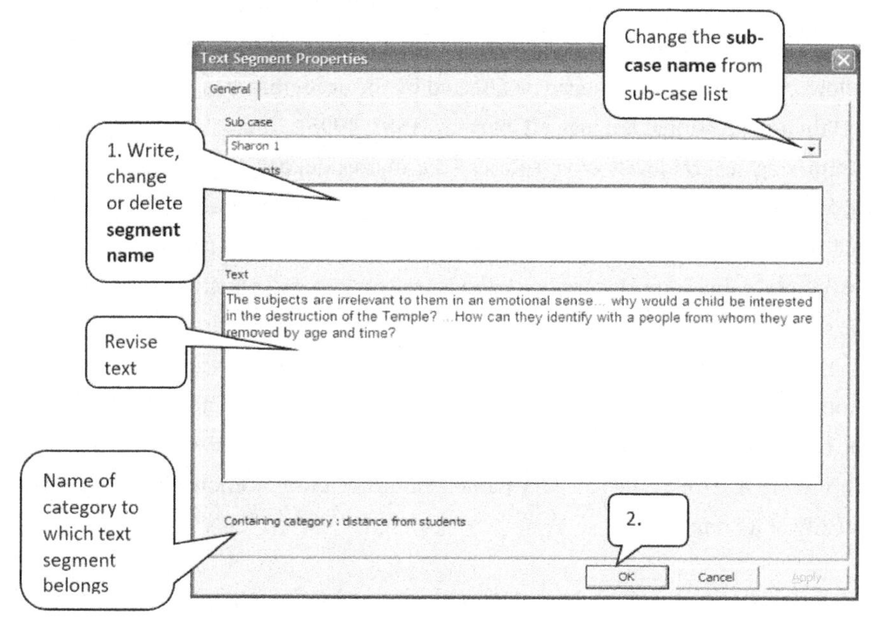

Figure 14R: Text Segment Properties Box

5. At any time, it is simple to change the location of any segment and transfer it to a different category by **dragging** the segment from the **Content pane** to the new category in the **Categories pane**, or by using **Cut and Paste**.

Category Names

Category names, like segment names in the initial analysis phase, can be taken directly from the natural language of the participants of the study, which Strauss and Corbin (1990) call "in-vivo categories." Although in the mapping analysis the researchers continue, as much as possible, to use a name taken from the language of the participants, the nature of the mapping analysis leads to a condition where researchers cannot remain completely close to the original words of the participants. The reason is that at the mapping analysis, researchers combine separate pieces of data segments that deal with the same topic into one unified category or into a "family" of categories. This is a process of conceptualization in which the researchers construct broader terms for groups of segments and categories. In this case, it is clear that at least some category names cannot be taken directly from the language of the participants, but from a conceptual language.

In the mapping analysis stage, and even more so in the stages of analysis that follow, category names can also be affected by the academic area of the researchers and their professional language (Charmaz, 1983, 1995). In fact, in a category array comprising several levels of categories--the main category, upper categories, sub-categories, sub-sub categories, and so on-- as we "go down" to lower levels, it is likely that the categories are more often close to the language of the participants, and as we "climb higher" in category level, the category names are likely to be affected by the researchers and research language. However, the category names, whether taken from the language of the participants or from professional-academic literature, should be as catchy as possible so that they immediately attract the attention and the hearts of the researchers and their fellow researchers. In the words of Strauss & Corbin (1990, p. 67), the category name "should be graphic enough to remind you quickly of its referent."

Expanding and extending the mapping category array

In order to expand and extend the mapping category level, please follow the instruction in chapters 9 and 10.

PART FIVE

ARRANGING THE ANALYSIS PRODUCTS

INTRODUCTION TO PART FIVE

The ultimate goal of any research is to present its findings to the readers and the public at large. The findings should be organized so as to be communicated to the potential readership (Lincoln & Guba, 1985). The mapping analysis process (as presented in previous chapters) is the process in when the researcher constructs meaning to all the data findings. However, once this goal is achieved, it should not be regarded as the final phase of the study. There is one additional important step, that of presenting a comprehensive view of the research picture. Organizing the analysis findings provides the means to create the "story line" of the study and to portray a significant depiction of the research results. The process of arranging the research findings aims not only to make the study accessible to potential readers, interact with them, and provoke their empathy for the study, but also to focus the readers' attention and interest upon specific defined issues and thus boost the contribution and impact of the research.

When researchers focus on presenting the research picture, they choose the objects of study from a certain estimation and consideration and not at random (Merrick, 1999; Peshkin, 1993; Sciarra, 1999). Qualitative researchers use a variety of options to present the data picture. In the analysis process, the researchers identify the data according to two main components: categories and sub-cases. Each data segment is identified by the categories that characterize it and the sub- case to which it belongs. At the final presentation of the research findings, the researchers must weave together all the components until an all-inclusive picture is attained.

Different combinations of components give the researchers a wide range of options for displaying the findings. As for the category component, researchers can refer to the entire final category array to present a comprehensive research picture, or alternatively selectively choose a certain representative group of categories. More emphasis may be placed on certain categories and less on others. Researchers can choose a specific order of categories to illustrate the research picture, present the descriptive and theoretical categories integrally or display them separately, and more. The same options apply to the focus on sub-cases: researchers may display all the sub-cases equally or accentuate certain sub-cases over others. They can choose several representative sub-cases and present them as reflecting on the overall picture. They can choose a certain order to present the sub-cases, either by dividing the sub-cases according study participant, to a temporary occurrence, to data types, or by one or more combinations of these parameters. They can arrange the sub-cases into groups according to these or other characteristics, and more. The combination of a reference to categories and sub-cases allows the researcher to choose from almost unlimited possibilities of final research presentations. A researcher's consideration is guided by the goal to present the findings in the best way while communicating successfully with the potential readership. This part of the book deals with the transition from the mapping analysis stage through several analysis steps up to the stage of writing the final report. The chapter will be devoted to the ways in which to organize the findings of the analysis as a basis for the final report

CHAPTER 15

THE FOCUSED AND THEORETICAL ANALYSIS STAGE

Focused analysis is a process through which the researchers focus the elements of data into a coherent explanation around a main category or several upper categories. This means that the researchers choose those categories from the upper categories that look meaningful to them and around which they wish to focus the entire study picture. While the mapping analysis ends with a broad category picture, the focused analysis phase scales down the perspective to specific upper categories which researchers consider to be sufficiently rich to present an interesting research picture.

Among the researchers who use the focused analysis, some are interested in a descriptive picture as the product of the study and may see this stage of the analysis as the final stage. Researchers who also want to display a theoretical picture and will execute the focused categorization with the theoretical categorization, as will be explained in the end sections of this chapter.

Constructing the Focused Analysis Category Array

Theoretically, focused categorization can begin relatively early. The sense of core categories begins early in the first stages of analysis. Even during data collection, researchers can generally sense which categories have the potential to become core categories. This sense is strengthened during the previous analysis stages. However, it is necessary to wait until the entire mapping picture is constructed and can be viewed. Following the mapping analysis, the researchers have an appropriate picture and can be certain that the main category or categories they chose are indeed the most suitable (Riessman, 1993).

In the focused analysis process, the researcher sometimes seeks to create a new main category instead of the main category of the mapping stage. This main category should be more focused and surrounded by a smaller amount of categories. In the process of constructing the main category, the researcher seeks a "theme", i.e., what appears as the main problem or issue and what contains enough data to justify its transformation into a main category of research. Strauss (1987) suggested several criteria for judging which category should serve as the core category: [1] It must be central, that is, related to as many other categories and their properties as possible. [2] It must appear frequently in the data. [3] It must relate easily to other categories. [4] It has clear implications for a more general explanation.

Technically, the researcher can use the mapping category tree as a means to see the overall picture and to find the appropriate main category. Using the table presented in the end of Chapter 10 may be much more efficient, because the table aggregates and summarizes the main points of the data and does not limit itself to presenting only category names. Often each of the upper categories of the mapping analysis (one "branch" of the mapping tree) can become a main category. Sometimes the researchers notice that a combination of two categories or more can lead to a significant main category.

The transition from mapping analysis to focused analysis will be demonstrated in Figure 15A. In this example, the researcher has concluded that one of the upper categories, "Knowledge," has a potential for becoming a main category of the research. Figure 15A illustrates the mapping tree, and highlights the categories that the researcher has decided to focus upon.

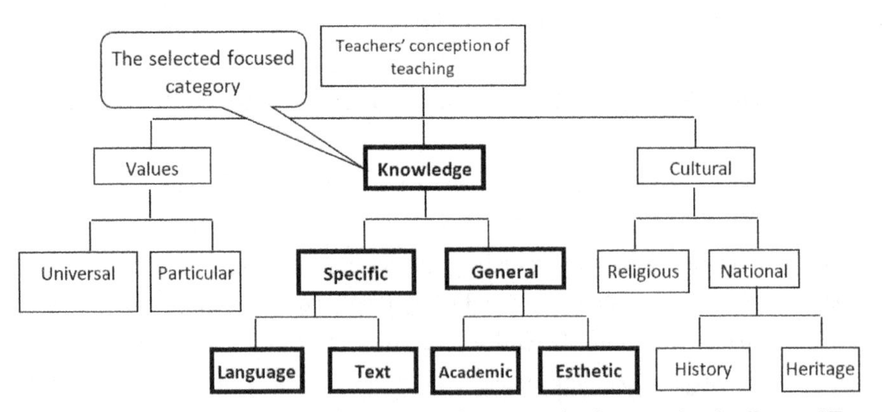

Figure **15A**: The Mapping Tree with the Potential Categories for Constructing the Focused Tree

As illustrated in Figure 15A, when the upper category "Knowledge" is identified as a potential for the main category, all the categories below continue to be connected to this category. But the focused tree cannot be the same as the original "branch" from which it was built. The transition from mapping analysis to focused analysis is not a technical change, because when the focus changes, the emphasis changes as well. The new main category and the categories beneath may change their names, and the sub-categories may split to one or more divided categories, as shown in Figure 15B. Moreover, in constructing the focused tree, it is not necessary for every relevant category from the mapping tree to "bring" all the categories beneath it to the focused tree. Choosing a new main category involves an addition of categories and examines the sorting of others. For each category, it is necessary to examine whether it can still be part of the new main category.

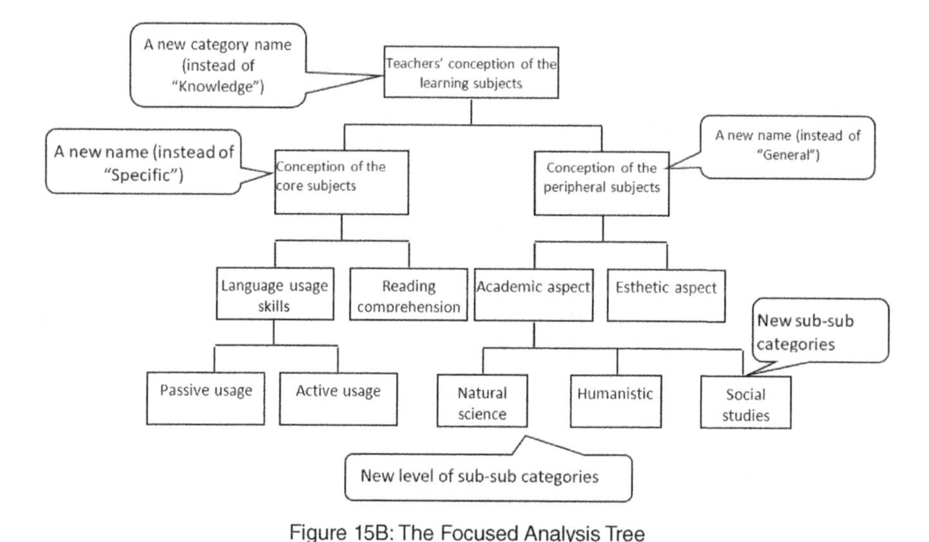

Figure 15B: The Focused Analysis Tree

Researchers conducting the focused analysis using a word processor should work on a new document. They will copy the relevant parts from a mapping analysis document, and make the necessary changes until receiving the appropriate focused analysis.

Constructing the Focused Analysis Category Array Using Narralizer

In order to make the transition from a mapping analysis document into a new focused analysis document, you must work with two parallel analysis documents. **Narralizer** offers two convenient ways to do this: Cascade operation and Tile operation.

A. To perform the Cascade operation:

1. Consecutively **open the Narralizer documents** you wish to work on. Open the mapping analysis documents and a new research document for the focused analysis.
2. Position the cursor in one of the documents and click **Cascade** in the **Window** menu, as shown in Figure 15C.

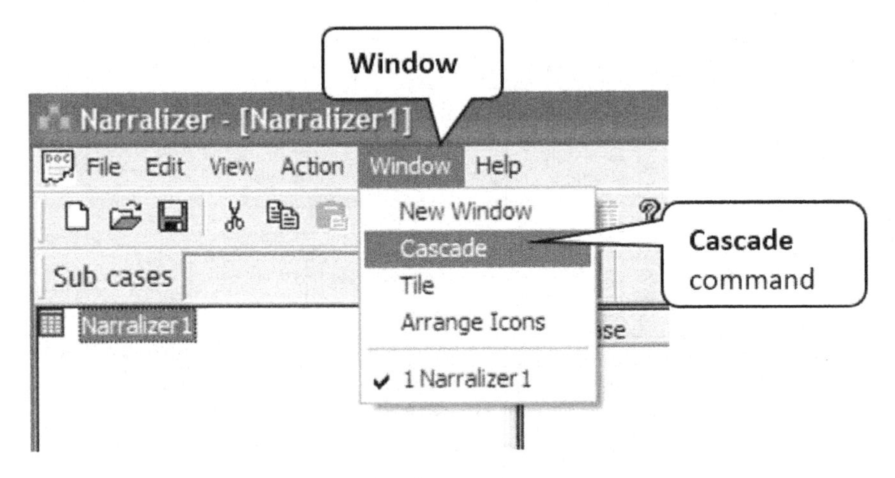

Figure 15C: Cascade Command

3. **Cascade** will arrange all of the open **Analysis documents**, one on top of the other (the example below shows three documents). The documents can be moved, maximized, minimized, and their order changed to suit the researcher's needs. In the case of transition categories and data from the mapping analysis array into a new focused analysis, you need just the two relevant documents. Thus, if other **Analysis documents** appear that are not needed, simply close them by clicking X in the top right corner of the superfluous document.

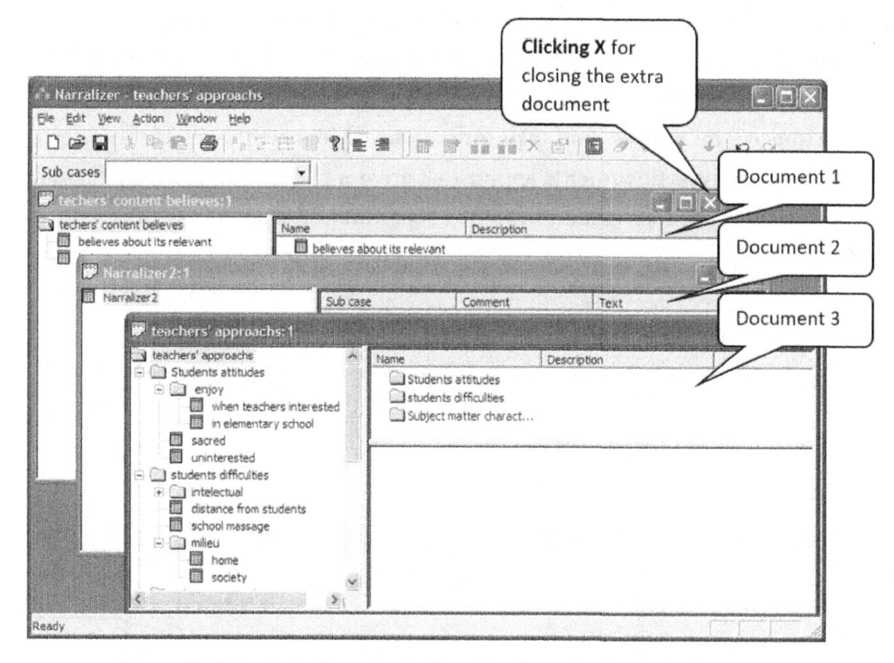

Figure 15 D: Analysis Documents Resulting from the Cascade Command

B. Tile Operation:

1. Consecutively, open the **Narralizer** documents you wish to work on. Open the mapping analysis documents and a new research document for the focused analysis.
2. Place the cursor on one of the documents and click **Tile** in the **Window** menu, as shown in Figure 15E.

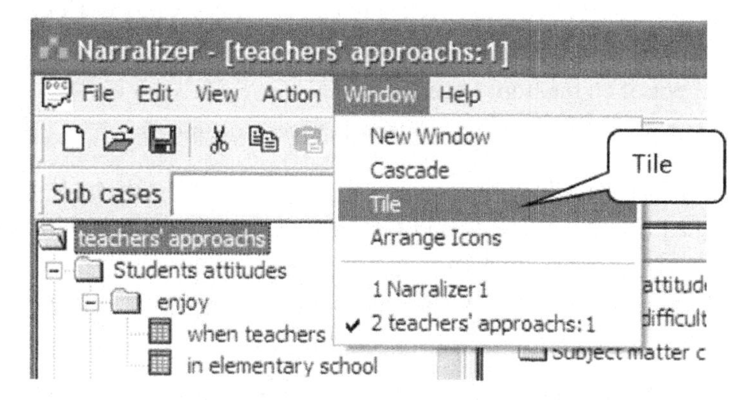

Figure 15E: Tile Command

3. Clicking the **Tile** command arranges all open **Analysis documents** one on top of the other. Unlike **Cascade**, the **Tile** option does not hide the lower documents. Instead, all documents are equally visible. The documents may be moved, maximized, minimized, and their order changed to meet the researcher's needs. If other **Analysis documents** appear that are not needed, simply close them by clicking X in the top right corner of the superfluous document.

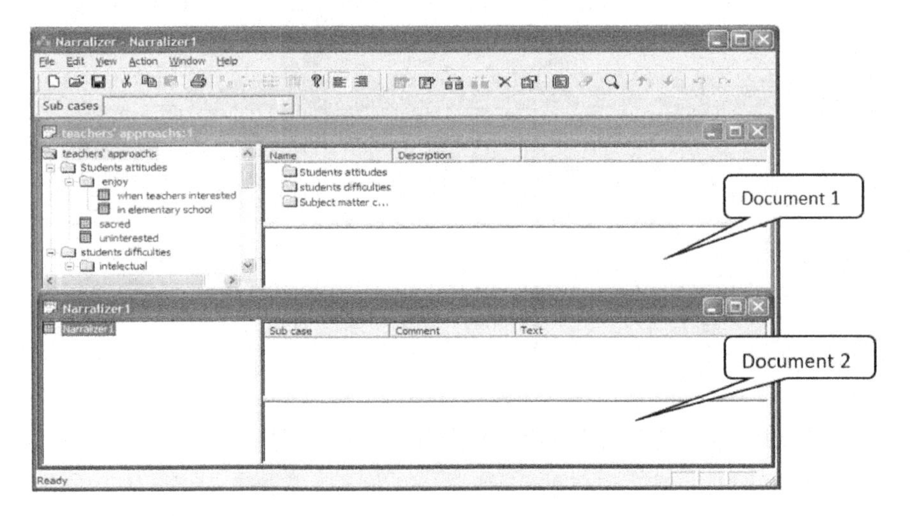

Figure 15F: The Analysis Documents Resulting from the Tile Command

Transferring Data Categories and Segments from One Document to Another Using **Narralizer**

The researcher can now work simultaneously on several documents and transfer data categories and/or segments between documents using **Copy** or **Cut & Paste,** or drag them when in the **Cascade** or **Tile** layout. Placing the cursor on each of the documents puts it on the front screen.

To move analysis categories or segments between documents:

1. **Open two analysis documents** side by side: the "source" document from which to **export** analysis segments, and the "new" document into which to **import** analysis segments, as illustrated in previous sections.
2. Name the "new" document by **Rename** or **Save as** commands, and then create the main categories for the new document (in Figure 15G, we split the main category into two upper categories using the Split categories command, neither of which derives from the "source" document).

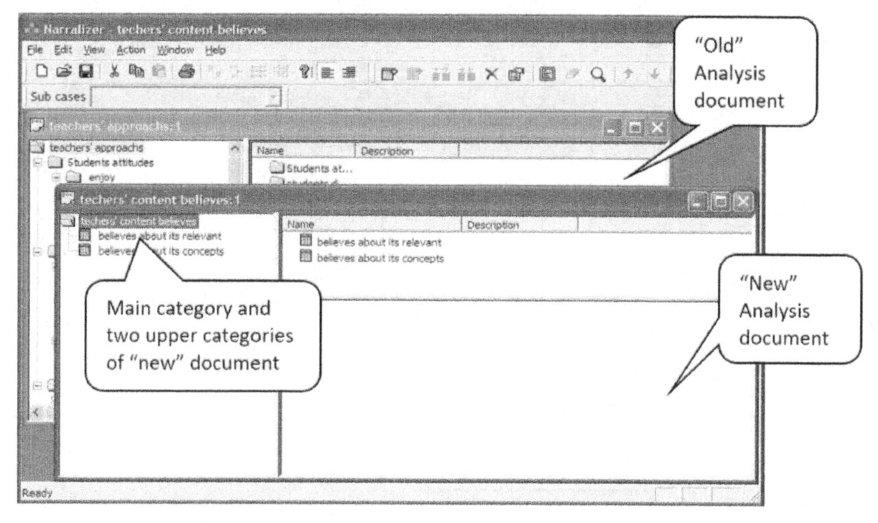

Figure 15G: Name of the Categories in the Focused Analysis Document

3. Now we want to move categories from the "source" (mapping analysis) document to the "destination" document (the "new" focused analysis document). In Figure 15G, the content category "distance from students" in the "source" document is **copied and pasted** into the "believes about relevant content" category, which is located in the "destination" document. This operation is performed by the following steps:

   A. In the **Categories pane** of the "source" document, highlight the category located one level above the category you wish to copy. The "distance from students" category will appear in the **Content pane** of the "source" document.

   B. **Highlight** and **Copy** the category, then position the cursor on the destination category in the "destination" document and **Paste**.

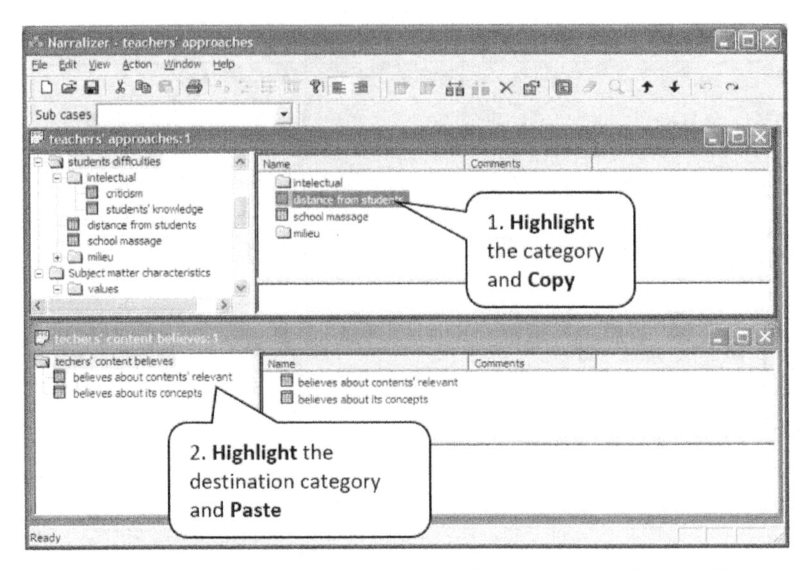

Figure 15H: Copying Categories from the Mapping Document to the Focused Document

Bear in mind: In this case, it is not advisable to **Cut** or **drag** the text since this will change the source document. It is important to leave the source document intact at a research stage we can return to at any time.

4. When a category is moved from the "source" document to a content category in the destination document, the content category in the destination document changes to an **indication category** and its icon changes from a hatched square to a folder. However, because we have only moved one content category (and an indication category must have at least two categories below it), the software will automatically create a temporary empty content category with the name of the indication category and a number 2, as shown in Figure 15-I.

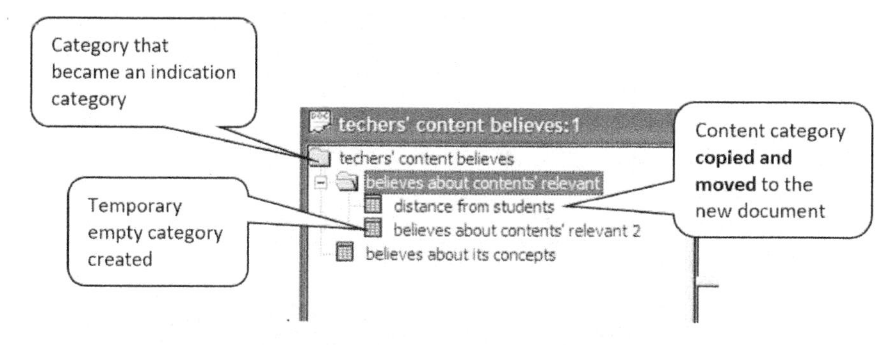

Figure 15-I: Content Category Changed to an Indication Category after Moving Category from Mapping Document

5. In the same way, you may copy not only single categories but also groups of categories from one document to another. Figure 18J illustrates how to move the entire group of categories below the prime category "knowledge" in the "source" document to the "believe about its concepts" category in the "destination" document. The procedure is the same as above.

   A. **Highlight** the "knowledge" category, and the indication categories beneath it will appear in the content pane.

   B. **Highlight** the categories in the **Content pane** that you wish to copy (using the **Ctrl** key).

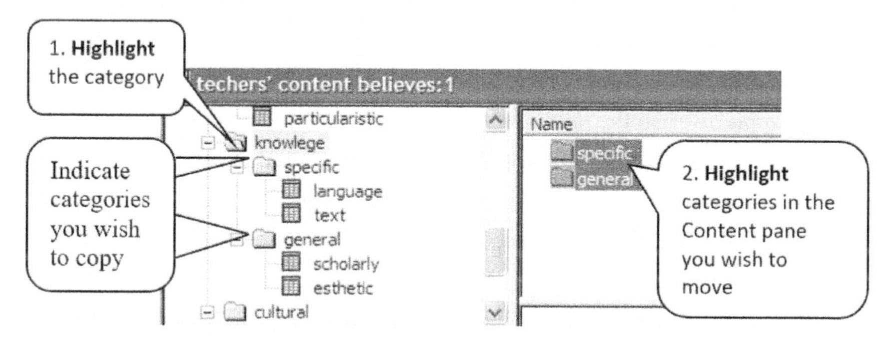

Figure 15J: Process of Moving Several Categories

C. **Copy** and **Paste** the highlighted categories in the **Content pane** of the "source" document into the "believe about its concepts" category in the "destination" document, as shown in Figure 15K.

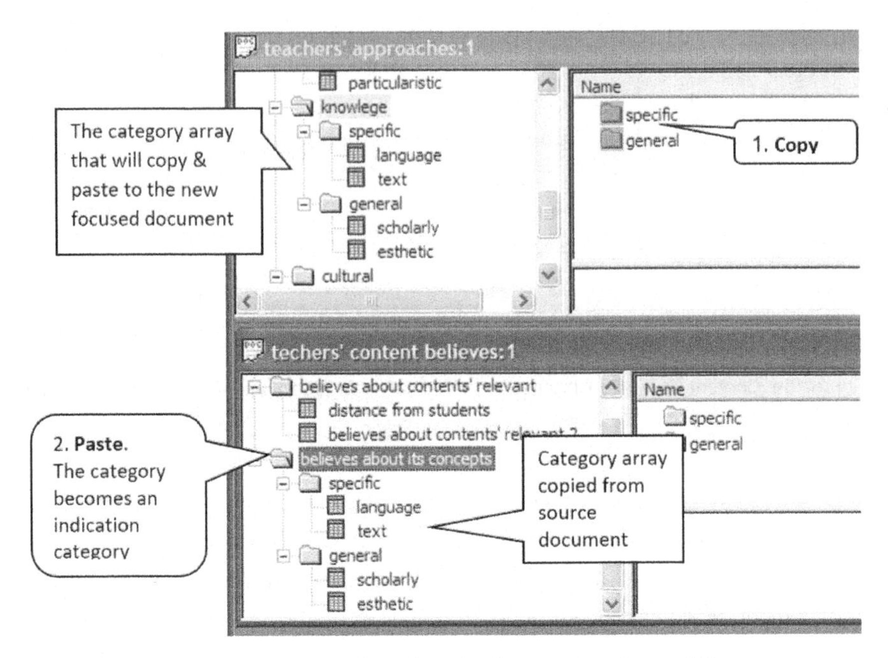

Figure 15K: Copy and Paste from Mapping Document to Focused Document

In a similar way, segments and categories (indication and content) can be moved from one Analysis document to another.

Constructing Focused Analysis with Significant Categorical Changes

In the example given in the previous section, we transferred an entire branch of the category array to a focused category array. However, often the creation of a focused category array involves significant structural changes in the original array of categories. In these cases, the focused category array is indeed based on the mapping categories and data segments, but its final categorical structure is completely different. Not only is its main category different, but the upper categories reflect a new categorical organization. However, in these circumstances as well, the entire process of creating the focused categories is based on the rich picture of the mapping stage, even if the end result shows a creation of new categories seemingly detached from the mapping analysis.

The example shown in Figure 15L demonstrates significant structural changes in the transition from the mapping category array to the focused array. In this case, we do not choose one or another branch of categories to relocate from one category array to another, but instead make structural changes that intersect the upper categories from the mapping category array. In Figure 18K, the researcher has marked several categories in the mapping analysis tree which are seen as relevant to the new focused category array. Note that the researcher did not choose all the categories under the original upper categories, but only those relevant to the new focused issue

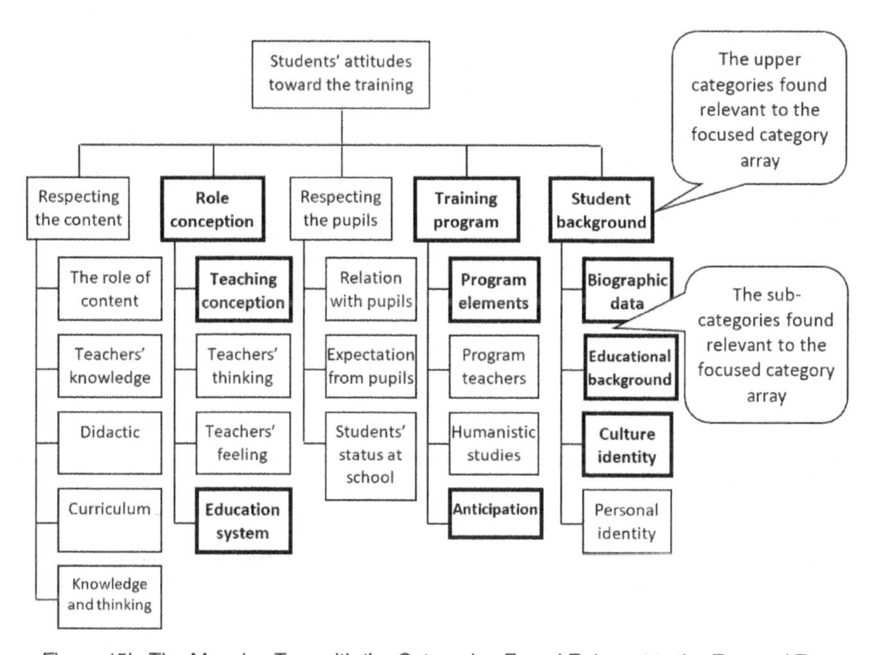

Figure 15L: The Mapping Tree with the Categories Found Relevant to the Focused Tree

In Figure 15M, the focused analysis array is based on the revision of the mapping analysis array, and the main category of the mapping tree ("Students' attitudes toward the training") has been changed to a more focused category: "Student motivation to join the teachers training program." None of the upper categories in the mapping tree relates directly to this issue, but a close look at the mapping categories and the data segments led the researchers to conclude that this issue evolved from the mapping analysis. The mapping analysis served as the resource and framework from which researchers formulated the new main category.

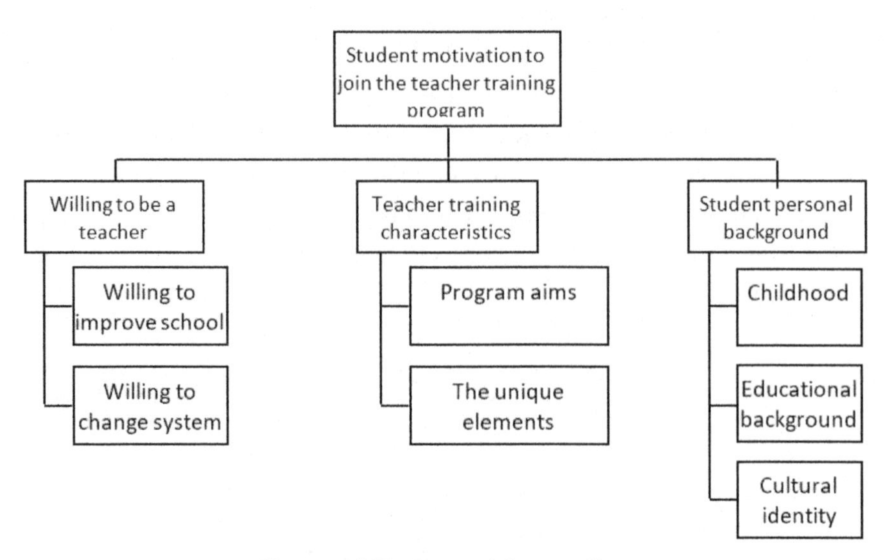

Figure 15M: The Focused Category Tree

It is assumed that following the change in the categories, some of the data segments associated with the mapping categories now become irrelevant to the new focused categories. On the other hand, it is possible that some data segments that originally existed in other categories have become relevant to the newly-created category array. It is important to locate these segments and attach them to the relevant categories.

It seems that in many cases, search activity may be beneficial. The researchers will mark several words or phrases which may reflect the relevant categories of the new focused categories. Using the Find command, you may quickly and effectively place these key words in the mapping analysis document or even in the raw data text to examine the relevance of such segments into the focused categories. If the segments are found as relevant, the researchers can attach them to categories by Copy & Paste, while ensuring that the identification details of the new segment will be added.

Constructing Focused Analysis with Significant Categorical Changes Using Narralizer

The procedure for constructing a new focused category array and moving data segments from the mapping category array is identical to the procedure of moving categories from the "source" (Word or **Narralizer**) document to the "destination" document explained in the above sections. There are two differences:

A. A new focused category array is constructed in the "destination" document before moving the data segments from the mapping array (the "source" document)

B. Following the same procedure of moving from document to document, the data segments and not the entire categories from the "source" document are mainly transferred, as seen in Figure 15N.

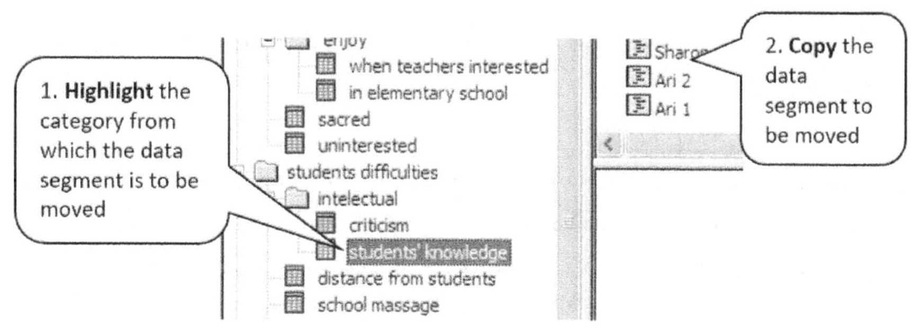

Figure 15N: Copy Relevant Segment in the Mapping Document to Move to Focused Category Array

Using the **Narralizer**, relevant segments may be found throughout the **Narralizer** mapping document by using the **Find** command for several key words. If the segments are found as relevant to the new focused categories, the researcher can attach them to categories by the **Add text segment** command, while ensuring that the identification details (the name of the sub-case) of the new segment is added. These procedures are explained in previous chapters.

Changing the Theoretical Framework of the Study Following the Focused Analysis

The focused analysis is done mainly in the methodologies that are not based on a-priory clear theoretical perspective. In these methodologies, the theoretical perspective becomes clear during the research process, and particularly through the analysis process. Naturally, in the focused analysis stage, which is an advanced stage, the theoretical perspective becomes more and more clear. Therefore, together with the focused analysis process, the researchers should try to formulate the theoretical perspective, and subsequently update their literature review. Updating the literature can help researchers to focus around an appropriate and rich main category. The two processes, the focused analysis and the clarification of the theoretical perspective, occur simultaneously and enrich each other. There should be a connection between the literature review and the definition of the research categories. This does not mean that literature is always the basis for defining the categories; the relationship between literature and research categories is more dynamic and less linear. This procedure will be explained in the following sections.

Translation of Descriptive Categories into Theoretical Categories

At this stage, researchers try to propose a match between the descriptive categories (containing the data segments) and the conceptual-theoretical categories. The researcher compares the descriptive categories (of the focused and mapping stages) with concepts taken from the theoretical literature and translates the descriptive categories found in the higher levels of the category tree to concepts taken from the theoretical arena (Araujo, 1995; Charmaz, 1990, 2005; Shkedi, 2005). By creating a significant link between the data and its theoretical translations, the categorization process becomes the primary means for developing a theoretical explanation (Charmaz, 1983, 2005). When researchers look at the focused categories and the data attached, they ask such questions as, "What do we see here?" "Which theoretical concepts characterize what we see?" When they try to discover, identify and ask questions about the conceptual-theoretical potential of the descriptive categories, they actually translate the descriptive categories to conceptual-theoretical categories.

Creating a theoretical explanation usually occurs around at least one main category and the upper categories attached to it (Strauss, 1987). In practical terms, as suggested above, at the theoretical analysis stage the researcher translates the main and upper categories from everyday terms to theoretical terms. The researcher identifies the features of all categories and relations between them as portrayed in the focused descriptive category array on the horizontal and vertical axes, and seeks to find theoretical explanations for these relations. As the upper categories in the analysis tree become termed as conceptual-theoretical, the category names as we move down the tree are characterized by more everyday expressions. Therefore the conceptual-theoretical terms of the "higher" categories (main and upper categories) reflect the theoretical explanations of the phenomenon under study.

The process of translating categories to theoretical terms is not a technical operation of translation (Charmaz, 1995). Thus, the researchers do not necessarily retain the focused category array created in previous phases. The process of translating the descriptive categories to theoretical categories does not involve only changes of terminology from everyday terms to theoretical terms. This process may also involve changes in the structure of the category array. The researcher reviews the data according to new theoretical insights, and as a consequence may change the categories and relations between them, divide or join categories, break up "families" of categories, and/or construct other categories. In this way, the original descriptive focused category array may change, entirely or partially.

Using Theoretical Literature

Researchers use professional literature throughout all stages of research. The integration of literature is characteristic not only to the methodological patterns based on external predetermined theoretical perspectives, but also to the methodological patterns that provide a greater expression to internal perspectives. The difference between the methodological patterns is not reflected in the degree of literature usage, but in the role that the researchers assign it, i.e., whether it will function as a source for predetermined categories or a source to arouse assumptions, ideas and ways of thinking. In methodological patterns that give expression to the internal perspective, researchers use theories in data collection and the descriptive analysis phase mainly as an intermediary to identify and focus their own conceptual perspective. They are careful not to force the theoretical expressions of the literature into the study. However, in the theoretical analysis stage, the picture changes completely, and the researchers use theoretical concepts from the literature. Now they refer to the data from the perspective of a "conceptual category, rather than merely as a descriptive topic or code." (Charmaz, 1990, p. 1168). Once the researchers create an idea, they may seek additional material from the professional literature in support of the idea. Researchers go back and forth from literature to data analysis in search of appropriate concepts and relationships.

It is possible that researchers can find existing theoretical literature that provides a basic, plausible theoretical explanation of the phenomenon under study and which is appropriate to the categories created in the first three descriptive stages of analysis. But in many cases the researchers will use existing theories and literature only as inspirations for formulating a theoretical explanation or a unique grounded theory. In that cases process the researcher translates the names of the categories located in the higher levels of the category tree into theoretical expressions, based on sources from the literature or relying on inspiration from literature, as illustrated in Figure 15-O. Sometimes researchers use the same term as appears in the literature; other times they may change the name and even generate an entirely new theoretical name.

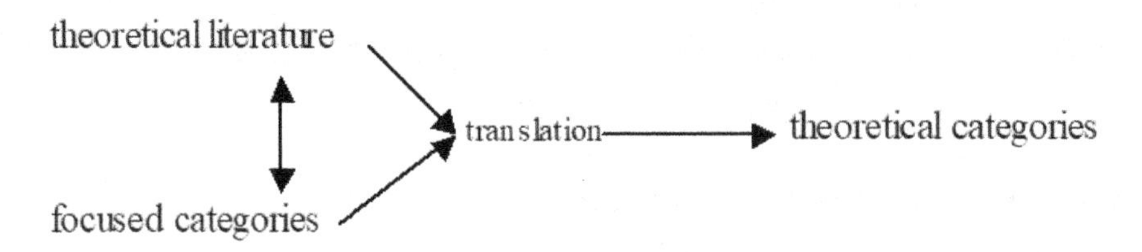

Figure 15-O: Using the Literature for Constructing Theoretical Categories

Figure 15P shows the process of the theoretical analysis with a focused category tree very similar to that presented in Figure 15M above. This tree is based on descriptive categories before translation to theoretical categories.

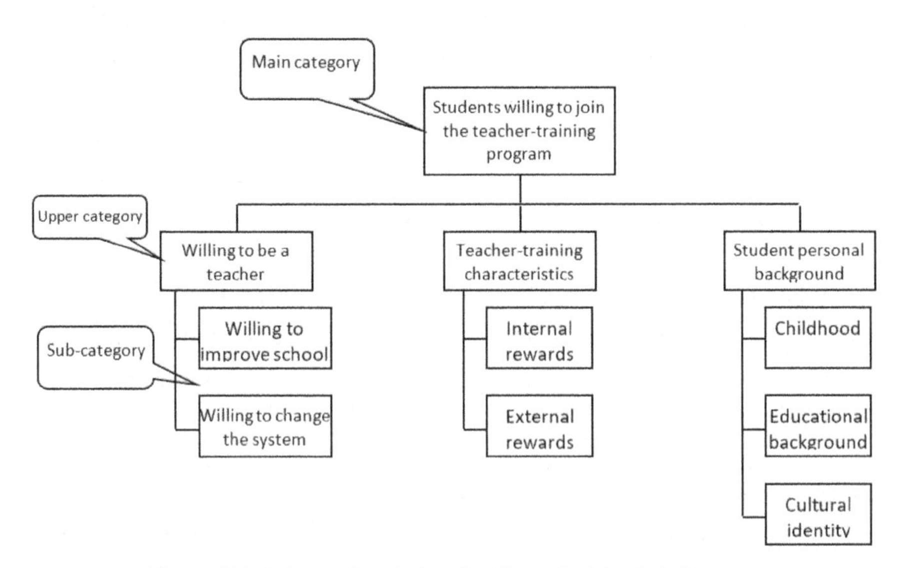

Figure 15P: Category Tree before the Theoretical Analysis Process

With the transition to theoretical analysis, we translate the descriptive categories to theoretical categories. This process is not performed precisely by translating each category to another, but in many cases by providing a new conceptual category which offers a theoretical explanation for some descriptive categories. As shown in Figure 15Q, the translation process of the descriptive categories (from Figure 19B above) is not one-way, but carried out as a "dialogue" between the descriptive picture (on the right side) and the theoretical concepts (left).

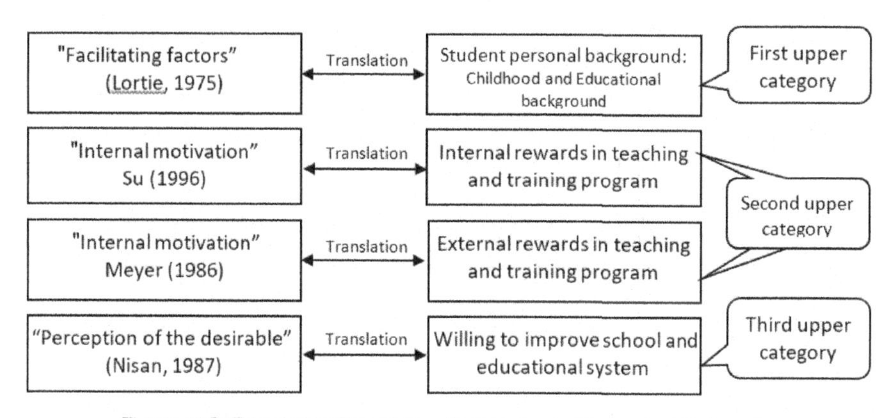

Figure 15Q: Translating Descriptive Categories to Theoretical Categories

The translation process of the descriptive categories to theoretical categories is, as mentioned, a two-way process: it operates not only from the descriptive category to the theoretical categories, but simultaneously from the theoretical categories into descriptive categories. Following this, the focused category array may change somewhat, namely, the translation process will affect the understanding of the descriptive category and its structure. Figure 15R illustrates the change that might occur in the descriptive category after the translation to theoretical concepts.

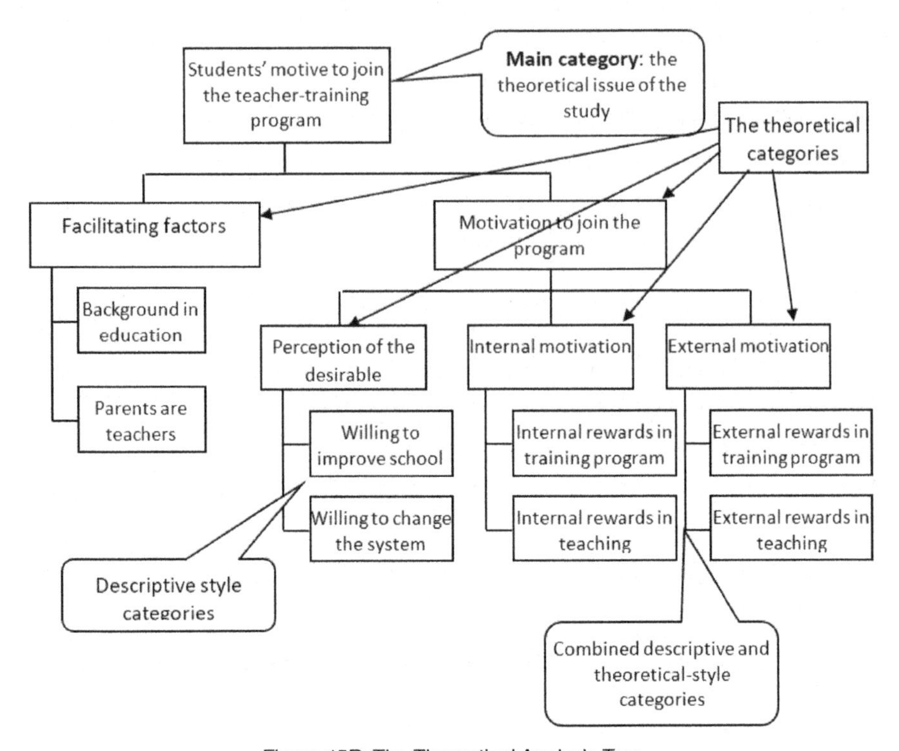

Figure 15R: The Theoretical Analysis Tree

After the researchers have designed the theoretical category array, they should insert it into a word processor or **Narralizer** document.

Constructing a Theoretical Category Array Using Narralizer

The procedure of constructing a theoretical category array is based on what has been presented in previous chapters and sections.

1. The theoretical category array is built on the basis of the focused category array (sometimes directly from the mapping category array). Thus, the first stage of the process is to open a focused analysis document.
2. Create a new data document (**New** command) and place the theoretical category array on its **categories pane** by clicking **Split Categories** and **Add category**, as

explained in previous chapters. At this stage this category array is still empty of data segments.

3. Using the **Cascade** or **Tile** commands (explained in Chapter 18), place the focused analysis document above the theoretical analysis document.
4. Mark in the content pane of the focused document (the "source" document) the corresponding data segments to copy. Transfer to the theoretical categories in the "target" document and perform **copy & paste** to the categories pane in the theoretical document. Similarly, consider other categories and segments in the focused analysis document and determine whether to copy them and to which theoretical categories they belong.

The Need for Second-Order Theoretical Analysis

The theoretical analysis process presented in the first part of the chapter is based on the translation of descriptive categories into theoretical categories. This process can also be called a "first-order theoretical analysis," since we are constructing the theoretical explanation directly from the data collected from the research participants and the descriptive categories which are based on the participants and the study site. However, there are occasions when it is impossible to point to a theoretical explanation that can be constructed through direct translation of the descriptive categories of theoretical categories. Still, we feel that our data has a potential theoretical explanation. In this case, the researchers need a more complex method. Using the second-order analysis process (Shkedi, 2005; 2004), we can connect different data parts and offer theoretical explanations based on this data. The second-order theoretical analysis, like the first-order, is based on the descriptive analysis stages, as explained in previous chapters.

The need for second-order analysis arises when the information received from participants is indeed of a sufficient quality to provide rich descriptions, but does not allow a theoretical interpretation to be derived by way of direct translation. The rich description and the theoretical clues found in the data challenge the researchers to construct theoretical explanations through the process of second-order theoretical analysis. A typical example of theoretical insights rarely acquired directly from the participants is causal relationships between different phenomena (Huberman & Miles, 1994). While we often receive only the description of phenomena, we can consider them in parallel, and with existing theories, by way of second-order analysis offering one or more possible causal explanations. While the theoretical analysis (of first-order) presented above is a direct interpretation of phenomena as subjects have presented them, the second-order theoretical analysis is an interpretation of interpretations (Gudmundsdottir, 1995).

The Process of Second-Order Theoretical Analysis

The second-order theoretical analysis utilizes the descriptive categories obtained in the first three stages of analysis to construct a theoretical explanation. While the first-order theoretical analysis described above is essentially a process of direct translation of the descriptive categories to theoretical concepts, the second-order theoretical categorization process is the process of conversion. In this process we do not translate categories from one language to another while preserving their original meaning, but convert one descriptive categorical array to a theoretical categorical array. Although based on components of an original category, the relationship between the elements is not a descriptive link, but one offered by existing theories or by the stimulus of existing theories. The difference between the first-order theoretical analysis and the second-order theoretical analysis is illustrated on the Figure 15S.

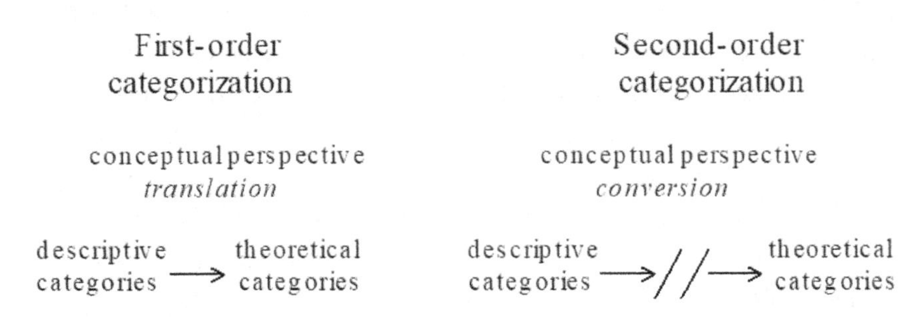

Figure 15S: The Processes of Two Types of Theoretical Categorization

Considering the entire analysis picture in second-order theoretical analysis, the researcher actually finds new core theoretical categories which have not emerged *directly* from the existing focused categories. The researcher might find several hints in in the data, but not massive evidence. As is the case with any other qualitative research, the second-order theoretical analysis is also based on the informants' narrative fragments, albeit not in a direct manner.

The main theoretical idea in a second-order analysis procedure is reflected in the choice of the main categories and the upper categories. Other categories are descriptive categories related to the main and upper theoretical categories. The main category is the category that reflects the theoretical idea and defines the theoretical phenomenon. The categories located under the main theoretical category represent the characteristics of the theoretical explanation. They include upper categories that describe the components and conditions that affect the phenomenon, the phenomenon's unique contexts, and other intervening conditions involved in the studied phenomenon.

In practice, the starting point for second-order theoretical analysis is the focused or mapping category array. Researchers will examine the array of categories to find categories with a potential for connection - a causal connection, suitable connection, or any other connection. In Figure 15T, for example, the researcher finds a causal connection between the upper categories "teachers' role conception" and the "students' background." Within each upper category, the researchers locate the relevant sub-category to the possible theoretical causal link.

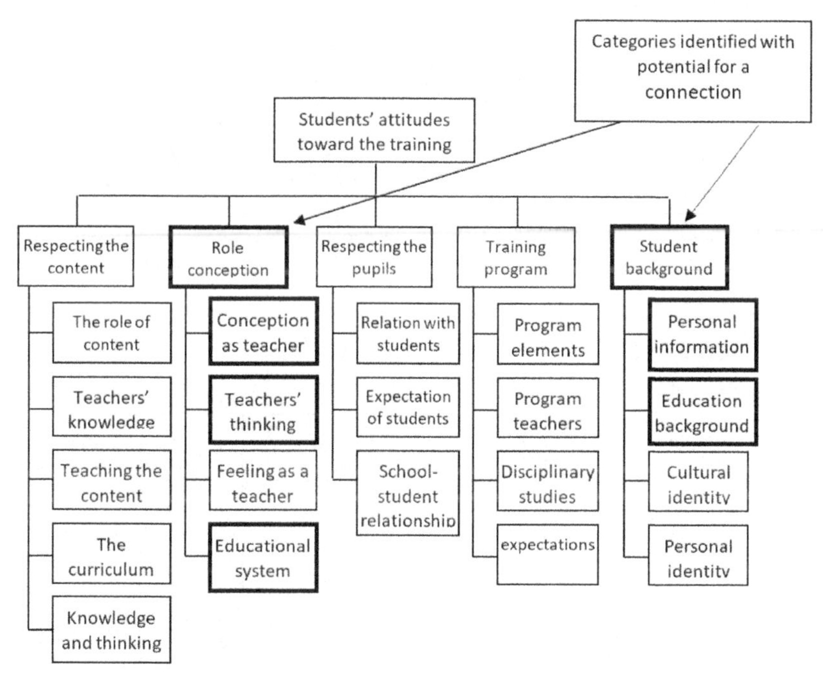

Figure 15T: Examining the Descriptive Category Tree Which Reveals a Possible Causal Relationship

The descriptive picture reflected in the category array in Figure 19F does not point directly to any causal connection between the two upper categories, but researchers find that examining the two categories in contrast to each other, assisted by theoretical literature, may suggest a causal relationship between the two. The literature dealing with teachers' conceptions introduces the idea that there is a relationship between teachers' conceptions of teaching and how they themselves experienced their studies at school (Lortie, 1975). This theoretical idea, revealed in many studies, raises the possibility of a possible causal relationship between these elements (upper categories), even though the participants did not indicate such a connection.

Once the theoretical categories are identified with all their connecting descriptive categories, the second-order theoretical category array is reorganized accordingly. Figure 15U illustrates the second-order theoretical array based on the causal relationship revealed in Figure 15T.

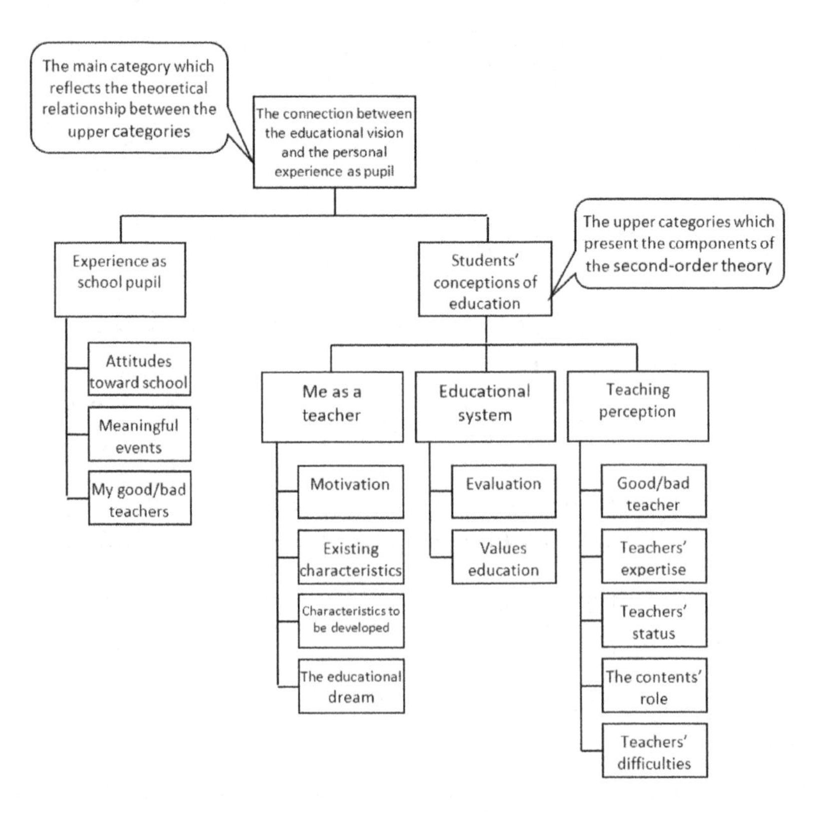

Figure 15U: The Second-Order Theoretical Analysis Tree

As illustrated in Figure 15T, the categories have changed their previous names according their function in the second-order theory. For example, the upper category "Student background" was sharpened in the second-order theoretical array to become "Experience as a school pupil," while the upper category "Students' conception of teacher's role" now bears the broad name "Students' conceptions of education." These two upper categories are located under the main theoretical category and related below to the descriptive sub-categories and sub-sub categories which specify the descriptive picture and give substance and validity to the theoretical explanation. The accepted theoretical picture of both first and second-order theoretical explanations is, in fact, a grounded theory.

Researchers who conduct the analysis using a word processor list the second-order category array on a new document and insert the relevant data segments into the categories, a process similar to that described in previous chapters in this part of the book.

Process of Second-Order Theoretical Analysis Using Narralizer

The process of implementing the second-order analysis using the **Narralizer** is identical to that of the first-order theoretical analysis, as explained in this chapter.

CHAPTER 16

ARRANGING THE ANALYSIS PRODUCTS: STRUCTURING THE COMPLETE PICTURE

The analysis process is the process in when the researcher constructs meaning to all the data findings. However, once this goal is achieved, it should not be regarded as the final phase of the study. There is one additional important step, that of presenting a comprehensive view of the research picture. Organizing the analysis findings provides the means to create the "story line" of the study and to portray a significant depiction of the research results.

The ultimate goal of any research is to present its findings to the readers and the public at large. The findings should be organized so as to be communicated to the potential readership (Lincoln & Guba, 1985). As emphasized in the first part of the book, the language of qualitative research is the language of words, which is characterized by its harmony with the epistemological world of the readers, and allows readers to be empathetic and participate in the story described in the study (Eisner, 1979; Stake, 1978). The process of arranging the research findings aims not only to make the study accessible to potential readers, interact with them, and provoke their empathy for the study, but also to focus the readers' attention and interest upon specific defined issues and thus boost the contribution and impact of the research.

When researchers focus on presenting the research picture, they choose the objects of study from a certain estimation and consideration and not at random (Merrick, 1999; Peshkin, 1993; Sciarra, 1999). Qualitative researchers use a variety of options to present the data picture. In the analysis process, the researchers identify the data according to two main components: categories and sub-cases. Each data segment is identified by the categories that characterize it and the sub- case to which it belongs. At the final presentation of the research findings, the researchers must weave together all the components until an all-inclusive picture is attained.

Different combinations of components give the researchers a wide range of options for displaying the findings. As for the category component, researchers can refer to the entire final category array to present a comprehensive research picture, or alternatively selectively choose a certain representative group of categories. More emphasis may be placed on certain categories and less on others. Researchers can choose a specific order of categories to illustrate the research picture, present the descriptive and theoretical categories integrally or display them separately, and more. The same options apply to the focus on sub-cases: researchers may display all the sub-cases equally or accentuate certain sub-cases over others. They can choose several representative sub-cases and present them as reflecting on the overall picture. They can choose a certain order to present the sub-cases, either by dividing the sub-cases according study participant, to a temporary occurrence, to data types, or by one or more combinations of these parameters. They can arrange the sub-cases into groups according to these or other characteristics, and more. The combination of a reference to categories and sub-cases allows the researcher to choose from almost unlimited possibilities of final research presentations. A researcher's consideration is guided by the goal to present the findings in the best way while communicating successfully with the potential readership. This chapter deals with the transition from the analysis stage to the stage of writing the final report. The chapter will be devoted to the ways in which to organize the findings of the analysis as a basis for the final report

The Category-Based Arrangement of the Analysis Products

An accurate display of the research picture may lead the researcher to present all the categories, while the desire to present the findings to the readers understandably and communicatively dictates the direction and choice of the researcher's decision. However, the organization of the categories in an array of categories and data segments, accompanied by certain sub-cases, allows the researcher great flexibility in the process of constructing the analysis findings.

The consideration of choosing the categories and how and what to present is a serious challenge. But the technical procedure is simple and convenient. Researchers consider the categories and the data in the analysis document, and decide which parts of the category array they wish to focus upon. Once the decision is made, the researchers can create a new document, separate from the original document. They will highlight the selected passages in the original document and transfer them to the new document. In this way, the researcher can create several parallel documents, each of which focuses on a different category branch, or a single document containing all relevant category branches with their associated data segments. Once the researchers organize the analysis picture in the new document/s, they can compare their respective places in the large research picture and determine what significance to assign to each in the final report.

The Category-based Arrangement of the Analysis Products Using the Narralizer

In order to obtain a data array for certain categories but not others, **mark the desired categories** and use the **Show sub-tree** function. When the box opens, click **Select all** to obtain a selective data array. **Clicking** an indication category (folder icon) will summon all data contained in the categories below it (See Chapter 16): Now you may print out the category and segment display using **the Print command** or export it to an rtf document using the **Export** command. You may work on the rtf document, introduce changes within it, adapt it to the final report, and more.

Arranging the Analysis Findings by Sub-Cases

The process of arranging the analysis findings by sub-cases is slightly more complex than the arrangement of categories. Generally, almost any analysis document contains several sub-cases. Each sub-case can refer to a specific event or to specific participants (e.g., "Joe"), to a particular type of data ("Joe, interview"), and to a specific event in reference to a specific time ("Joe, Interview, 1st year"). As explained in previous sections, the division to sub-cases is very efficient, and allows different combinations and then various distinctions as needed. For example, it is possible to centralize the entire sub-case that deals with a specific participant ("Joe" or "Cindy," etc.) or with a specific group of participants (men/women etc.), or all sub-cases that deal with a particular unit of time (" 1st year" or "2nd year," etc.), that are based on interviews, or any other relevant combination. In the upcoming sections we will introduce the procedure of organization by sub-cases, using a word-processor and/ or **Narralizer.**

Changing the Names of the Sub-Cases

Sometimes, as the study progresses and especially when the analysis findings are spread out before the researchers, they may conclude that the name of a sub-case must be changed to a more representative title. For example, if a sub-case was originally named for the participant, the researcher may later conclude that the focus should be upon specific events rather than on the participants. In a study of social workers, for instance, the researcher may conclude that the focus is on the type of treatment provided by the social workers rather than on the specific social workers themselves, thus changing the names of the sub-cases from "Mary," "John," and the like to such titles as "working with Ethiopian immigrants," "working with Russian immigrants," etc. It is possible, of course, to create a new name that includes both characteristics, for example, "Mary, working with Russian immigrants," or any other such combination.

Using the word processor, renaming the sub-case is relatively simple by using the **Replace** command, one of the effective tools of word processing. Of course, researchers need to vouchsafe that the changing name will remain constant in all relevant documents, or the picture may become quite confused. It should be noted that the "replace" operation of the word processor will affect only the words typed accurately. A small, unnoticed typo may cause a missed exchange. It is also preferable to avoid the command "Replace All," and use the command "Replace," instead which allows the researcher to individually examine the place to enact the command.

Changing the Names of the Sub-Cases using the Narralizer

1. Highlight the main category in the **Categories pane to** click on **Rename sub-cases** in the **File** menu (see Figure 16A). Alternatively, place the cursor on the main category in the **Categories pane.**

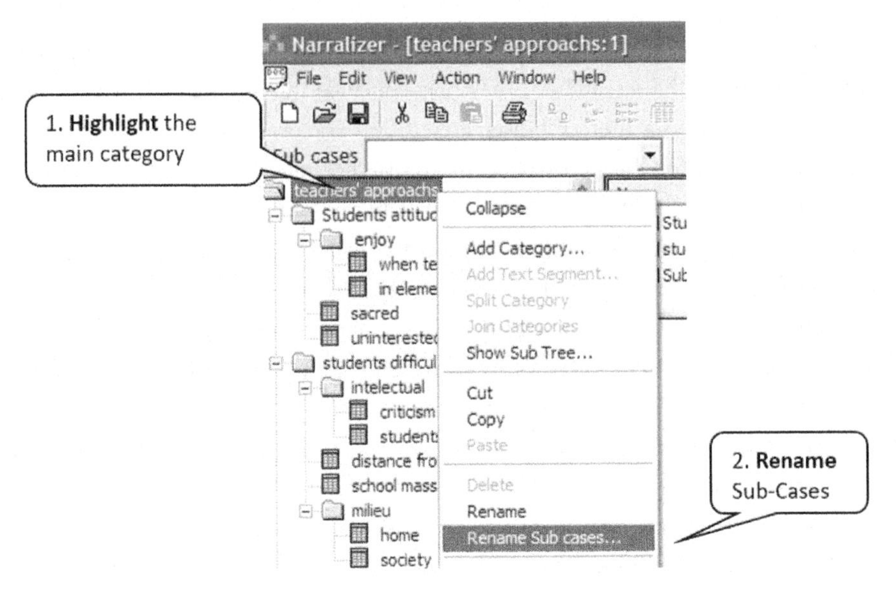

Figure 16A: Rename Sub-Case Command

2. Clicking the **Rename sub-cases** command opens a dialog box with this name.
3. Find the name of the sub-case ("Sharon 1" in Figure 16B) to rename using the arrow in the upper pane.
4. Write the new name of the sub-case in the bottom pane ("Teacher 1" in Figure 16B) and Click **OK** to replace the old name.

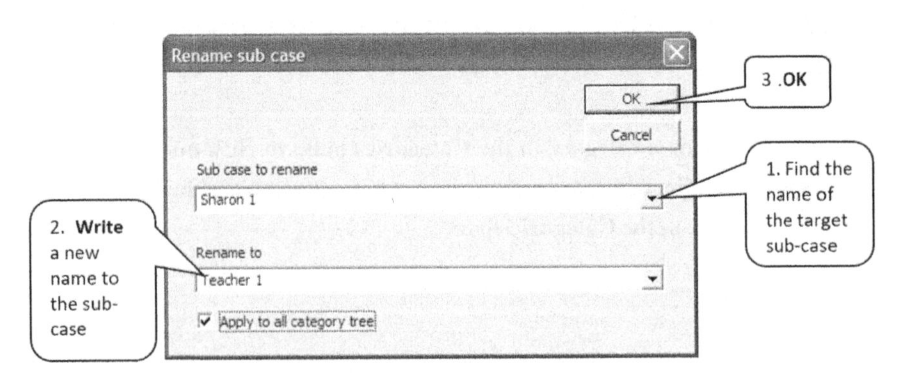

Figure 16B: Rename Sub-Case Box

If the sub-case you wish to rename appears in several Analysis Documents, be sure to rename it in all the **Narralizer** analysis documents.

Bear in Mind: At the bottom of the dialog box is the indication: "Apply to all category tree" and next to it is the symbol V. This symbol appears automatically, which means that changing the name of the sub-case will be made throughout the entire document. To change the name of a sub-case only in a particular category, or not throughout the entire document, first mark the specific category in the category pane before giving the Rename Sub-Case command, and remove the V sign in the box.

Unification/Merger of the Sub–Cases

In Chapter 12, it was recommended to place every source of information (interview, observation, document, etc.) as a separate sub-case, assuming that the researcher can always merge several sub-cases (although dividing sub-cases is much more complex). At the end of the analysis process (or even in the earlier stages), it is easy to merge while maintaining all the analysis actions taken to that point. Before performing the sub-case unification, it is important to keep the original documents with the separate sub-cases. In this way, you will not lose the original sub-cases, and will have them available for additional or parallel combinations of sub-cases. Note that you may use the same sub-case in different combinations, each of which will highlight a different aspect of the analysis findings.

Unification of several sub-cases using a word processor is quite complex, requires a great deal of effort, and may not be particularly efficient. The goal is the creation of a new united sub-case, in which all the segments "belonging" to the same category will appear under this category. Using a word processor, researchers must create a new document, name it, and copy all the pieces of data one by one from the source documents to the new document so that all sections belonging to the same category will appear in the new document under this category. This is done using the Copy & Paste commands. Of course, you must make certain that the identification name will appear on each segment.

Unification of the Sub–Cases Using the Narralizer

As illustrated above, unifying the sub-cases using a word processor is possible, but quite tedious and error-prone. It seems that the use of the **Narralizer,** however, may greatly facilitate the operation. Merging several sub-cases by using the **Narralizer** is actually the combining of one sub-case into another. This is basically identical to the operation of **the change of the sub-case name** explained above. In principle, there may be two options:

A. Keep the name of one of the existing sub-cases and change the names of the other sub-cases to that name.

B. Give each of the sub-cases a new mutual name.

In both cases, the operation is carried out by the **Rename Sub-cases** command which opens the Rename sub-case box. In the lower pane of the Rename sub-case box, type the name of the mutual sub-case (see Figure 20B). The rename operation will join the relevant sub-cases in all the categories of the analysis documents.

> Bear in Mind: When working with several documents simultaneously in the same analysis research project, the unification operation of sub-cases (or changing the sub–case name) is only valid for the document in which "Rename sub-cases" was carried out. Be sure to perform this action in the remainder of the analysis documents of the study.

Unification of the Sub–Cases While Preserving the Identity of the Original Sub-Cases

In many cases we want to unify several sub-cases while simultaneously maintaining the separate identity of each of the original sub-cases. This kind of combination is necessary especially when the original sub-cases are desired to appear in several combinations of sub-cases. For example, in comparing men and women, we want to unify all the sub-cases of men and in parallel all the sub-cases of women. However, we can organize other merges by men and women's age, such as "aged 21-30," "31-40," and so on. Keeping the original sub-cases allows the creation of countless combinations of sub-cases to compare and to examine the studied phenomenon from many different angles.

Using a word processor to create sub-case groups while preserving the original sub-cases is possible, but requires long, hard work to copy and paste from document to document. The greater the number of the sub-cases and combinations to be executed, the more complicated the work will become.

Using the Narralizer for Unification and Preserving Sub-Cases

The Narralizer facilitates the work of unification and preserving sub-cases through the possible distinction between separate sub-cases and sub-case groups. It provides an easy unification of separate sub-cases that preserves the separate existence of each in parallel. To create **sub-case groups:**

1. Click on **Sub-cases groups** in the **File** menu to open the **Sub-case Groups box** with four panes. The lower pane, called **Non-group members**, lists all sub-cases in the **Analysis Document** (Figure 16C).

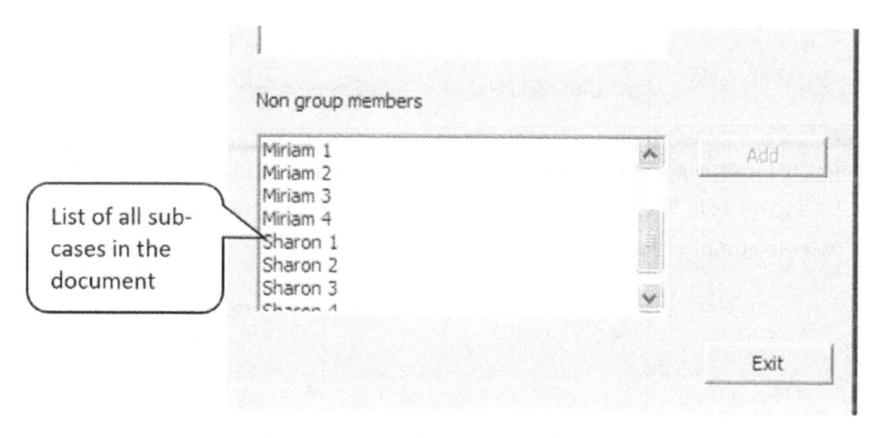

Figure 16C: The Lower Pane of the Sub-Case Groups Box

2. From the list of sub-cases, identify the sub-cases to be placed in a group and type the name of the group in the upper pane of the box (in Figure 16D, "Sharon"). Click the **Add** button to add the new name to the list of groups.

Figure 16D: Typing the Name of the Sub-Case Group

3. Add names of groups as required (in Figure 16E, "Miriam" was added to "Sharon" which already existed in the list of groups.)

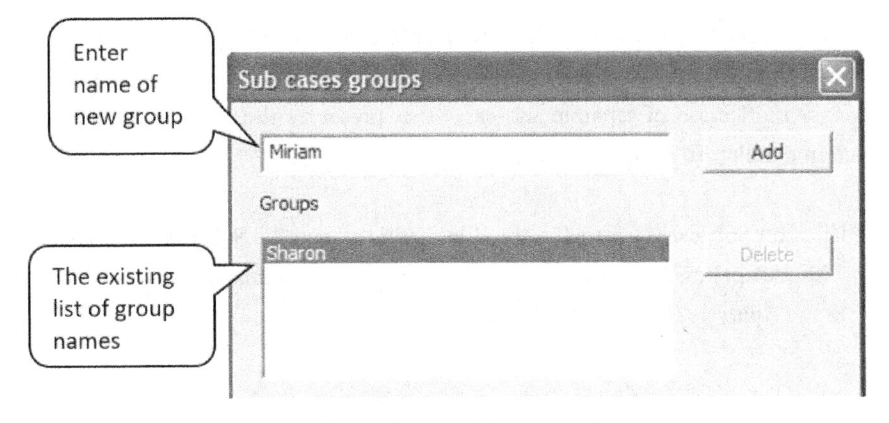

Figure 16E: Add Names of Sub-Cases Groups

4. In the **Groups** pane, highlight the name of the desired group to insert sub-cases (in Figure 16F, "Sharon"). In the **Non-group members** pane, highlight the sub-cases to combine as a group (in Figure 20F: "Sharon 1, 2, 3, 4").

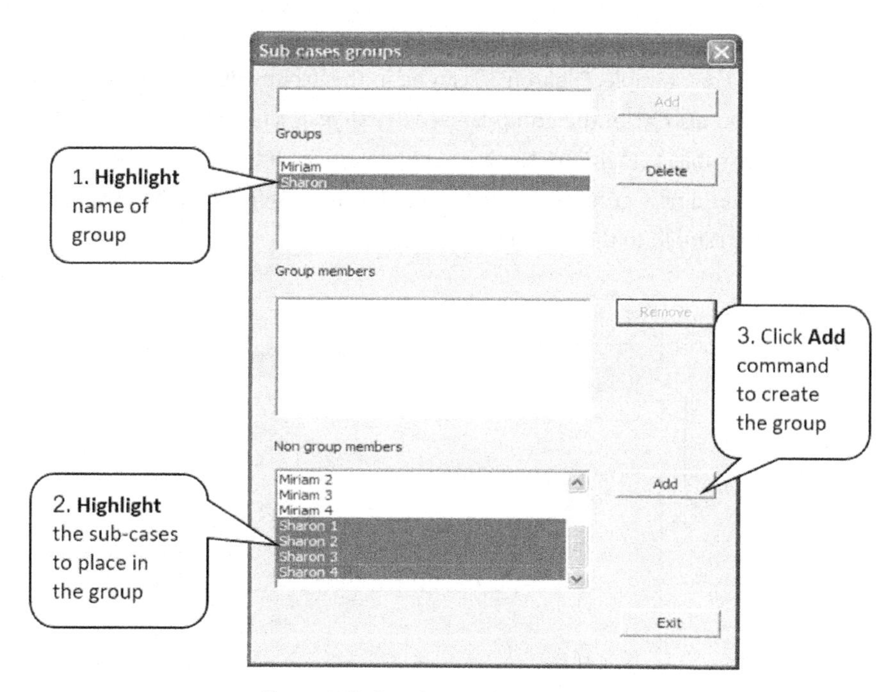

Figure 16F: Creating the Sub-Case Group

The **Add** command forms the group and transfers the names of the sub-cases in the group to the **Group members** pane, as shown in Figure 16G.

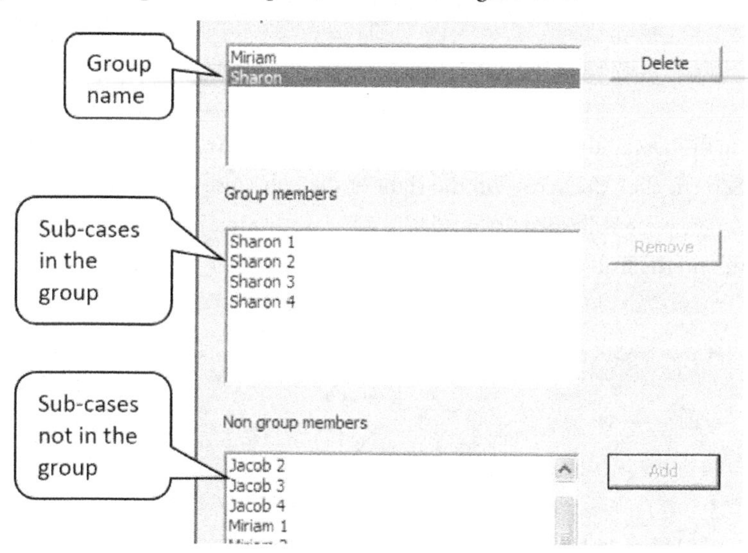

Figure 16G: Revealing the Sub-Cases in the Group

Note: A sub-case can be in several groups at once depending on the research requirements. For example, "Sharon 1" can be in the group with the other Sharon sub-cases; it can also be in the group labeled "First year" containing the sub-cases from the other subjects (Miriam, Jacob, etc.) in the research for "Year 1" (see Figure 16H.) Whenever a new group is created using the above steps, all sub-cases in the document are restored to the **Non-group members** pane.

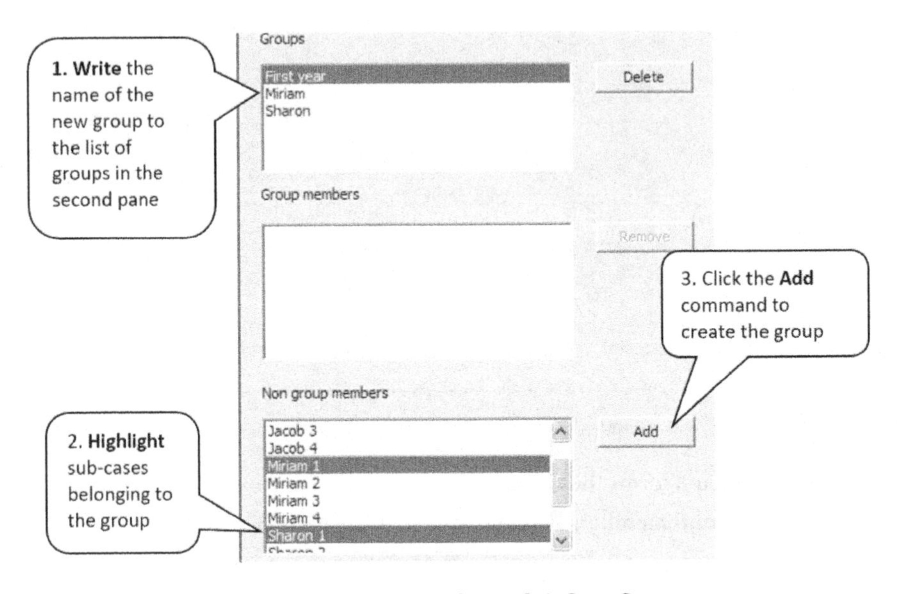

Figure 16H: Continuing to Create Sub-Case Groups

In order to display the list of Sub-Case groups in the Analysis Document on the Work Screen, click the arrow on the right of the Sub-Case groups pane (see Figure 16-I). A new pane will open listing all the sub-case groups. Use the scroll bar to view all names on the list.

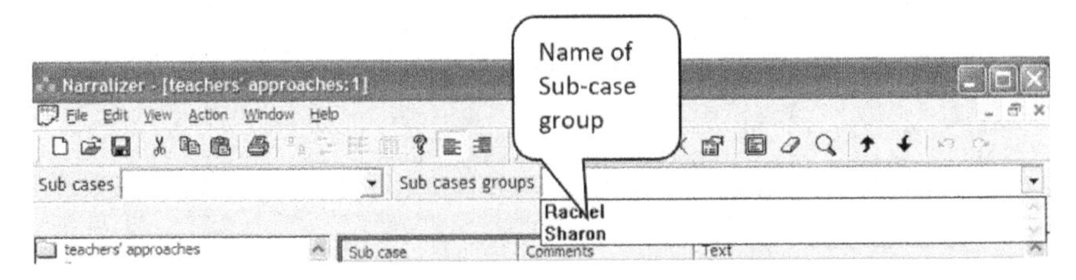

Figure 16-I: Sub-Case Group Pane of the **Narralizer** Work Screen.

Pattern Organization of Research Findings

The previous section described how to unify several separate sub-cases to one united group. This type of arrangement can be called a "research pattern." Indeed, by referring simultaneously to a categorical dimension and to the sub-case dimension, there is a potential to represent the research findings by focusing on specific features which are typical of certain participants, for a specified time or a particular context, and to distinguish those subjects from other participants, periods or contexts. Highlighting similarities and differences between sub-case groups constitutes a deeper recognition of the phenomenon under investigation. The combination of the categorical point of view and the perspective of the sub-cases allow access to almost limitless combinations, each with a certain pattern. These patterns may be significant, and organize the findings with an emphasis on insights gained during the analysis process. As mentioned, since qualitative research deals with an enormous amount both raw data and analysis findings, there is a danger that the wealth of details may cloud the entire picture. In this stage that follows the analysis process, the researcher should consider ways to focus the research picture simultaneously upon the details and the entire picture. Organizing the sub-cases around patterns is an effective means that also grants significance to the entire research. Creating patterns highlights particular characteristics of the investigated phenomenon.

The organization the sub-cases by pattern according to selected categories is mainly based on suitability. The term "suitability" in this context means that we can gather a certain number of sub-cases into a group (pattern) when they are characterized by similarities and/or identified with respect to the an upper category, or with respect to several categories and/or to several dimensions of categories. Clearly, as the research poses a greater number of sub-cases, the issue of the pattern arrangement becomes more important.

Figure 16J is an example of organizing the analysis findings around four patterns. Each of the four distinctive patterns gives a different expression to the issue of research formulated in the main category "The connection between the educational vision and experience as pupils." Each participant in this study (teacher-students) is identified as a sub-case. Upon completing the analysis process, the researchers compared the perceptions of all participants, and focused on several categories that they considered most significant. The researchers noticed that they could classify all the sub-cases in four patterns that express the relationship between the participants' perceptions. As such, the researchers organized the analysis findings according to patterns. Each pattern refers to the same categories but gives them a different expression, based on the differences between them. Consequently and purposefully, the sub-categories and the sub-sub-categories of the four patterns in the category array are identical but the content of each category is different. The structure of the category array allows the comparison of the sub-cases over a common theme and the construction of the patterns.

A. Pattern of **similarity and full continuity** between the vision of teaching and the experiences as school pupils

B. Pattern of **corrective continuity** between the vision of teaching and the experiences as school pupils

C. Pattern of **continuity with change** between the vision of teaching and the experiences as school pupils

D. Pattern of **renovation** of the vision of teaching in light of the experiences as school pupils.

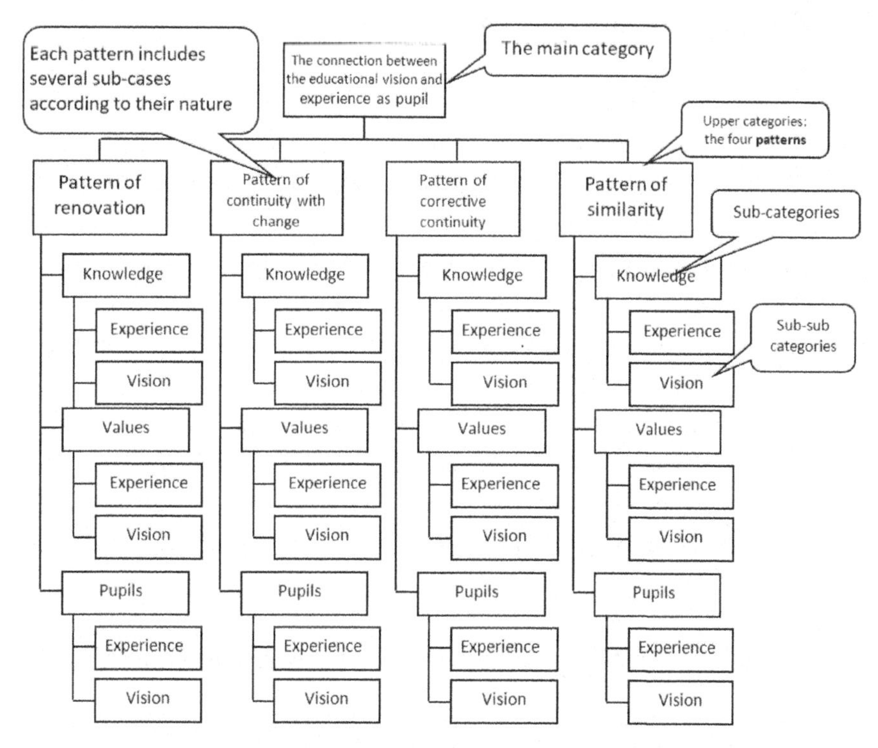

Figure 16J: Category Array Created by the Pattern Construction

Pattern Organization of Research Findings Using the Narralizer

As mentioned above, the **Narralizer** does not make the analytical decision for the researchers, but help them to perform the process, providing particular assistance in determining sub-case groups and constructing patterns. Using the **Narralizer**, the researcher can easily merge sub–cases into a pattern (a sub-case group), as explained in previous sections. On the basis of such patterns (sub-cases groups), the researcher can construct a new research document with the appropriate category array, as explained in Chapter 15 (dealing with focused analysis and theoretical analysis).

Splitting the Analysis Document

At this stage of the research, after collecting and analyzing the data, the researcher may possibly conclude that the current findings are too rich to include them all in one study since the plethora of material may prevent the focus on specific significant issues. This situation is very typical of qualitative research, where data is primarily verbal and often very rich. In this situation, the researcher may decide to split the research and focus only on certain sub-cases (for example, only the sub-cases dealing with participants of a particular profession, or with some professional experience, or a certain age, gender, and the like).

In this situation, the other sub-cases may be saved for a parallel study either now or in future. When researchers decide to focus only on certain sub-cases, those working with a word processor separate the sub-cases by using operations that the word processor allows, in full accordance to what has been explained in the previous sections.

Splitting the Analysis Document Using the Narralizer

1. On the document to be split, click **Divide** in the **File** menu, as shown in Figure 16K.

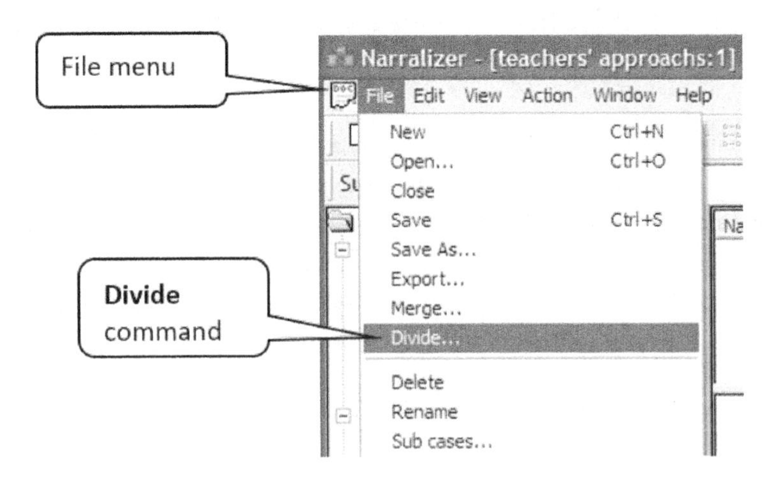

Figure 16K: **Divide** Command

2. The **Select Sub-cases box** is opened. This is the same box used in Show Sub-tree. The left side of the box lists all the Sub-cases and Sub-case groups in the research document to be split. Highlight the Sub-cases or Sub-case groups to be placed in a different document and use the top arrow between the two panes to move the highlighted Sub-cases into the right hand pane, as shown in Figure 16L.

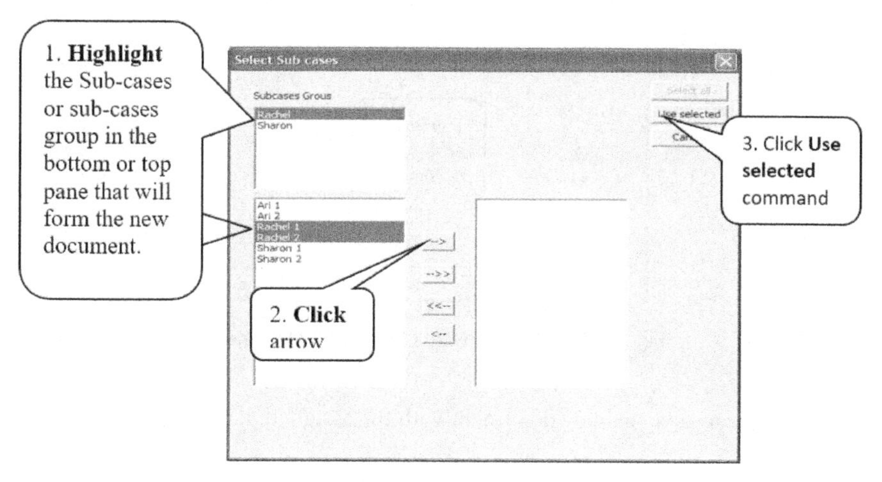

Figure 16L: Select Sub-Cases Box to Divide Documents

3. Click **Use selected** to open the **Save as** box (Figure 16M). Type the name of the new document in the box (there is no need to add the suffix ***nbr,** as this is added automatically) and place the document in the appropriate folder.

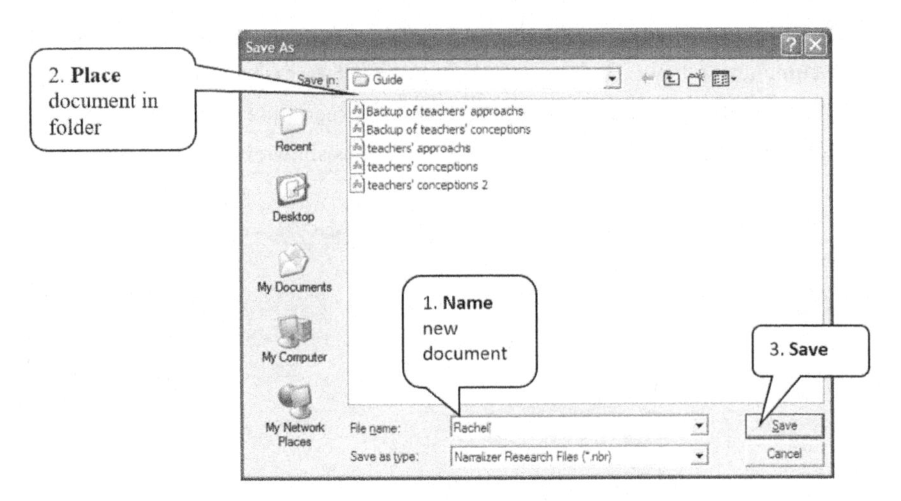

Figure 16M: Giving a Name to the New Document

4. Click **Save** to save the new document with the assigned name.

Bear in mind: Creating a new document does not eliminate the original document, which retains all its components, including those sub-cases that now make up the new document. We now have two documents: the original document and the new document.

Merging Analysis Documents Containing Varied Sub-Cases

Researchers may sometimes carry out the analysis process on several parallel documents, or conduct the study with separate analysis documents in partnership with researchers who are simultaneously investigating the phenomenon. The stage of organizing the analysis findings is the best time to create merged analysis documents. These separate analysis documents have an identical category array, but each deals with one or more different sub-cases. Merging means that all sub-cases will combine within a single document, allowing the researcher to see the big picture and to compare the sub-cases. Merging analysis documents using word processor tools requires such multiple efforts as creating a new document, copying section by section from one document and pasting it the appropriate category in the merged document, and ensuring that the identification data of each segment will be properly linked.

It is impractical to merge separate analysis documents with different category arrays, even if this is technically possible. The ultimate condition for merging is a common category array. Technically, two sets of different category arrays may be placed side by side in one list, but it would lack analytical significance. However, if the difference between analysis documents appears only at the lower levels of categories, especially the content category, a significant merge can be performed, even if not every sub-case will express all the categories at the lower levels.

Merging Analysis Documents Using the Narralizer

The **Narralizer** allows us to merge several Analysis documents into one. In other words, we can import data from several Analysis documents into one Analysis document.

> **Bear in mind** that the method of merging Analysis documents only applies when the two documents being merged have the same category array or at least an array that is very similar.

1. In the Analysis document to which another document is to be added, click the **Merge** command in the **File** menu (Figure 16N).

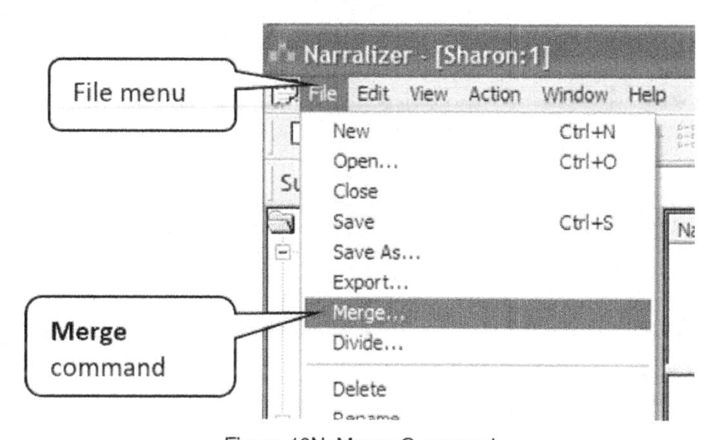

Figure 16N: Merge Command

2. An **Open** box will appear. Use the **Look in** pane in the upper section of the box to find the Analysis document you wish to add to the current Analysis document in use, and **highlight** it (Figure 16-O). The document name appears automatically

in the bottom section of the box in the **File name** pane.

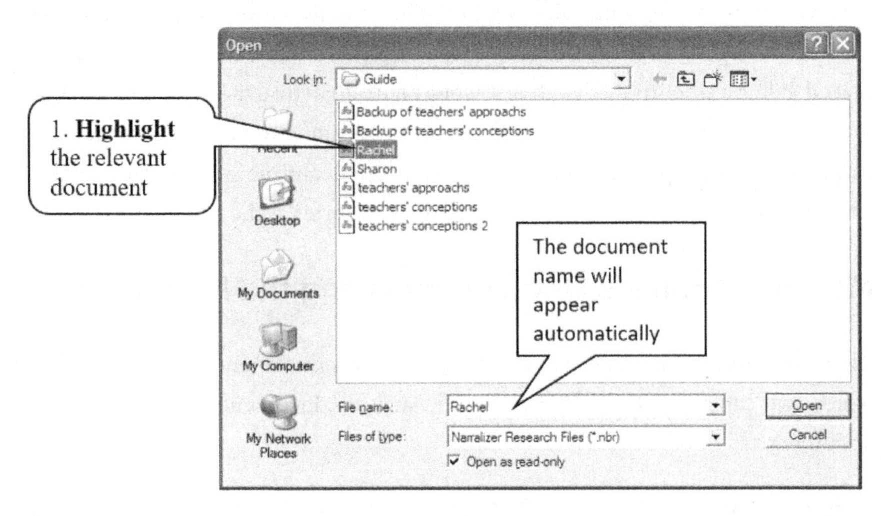

Figure 16-O: Open Box

3. Click **Open** to open the information and the Merge question box (Figure 16P)

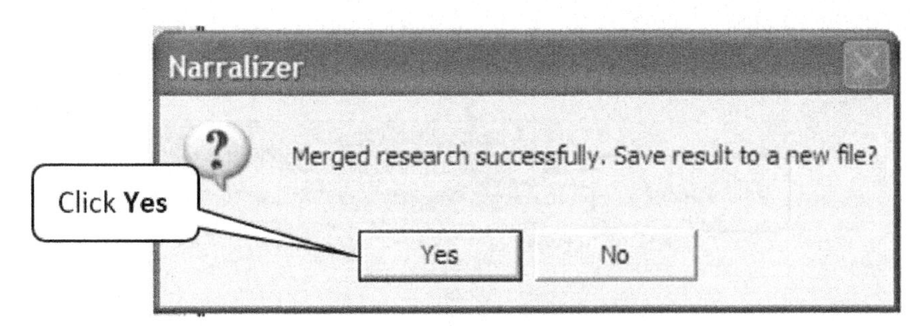

Figure 16P: Merge Question Box

4. Click **Yes** and the **Save As** box opens automatically (Figure 16Q). **Type the name** of the merged document, and **assign** it to the appropriate folder.

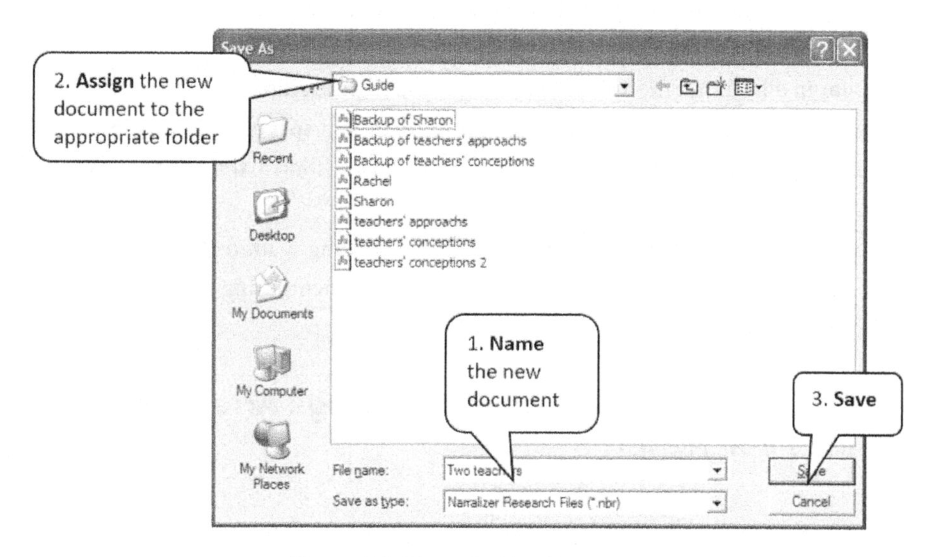

Figure 16Q: Name the New Document

5. Click **Save** (Figure 16Q) to merge the two Analysis documents into a single Analysis document (the source documents are also saved as separate files).
6. Where a difference is found between the category arrays of the two documents, the software identifies the categories that appear in the "added document" but not in the "recipient document" and displays a question box (in Figure 16R, the question is whether to add the "school message" category to the prime category "student difficulties"):

Affirmative answer – category is added
Negative answer – ignores category (does not add it)
Cancel – exits Merge sequence

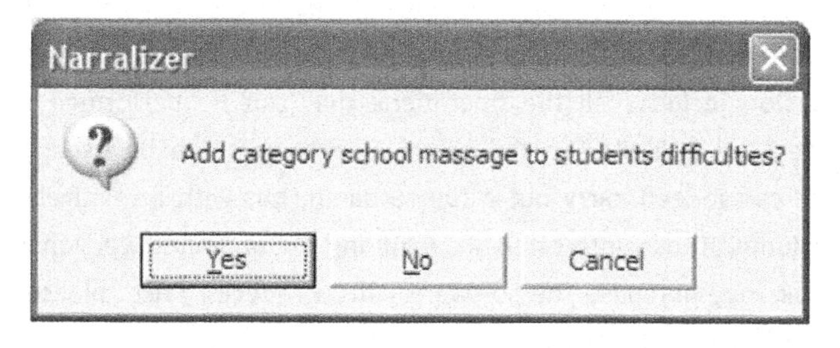

Figure 16R: Question Box: Whether to Add the Category

A similar question box is opened each time the **Narralizer** identifies a difference.

> **Bear in mind:** that the software identifies the slightest difference in category names (e.g., a different letter, additional comma, a space) as a completely different category. So, it is important to read the question in the box carefully.

After attaching two documents analyzing and creating a unified document, we can attach additional documents to the merged document, using the same process described above.

Summary: From the Analysis Findings Arrangement to the Final Report

The arrangement of analytical findings is the interim stage between the data analysis and the final report writing. At the analysis findings arrangement stage, more than at any other stage, the researcher sees the entire picture. This creates an opportunity for the researcher to examine the process of analysis, and often to return to the previous stages and modify the analysis if its results do not seem convincing or if it raises questions. The fact that the researcher conducted a transparent process by carefully preserving the analysis documents from previous steps allows the researcher to detect any inadequacies. On the other hand, when researchers arrange the analysis findings into a "story" as they write the final report, they often go back to re-examine the arrangement of the research outcome and arrange the material in a different way. This last step of research now opens the way for writing the final report.

> **Bear in mind:** As mentioned in various places in the book, the attached Narralizer educational software contains a limited version of the Narralizer. This version includes all the operations that can be performed by the Narralizer, but is limited in the scope of data that can be used. In this way, we can indeed carry out a full research, but with a relatively small data volume. Those interested in acquiring an unlimited version of the Narralizer may purchase the software with a reduced price, please go to **www.narralizer.com/Software/BuyNow.aspx**

PART SIX

MIXED METHODOLOGIES AND METHODOLOGIES WITH VARIETY OF APPROACHES

INTRODUCTION TO PART SIX

In this part of the book, we add two more methodologies: structural-focused methodology and the self-focused methodology, each of them characterized by a variety of approaches, some objectivistic (positivistic or post-positivistic) and others interpretive-constructivist or critical-constructivist.

In terms of the philosophical research approaches, presenting by the four approaches: positivism, post-positivism, interpretative-constructivism and critical-constructivism, the two qualitative methodologies are located in the "space" of qualitative research. In some of the research types, there are more analytical skills expressed; in others there is a dominance of the intuitive research skills, some are closer to the objective approaches, and others closer to the constructivist approaches (See figure 5-1).

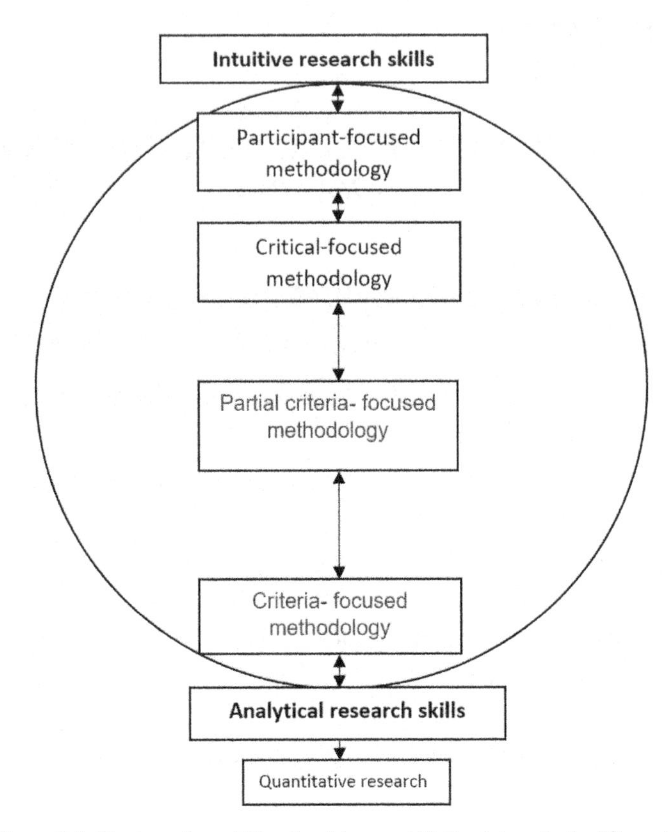

Figure 5-1: The Location of Structural-focused Methodology in the "Space" of Qualitative Research

In first part of the book, qualitative research is defined by these three characteristics:

A. Research utilizing the language of words, a natural human language in the context of natural human life

B. Research based on intuitive human research skills, focused on closeness, participation, and empathy with the phenomena investigated

C. Using analytic human research skills focused on distancing, reflection and control of the process

The first component, the language of words, is typical to all qualitative research methodologies and is perhaps the most prominent marker of qualitative research, although it may be expressed differently in each methodological pattern. This component significantly distinguishes between qualitative research and quantitative research, which bases its language of study on the language of mathematic. Therefore, most of the literature addressing the issue of mixed methodologies relates to the combination of quantitative research with qualitative research. The combination of the two is perceived as both a meeting of the weaknesses of each of the two approaches, and as a recruiter of the strengths of each approach into a joint research work. The combination of qualitative with quantitative methods is also considered to extend the scope of qualitative research (Denzin & Lincoln, 2005a).

CHAPTER 17

STRUCTURAL-FOCUSED METHODOLOGY

Within the qualitative research methodologies we can identify two traditions, the content tradition and the structural tradition. The content traditional sees the spoken and written word as a window that reflects the human experience, and thus utilizes what is said or written to learn about the world of the participants and the phenomenon under investigation. Research adhering to this tradition is focused on content. In other words, the content tradition deals with "what" is written or said (Ryan & Bernard, 2000). The structural tradition, however, asks "how" things are said or written (Lieblich, et al., 2010) and aims to understand the behavior of people by understanding how they use their language (Seidman, 1991). Research based on this tradition is focused on structures (Ryan & Bernard, 2000).

All the methodologies presented thus far have belonged to the content tradition. This chapter presents the qualitative research studies which belong to the structural tradition, called "structural-focused methodology." The structures, which reflect the way things are said, turn our attention to linguistic aspects of the phenomena under study. "It is important to study from literature and media researchers about the importance of the way things are delivered to the researcher. Sometimes we can ignore the content embodied in the text and engage in the structure of the plot, the order in which things are presented, the relationship between the story and the time dimension of the story, the extent of complexity or coherence, the speaking style, the choice of certain metaphors, questions, active or passive spoken form, singular or plural, first person, second or third, and much more. Examination of the structure of the story may yield significant information about the narrator or the problem of the research." (Lieblich et al., 2010, pp. 24-25)

Between Content and Structure

We perceive language as a means of communication, and this is certainly one of its characteristics---but there is also much more. Language is a means of conceptualizing and of data compression which allows us to represent complex information in compact codes. The uniqueness of the human language does not lie in its vast vocabulary that people can learn and retain, but in its innate natural grammar. A language based on grammatical rules is built from the connection of sounds into different variations of the same word, and the connection of different words into an infinite number of sentences with different meanings. The innate grammar allows us to build an infinite number of combinations of words from a finite number of words (Diamond, 1992). Humans also possess a unique ability to modulate the sound while adding accents and intonations that help to convey emotional meaning (Ratey, 2002).

It seems that language shapes the way we understand reality and ourselves in the world. Language allows us to arrange and rearrange our ideas. Language is a flexible tool, capable of creating new patterns, and it is easily adapted to structural changes (Goldberg, 2007). Language lets us create symbolic models, not just of the world as it is, but also as the world we want it to be. The ability to use language for guidance in planning future operations is in the heart of humanity (Harari, 2011).

Socio-linguistics researchers such as Wittgenstein refer to language not only as reflecting reality, but also as a designer of reality (Phillips & Hardly, 2002). This view characterizes many constructivist researchers, especially those of the critical school (Vardi-Rat & Bloom-Kulka, 2005). Language is rooted deeply in our existence, so that thinking with language and human existence are seen as one element. Life experience appears as a linguistic structure; therefore we can talk about the human experience and human interaction as a text (Van Manen, 1990). Consequently, some researchers refer to consciousness not as a mental entity, but as a discourse (Potter, 1996). This view "is a rejection of the idea that language is simply a natural means of reflecting or describing the world and a conviction of the central importance of discourse in constructing social life" (Gill, 1996, p. 141).

Language is seen not only as reflecting the world of people and the social relationship between them, but also as a creator of meanings and social realities (Richardson & St. Pierre, 2005). According to this view, in structural-linguistic study there is a significant expression to "analysis of the social context and the discourse that supports it" (Phillips & Hardly, 2002, p. 23). It seems that the human language had a significant function in the development and fortification of the social foundations of mankind. In the absence of language, social information and messages would not have passed from one to another (Diamond, 1999). British psychologist Robin Dunbar (1997) argues that language plays a role in supporting and developing social relationships. In many studies, Dunbar and colleagues found that language is used not only to provide useful information, but mainly for social information (gossip), designed to help people to function in the social framework in which they live. Each unique language of each human community is used to create social cohesion.

What distinguishes the human is the powerful ability to transfer through culture the selection of specific knowledge patterns from person to person and from generation to generation. We are not required in every generation "to reinvent the wheel." We utilize the impact of knowledge gradually stored in the society over thousands of years (Jacob, 2004; Blackmore, 1999). Thus, the structure of language, more than its vocabulary, gives the potential and uniqueness to human language.

Discourse, Narrative and Story

The object of structural-focused research methodology is the human's written and spoken text. In this context, there are three key concepts prevalent in literature relating to the text: discourse, narrative and story. In the relevant literature for structural-focused methodology, these concepts are often interchanged. This is due to a lack of agreement among researchers on the definitions of these concepts, the relations between them, the differences between them and the division of each to sub-concepts. In the opening discussion on the methodology, distinct definitions will be offered, even if they have not gained the consent of all researchers in the field.

The term "discourse" is a common idiom, yet vague. In general, we can say that it usually refers to a specific social-cultural context. (Bar-On & Sheinberg-Taz, 2010). The concept "discourse" is a linguistic continuity that is larger than one sentence. It seems that most researchers would agree to regard discourse as a top meta-concept which can be divided into sub-concepts including "narrative" and "story." "'Discourse' refers to both talk and text, and it captures the social and constructive element of interaction. That is, 'language' may implicate a linguistic emphasis, with a focus on grammar punctuation or other technical aspects of the language system." (Wiggins & Riley, 2010, p. 138) It thus seems that discourse is perceived as a concept that includes many linguistic forms of expression. Viewing discourse as the top overall concept, Catherine Riessman and Jane Speedy (2007) see discourse as including chronicles, reports, arguments, questions and answers and more. Martin Cortazzi (1993) suggests a division into four genres of discourse: "procedural discourse," referring to texts or speech dealing with the issues of processes of action; "expository discourse," referring to ideas and explanations; "hortatory discourse," referring to a text focused on persuasion and arguments, and the "narrative." The term "discourse" may also include such terms as "discussion," "episode" and more. Researchers often talk about discourse in general, without distinguishing between narrative and other genres of discourse, and sometimes not even between discourse and narrative.

Many scholars relate the concept of "narrative" with the "story," and regard them as synonymous. The concept of "narrative" (or "story") is a comprehensive reference among qualitative researchers, but it seems that there is no definition that everyone agrees upon. In everyday language, the concept "narrative" is very common, and is attributed not only to the discourse of people in its various forms, but also to works of art, music, photography and the like. Tuval-Mashiah & Spector-Marzel (2010) indicate that there are researchers that refer to all non-numerical data literally as a "narrative." This is definitely the broadest possible definition, and makes the narrative an even wider concept of "discourse." At the other end of the spectrum, there are those who have greatly reduced the term "narrative" and see the narrative/story as a unit consisting of at least two sentences linked by units of time (Labov, 1972),.

Polanyi (1967) defines "narrative" and "story" as a form of discourse which is located on an exact timeline and consists of distinct units that form the description. Most researchers agree that the narrative/story woven around the plot and the units of time constitutes the main element. The plot consists of a sequence of events in units of time (Riessman, 1993). Thus, a minimal narrative/story includes at least three units of time: beginning, middle and end. Causality is the second component of the units of the narrative/story connecting several pieces of the narrative. The third element of the narrative/story is the human interest that determines which events and causal relations will be integrated into the plot (Cortazzi, 1993). Although most researchers refer to the narrative/story as containing events that have occurred in the dimension of time, there are those who do not restrict the narrative/story to the time dimension, but rather refer to it as containing events organized in a particular context, according to the place of occurrence (Dargish & Sabar-Ben Yehushua, 2001). In conclusion, the central elements of the narrative/story are plots, units of time, and events and components with a connection and/or causal relationship.

While the above definitions refer to the narrative and story as synonymous, other researchers distinguish between them and see the story as a product of the narrative. According to this distinction, if we define the story in a narrow way, then any story is a narrative, but not all narratives are a story. The narrative is perceived as an overall concept that also includes other forms of expression beyond the story. "Personal narrative encompasses long sections of talk – extended accounts of lives in context that develop over the course of single or multiple interviews (or therapeutic conversations) (Riessman & Speedy, 2007, p. 430). Based on this definition, any units of speech and text can be seen as a narrative, but not necessarily as a story. Under the heading "stories" (as a product) there are different genres: a novella, epos, history, drama, comedy, and so on.

The Narrative as a Process of Thinking

Regarding the narrative as a process (as opposed to viewing the story as a product) focuses attention to the way things are said or written, "how" they are said, not "what" is said. This corresponds to Bruner's (1985, 1996) conception of two ways of human knowledge and thinking: one paradigmatic, and the other narrative. Paradigmatic thinking is positivistic in nature and takes place according to orderly, linear, logical-scientific, empirical, and universal processes, free of values and unrelated to the context. The expression of paradigmatic thinking is focused on concise descriptions and explanations, and reflected in the language of argumentative discourse. In contrast, narrative thinking is constructivist in nature. Narrative thinking is interpretive, holistic, and dependent upon the context, personal values, and the ways in which people experience the world and give it meaning. Narrative thinking focuses on the unique rather than the universal, emphasizing dynamic tension, dilemmas and situations of ambiguity. Narrative thinking is a means by which people organize separate facts into an order that gives meaning to things (Clandinin & Connelly, 2000; Gudmundsdottir, 1995, 1996). Narrative thinking is reflected in the language of words by narrative outcomes (Lyons, 2007). These can be partial stories or a fragmented narrative, but not necessarily "complete" stories.

Elliot G. Mishler (1986) refers to the narrative as a basic structure common in the mind of people. He views the characteristics of the narrative as mental skills that allow people to find meaning in the world. These skills develop naturally and relatively quickly in children, without any external guidance. One of the main purposes of the mind is to weave our lives into a coherent story. It does so by producing explanations for our behavior on the basis of our self-image, our memories of the past, expectations for the future, present social and physical circumstances, and the surrounding physical environment where we act (LeDoux, 1998). Narrative is a landscape in which humans function and find meaning in their functioning (Clandinin & Connelly, 2000). People use narratives as a heuristic device by which they arrange the relevant facts and organize them in some logical order (Gudmundsdottir, 1995). It is easier to remember narratives, because narratives are in many respects the way we organize our memory. The narrative is our primary means to look into the future, predict, plan and explain (Turner, 1996). The meaning of life does not exist independently of our narratives about life. Narratives describe life, change it, and give it a special structure (Grimmet & Mackinnon, 1992; Widdershoven, 1993).

We live our lives as a lifelong study of ourselves, and thereby form a hermeneutic circle, the circle of understanding between the ego and our self (Fontana & Frey, 2005). The relationship between life and narrative is a hermeneutic circle: the narrative is based on a prior understanding of life, and then changes to a more fully-developed comprehension (Widdershoven, 1993). The "truth" of our narratives is not a scientific or historical truth, but what might be called a "narrative truth" (Bruner, 1990, p. 111; Freeman, 2007, p. 136). Guy Widdershoven (1993) argues that the relationship between life and narrative is usually reflected in one of two ways: either life seems like something that might be described by narratives, or the narratives present ideals whose nature we try to portray in our life. Thus, one can say that life experiences and narrative interact.

At the end of this section, we emphasize that the analysis of discourse, narrative and story is not unique to the structural-focused methodology, which views the world from a structural perspective. These elements are also found in methodologies that examine things from a content perspective, such as those described in previous chapters.

Structural Analysis Units

The linguistic-structural analysis of the texts can be arranged on a continuum between two dimensions: a holistic reference at one pole and a categorical reference at its opposite (Lieblich et al., 2010; 1998). The holistic analysis focuses on large textual units, up to the entire text, whereas the categorical analysis focuses on small units, up to the reference of a single word as a unit. The character of the analysis units stems from the linguistic-structural theory. If researchers adopt a theory that focuses on the story as the unit of reference, for example, the tendency will be towards a holistic reference and the analysis units will be part of the whole story (e.g. Labov, 1972). If the unit of reference is the words and defined concepts, then the analysis units will be words, and the context and general structure of the text will be of secondary importance, if any. More specifically, we can say that there are two main types of analysis: "the text is segmented into its most basic meaningful components: words. In the other, meanings are found in large blocks of text." (Ryan & Bernard, 2000, p. 775)

In the holistic analysis, referring to large units of text and the text as a whole, researchers focus on the way in which the author tells the story: the structure of the time sequence; the location of the plot; the relationship between the parts of the text; the relationship between events and characters, if they are linked or disconnected; the causality relationship, if the plot structure expresses progression or regression; and the genre of the text: tragedy, comedy, novel and so on. In analysis focused on small units, researchers face plentiful options, from an examination of short pieces to an examination of a separate word. Such an analysis might examine the use of certain words, using questions and expressions of doubts as opposed to using definite expressions; speech in the first person versus second or third person style, and more. Using or deleting certain words or phrases may teach about the personality of the speaker or writer. We can also use the quantitative analysis, particularly when focused on individual words or phrases (Lieblich et al, 2010). Intermediate analysis units between the whole and its parts enable the identification of different types of text expression, such as the distinction offered by Rosenthal (1993) between story, description, reporting and argument, and revealing and interpreting metaphors in the text (Fielding & Lee, 1998).

Another feature commonly seen as central is the story's coherence characteristic, the structural interactions between different parts of the story. The coherence of the story will be judged according to the degree to which and the way the content is displayed, as well as the way in which the content of the story and its background are intertwined (Fielding & Lee, 1998). The relationship between the parts of the story and its internal logical sequence will also be examined (Lieblich et al., 2010). It is possible to refer to three types of coherence: "(a) global coherence, or how a particular utterance is related to a speaker's overall plan, intent, or goal for the conversation; (b) local coherence, which refers more narrowly to relations between utterances and parts of the text; [...] and (c) themal coherence, or how utterances express a speaker's recurrent assumption, beliefs and goals, or 'cognitive world.'" (Mishler, 1986, p. 89)

Various types of structural-focused methodologies also differ from each other in the techniques by which the segments of the text are examined. There are those who scrutinize the text based on the appearance of certain words, while others relate to the syntax of the text, grammar or structure. There are various techniques for analyzing words in the text, each of which reflects a different research approach. One technique is to focus on keywords in the text with reference to their context. This technique is reflected in finding all occurrences where a particular word or phrase appears. Another technique is to count words, which means denoting meaning to the number of occurrences of certain words in the text (Ryan & Bernard, 2000).

Philosophical Approaches to the Structural Analysis

Structural-focused methodology includes a large number of different study types. Various researchers anchor their research in a different research approach (research paradigm) expressing different assumptions about the status of the phenomena under study and about the relationship between the researchers and the phenomena (Guba & Lincoln, 2005). We can sort the paradigms that guide the structural-focused methodology into the objective, interpretive-constructivist, and critical approaches. Each of these theory groups implies a unique methodology.

The **objective approaches** (formal approaches) applied in structural-focused methodology are based on positivistic or post-positivistic assumptions, which view the text structure and its components without any necessary connection to the context of the text. The text is understood as an expression of the existing reality outside the text. In contrast, the **interpretive-constructivist approaches** assume that the understanding of human discourse cannot be reduced to a set of properties isolated from one another, but rather view the text in its cultural-social context and view each unit of the text in the context of the entire text (Bruner, 1985). While the objective approach bases its research on an external defined theory and an analysis model based upon this theory, the interpretive-constructivist approach seeks to select the structural analysis model which is also based on the personal experience of those who stand behind the text. "In formalist inquiry, people, if they are identified at all, are looked at as examples of a form - of an idea, a theory, a social category." In the interpretive-constructivist approach, however, "people are looked at as embodiments of living stories" (Clandinin & Connelly, 2000, p. 43). The **critical approaches** relate to the text as a factor that constructs the world, and each human discourse is perceived as a social practice. The overt meaning of the text does not necessarily reflect the world that the text represents. The researcher's role is to listen to what the text says and try to understand it in light of a cultural, social, and political context. According to this approach, we cannot understand the verbal expression of the human in a social vacuum and in an isolated context. Discourse analysis in the critical approach simultaneously combines the text analysis with an analysis of the context in which the text was constructed (Gill, 1996).

If we consider the three paradigms that guide structural-focused methodology in terminology taken from the field of hermeneutics, we can identify each of the paradigms with a hermeneutic approach. The first approach, identified on the objective paradigm (positivist and post-positivistic), deems that the ability exists to reveal the original intention of the text writer. This is a hermeneutic of reproduction of the text (Gallagher, 1992). This hermeneutic concept is based on a model that examines the elements of the text in an objective way, assuming that each expression has an objective meaning, and explains these meanings to allow researchers to reconstruct the objective reality reflected by the text. The second hermeneutical approach seeks to establish a dialogue with the original meaning of the text. This approach assumes that there is no objectivity, and that any study of the text is actually a dialogue between the researchers (and readers) and the texts, which converges at the horizon of the text and the researchers or readers. In this approach, researchers attempt to see the text in its original context and hold a dialogue with it from their own perspective and context. This approach is associated mainly with Gadamer (1975) and corresponds to the interpretive-constructivist approach.

The third hermeneutic approach seeks to move the text to a new context. This is a constructive approach, but in contrast to the interpretative-constructivist approach, this approach does not believe that we can communicate with the original context of the text. "The transfer of the text just means that new relations are created and new meaning is produced. A text is fundamentally open to new interpretations and can be infinitely transferred to new contexts" (Widdershoven, 1993, p. 13). This approach, also called the "hermeneutics of suspicion" (Josselson, 2004), seeks to explore what lies in deeper layers of text, invisible to the reader. This approach connects to deconstruction and post-structuralism, associated with Derrida (1976,

1978) and Foucault (Mills, 2003), and negates that any particular text will have one single literal meaning. This approach focuses on the role of discourse in creating non-balanced power relations, and seeks to describe and explain how the abuse of power obtains its legitimacy through speech and texts of the hegemonic group (Phillips & Hardly, 2002). This hermeneutic approach aims to emancipate the mind from social, cultural and political constraints that serve the dominant group (Travers, 2001).

Structural Research Methods

Following are several common methods employing structural-focused methodology. Although this is by no means a complete list of all the standard methods, the six methods presented are all prominent in their use of structural-focused methodology.

Content analysis – a method for categorizing and systematically encoding text that allows the analysis of a large amount of text in order to identify trends and patterns of word use, word frequency, the relationship between them, and the structure of communication and dialogue with words. This method can be applied to analyze various types of texts, including media, documents, protocols, visual images and others. Researchers choose a specific coding system to highlight certain aspects of the text. Content analysis is accompanied by a significant quantitative dimension, and some see this as the advantage of this method. Content analysis is primarily used in positivistic and post-positivistic objective-formal approaches (Grbich, 2007).

Socio-linguistic analysis of the story - a method of analysis associated largely with the work of the linguist Labov (1972), which sees the story as a unit with a unique structure (Grbich, 2007). A story structure includes a sequence of at least two sentences, arranged in chronological order. The chronological order actually determines the meaning of the story. Every story has certain elements that stand out: an abstract, which appears at the beginning of the story and briefly summarizes; an orientation, which relates to the time, location, and main characters; a complicating action which relates to the sequence of events on a timeline that reflects a problem or describes a problematic situation; a resolution, how the problem is resolved or the unexpected situation is explained; an evaluation, the means by which the narrator indicates his point of view and its positions; a coda, the last part of the story, in which the narrator concludes the story by returning its perspective to the present time (Kuperberg & Gilat, 2002; Cortazzi, 1993; Hollingsworth & Dybdahl, 2007; Riessman, 1993). The assumption of this method is that we can break each story into units of meaning and map them in a way that exposes their characteristics. The research approach that characterizes the socio-linguistic analysis method is the positivistic or post-positivistic objective-formal approach which relates objectively to the text (Grbich, 2007).

Discourse analysis extends across a wide area, from a formal-objective linguistic approach to a critical analysis of culture and communication, and is employed in many disciplines including linguistics, psychology, education, sociology and others (Grbich, 2007). "Discourse analysis is [...] an epistemology that explains how we know the social world, as well as a set of methods for studying" (Phillips & Hardly, 2002, p. 3). The assumption is that social reality is created through discourse, and no one can understand the world and social interaction without the analysis of discourse. Discourse analysis researchers seek to find the relationship between discourse and reality. Discourse analysis is "interested in how and why the social world comes to have the meanings that it does. Discourse analysis provides such a methodology because it is grounded in an explicitly constructionist epistemology that sees language as constitutive and constructive rather than reflective and representative." (Phillips & Hardly, 2002, p. 14).

Discourse analysis focuses on the structure of the text and the discourse elements that subjects have used to relate to the content and rhetorical organization of the speech or text, and examines how all the elements of the text are organized in order to render it as convincing (Gill, 1996; Potter, 1996). "Discourse analysis involves changing the way in which linguistic material tends to be seen, so that instead of seeing discourse as reflecting underlying social or psychological realities, the focus of interest shifts to the ways in which accounts are constructed and to the functions that they perform. Doing discourse analysis fundamentally changes the way in which you see and hear language and social realities." (Gill, 1996, p. 144). As mentioned, there are different types of discourse analysis, which reflect different epistemological approaches. (Grbich, 2007; Wiggins & Riley, 2010).

Conversation analysis: Conversation with other people is a natural human activity, carried out almost without special attention. But conversation is not only talk; it is speaking in reference to interactive purposes. The intention of conversation analysis is to provide a detailed and systematic description of interactive speech, and to clarify the meaning of the conversation. Conversation analysis is associated with the concept of "ethnomethodology" which is "a study of the ways in which people make sense of what other people do in the processes of social interaction" (Grbich, 2007, p. 137). This means focusing on the way people themselves create and identify the practices that give sense to the conversation and allow its development (Forrester, 2010). The conversation's speakers actually provide their interpretation of what was said and what is happening, and their response to the words of the other person (Potter, 1996).

It seems that participants in the conversation have an intuitive awareness of their role, their contribution to the development of the conversation and their responsibility to the interaction of the conversation. The narrative generated is a joint contribution of the participants. The extraordinary nature of the conversation creates a unique investigation that takes into account the characteristics of the conversation, which differ from those of other texts (Cortazzi, 1993). Conversation analysis is focused on the structure of speech and the development of the relationship between the participants. The analysis is based on a careful examination, line by line, of the course of conversation (Forrester, 2010). However, in conversation analysis, like other structural methods, the meaning assigned to the conversation will be determined by the approach that guides the researchers, as mentioned above.

Structural semiotic analysis: The semiotic method, called "the science of signs," (Fielding & Lee, 1998, p. 50) argues that people communicate with others using signs, and that we think and create meaning only through a system of signs. The system of signs includes words, but also sounds, shapes, smells, textures, behaviors, objects, etc. (Bar-On & Sheinberg-Taz, 2010). The purpose of this method is to explain how the meanings of objects, behaviors, speech or written text are transferred or distributed. According to Ferdinand de Saussure, a pioneer of the modern semiotic linguistics branch which developed in the early 20th century, the sign is comprised of two parts: a "signifier" and "signified," and the combination of signifier and signified creates the sign (signifier +signified = sign). For example, the signifier, which is a sound or a graphic symbol (e.g., "home"), and the signified will be marked as the abstract meaning associated with the sound or graphic symbol (in this case: "the place we live in"). The relationship we make between the signifier "home" and the signified "the place we live in" creates the sign, which is the uniqueness of the word "home." But it is clear that the sign is arbitrary. There is no natural link between the signifier and the signified, and in other cultural contexts it could be called the same physical entity (signifier) in another expression (signified). Moreover, the meaning of signs is not perceived as being connected to the signifiers, but understood from the relationship with other signs (Grbich, 2007).

The assumption of the semiotic approach is that we create meanings only through a system of signs. In an unconscious process, we associate things to a familiar system of conventions (Avriel-Avni & Keiny 2010). For example, the signified "walking through a minefield" makes contact with the signifier "courage," which reflects the significance we attach to the phenomenon. But from a different context linking, there are those who connote this signified with the signifiers "stupid" or "irresponsible," which allocate a different meaning to the phenomenon. The semiotic approach to text analysis seeks the meaning of signs, and clarifies the concept of the individual or group in a specific culture and society.

We can identify two approaches for semiotic analysis: the structural approach, which assumes that signs have limited meaning within an identifiable range, and the post-structuralist approach, which believes that the semiotic structuralist analysis offers a superficial meaning, and that many alternative meanings can always be uncovered (Grbich, 2007). It is the second approach, deconstruction post-structuralism, which will be illustrated here.

Deconstruction post-structuralism: Derrida (1976, 1978) challenges the concept of de Saussure in his claims that the "signifier" (i.e. the sound or graphic sign to which we refer) does not refer to reality or to the "signified" (i.e., the abstract meaning associated with the sound or graphic symbol) in a way that gives specific meaning to the "sign" (i.e., the meaning that we assign to the word). He argues that there are many possibilities of connections and meanings, as expressed in his famous phrase: "There is nothing outside the text," connoting that things have no meaning in themselves, but the meaning is grounded and created in the text (Derrida, 1976, 1978). A similar position was expressed by Roland Barthes (1977) in his famous essay "The Death of the Author," where he claimed that the author has no more absolute control over textual meaning than the reader. Derrida offers the deconstruction method as an analytical method adhering to the concept of post-structuralism. From this, there are infinite possibilities to analyze any text which is not visible to the eyes of the speaker, writer, listener, reader or the text analyzer. To understand the meaning we have to deconstruct (dismantle) the structure of the text and comprehend its meaning in the specific context under study. The multiple possibilities for understanding the text do not allow suggesting one ultimate proper methodology for analyzing texts, but the analysis method should be adapted to the specific texts and their context (Grbich, 2007). "The objective is to show not only that a text can be interpreted in different ways, but that one can never establish a final meaning." (Travers, 2001, p. 155).

The Study's Theoretical Perspective

Structural-focused methodology deals first and foremost with language structures, the structure of discourse, or "how" things are said rather than "what" is said. As such, it is clear that the methodology should be based on a theoretical system that explains the construction of human discourse. However, discourse structural analysis is not in itself the ultimate goal of the study. The assumption of structural-focused methodology is that the way things are being said or written reflects the meaning in a significant way, albeit not necessarily visible to the speaker or writer, (Lieblich et al., 2010). According to this tradition, understanding people's behavior means understanding how they use language (Seidman, 1991). Hence the conclusion that when we speak of a theoretical perspective in the structural-focused methodology, we are dealing, in fact, with two groups of theories: one regarding the structural issues and seeking origins in the study of language, linguistics, literature, etc., and the other group studying the human and his society, seeking to offer meaning to the linguistic structure and its components in the areas of social sciences, sociology, psychology, and the like.

For example, the structural analysis may indicate a significant presence of the word "I" in the text under investigation. This may correspond to the realm of social science theory, which offers an explanation for the frequent use of "I" as an indication of certain personal characteristics. The same linguistic-structural analysis may yield different social science interpretations, with each based on different theoretical social science perspectives. Common to all these studies is their analysis of the text on the basis of structural theory, although different meanings may be attributed to the same text. In most cases, researchers come to the structural analysis with a given structural theory and method. Sometimes, researchers do not decide which theory and research model is appropriate for structural analysis, and thus construct the structural analysis model during the process of the analysis itself.

As will be explained, we can sort structural-focused methodology research into various theoretical perspectives. Some are focused on formal-objective concepts, others focus more on interpretive-constructivist perceptions, and still others place the main focus on critical approaches (Phillips & Hardly, 2002). Each of these theoretical perspectives is a "dialogue" between the structural-linguistic theory and the social–cultural theory, which interact to create a whole conception. Because of the different characteristics of each of the three groups of theoretical perspectives, research employing these theories will take place at different stages of construction of the theoretical perspective of the study.

In an objectivist approach grounded in the positivist or post-positivistic philosophy, the setting of the theoretical perspective and its implication upon the analysis model will constitute the initial stages of research, sometimes even before the researchers enter into the study field. One example can occur if the researcher is interested in examining the social interactions in the classroom. Based on objectivist assumptions, he may assume that there are universal processes that occur in each class, regardless of the contexts of each school culture, or even of the elements related to the specific content discipline. These researchers may come to the research field equipped with a theoretical perspective that explains classroom social interactions, such as that proposed by Flanders (1970). This theoretical perspective should guide all phases of the research: selecting a field of research and the participants, data collection and analysis, and others (Carr & Kemmis, 1986; Clandinin & Connelly, 2000).

In research conducted under the interpretive-constructivist approach, the researcher's aim is to set the theoretical perspective and the analysis method during and after the stage of data collecting and before the stage of data analysis. Sometimes, when the researcher cannot decide during the data collecting stage which model for structural analysis is appropriate, he will construct the structural analysis model during the process of the data analysis itself (by a procedure of initial analysis, to be explained in Chapter 14). In the interpretive-constructivist approach, the researcher is attentive to the unique context of the study subjects and texts, seeking to offer the theory and the structural analysis model respectively (Creswell, 1998).

Research adhering to the critical approach will be attentive, like the interpretive-constructivist approach, to the context of the studied texts. But the process of the critical approach will be more complex, with researchers approaching the study with a characteristic "suspicion" toward the studied texts. The critical researcher seeks not only to identify the context of the text, but also to reconstruct the text and move it to a new context. Here the researcher reconstructs a theoretical perspective, a process that will be clarified only in the course of the text analysis stage. The critical approach seeks to offer a theoretical perspective that could not only guide and explain the text, but also emancipate the text/discourse from the social forces that bind them (Travers, 2001).

As with any qualitative study, researchers in the structural-focused methodology begin the process of writing the literature review at an early stage of the research process. But the dynamics of writing this review will be different in each approach. In the objectivist approach, the writing will be completed in the early stages, before entering into the research field. This theoretical perspective, both of structural-linguistic and social science aspects, will serve as the guideline for all phases of the research. Researchers of the interpretative-constructivist and critical approaches begin the writing of the tentative literature review at an early stage of the research. But unlike the work of their objectivist approach counterparts, this review will be preliminary and it will be revised as the study advances and the theoretical perspectives become clear.

Formulation of Research Questions and Assumptions

Each of the three research approaches mentioned above leads to a different formulation of research assumptions. In the objectivist approach in which the theoretical perspective is determined before entering the study field, the research assumptions and questions are articulated accordingly and will accompany the study throughout, as illustrated in Figure 17A.

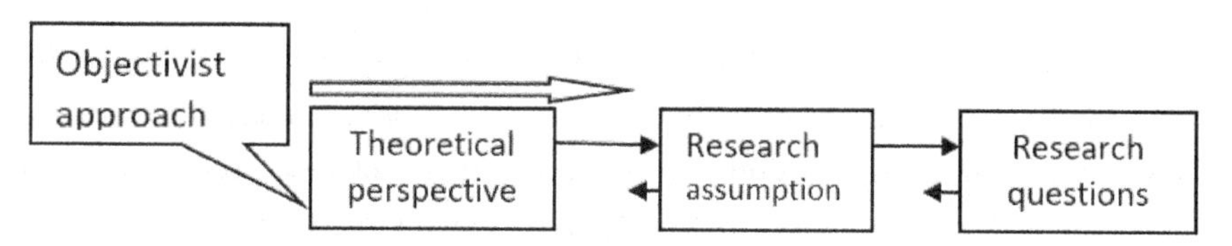

Figure 17A The Connections between the Research Components in the Objectivist Approach

The logic of the research process in this approach is essentially linear, leading from one stage to another. But as in all qualitative research which "recruits" not only the researcher's analytical skills but also intuitive skills, there exists a certain degree of repetition, review and change, as shown in Figure 17A by the short arrows from right to left.

In the interpretive-constructivist approach and the critical approach, the process of formulating the assumptions and research questions, which relate to the theoretical perspective, is relatively late. Researchers begin by drafting assumptions and researching primary questions that may change with the formation and clarification of the theoretical perspective. Figure 17B illustrates the relationship between the components of these approaches.

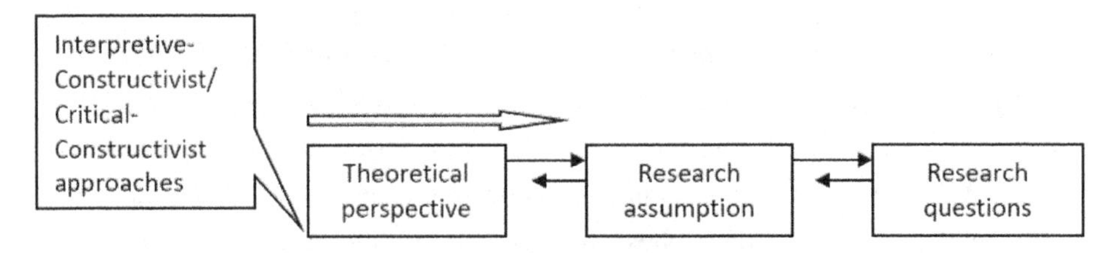

Figure 17B: The Connections between the Research Components in Interpretive-Constructivist and Critical-Constructivist Approaches

Note that although the study "flows" from the theoretical perspective to the research assumptions and research questions, there is a significant "flow" in the return process (arrows from right to left). This is due to the nature of these two approaches, which attaches significant importance to the cultural-social context, the unique features of the phenomenon under investigation, and the texts that reflect it. Therefore the components of the research, especially the theoretical perspective, are designed in the advanced stages of the research.

In order to understand the process of formulating the research questions in critical-focused methodology, it is necessary to reiterate the fact that this methodology combines two groups of interrelated theories, one which seeks to describe the linguistic- structural phenomenon, and the other that seeks to give meaning to the described phenomenon (Guba & Lincoln, 1998; Roberts, 2002). Qualitative researchers who adopt the structural-focused methodology do not intend to study the linguistic aspects alone (that task they leave to linguists and language scholars of all varieties), but rather to examine the social-cultural meaning arising from the discourse of people. Therefore, understanding the structure of the language when necessary, is only one step toward understanding the meaning of the discourse (Cortazzi, 1993)

Following is an example of a research question:

How many plot cycles appear in the text, and what is their meaning in the story?

This question has two parts. The first concerns the linguistic-structural aspect, and relies on a linguistic-structural theory dealing with the components of the story (Labov, 1972). The second part of the question concerns the field of social science theories, and seeks to explain the identified structural phenomenon based on the assumption that linguistic structural aspects teach us sociological, psychological and other social science aspects.

It seems the two parts of the same question (sometimes indeed formulated as two different questions) suggest a distinction between questions of the first order and questions of the second order (Miles & Huberman, 1994). First-order questions are those which characterize a situation in which the answers may be studied directly from the text. If we can identify, for example, the "plot" and the "plot cycle" in the text, then the answer to the structural part of the question will be found directly in the text. It can be determined that questions (or parts of questions) dealing with the structural aspect of the phenomenon under study will be defined as first-order questions. (Jovchelovitch & Bauer, 2000). In contrast, questions or parts of questions that refer to explanations and meanings are identified as second-order questions, since these questions have no answer in the body of the text, but are implied.

There is a connection between the linguistic-structural element of the questions and the social-cultural component, which probably influence each other. Researchers in each of the three theoretical approaches could be interested in other components of the text's structure in order to understand its meaning. But in principle (and in many cases in actuality), a question (or part of a question) from the first-order that deals with the structural dimension of the text may be the same in all three theoretical approaches. But a question (or part of a question) dealing with the social-cultural dimension will be completely different in each of the theoretical approaches, and must be formulated differently. For example, let us take a research question based on the life stories of community members found in a process of privatization. In a research study adopting the objectivist approach, the question would be formulated as follows:

How many times in the text does the word "I" appear, how many times does the word "we" appear, and what are the implications regarding the individual dimension and the collective dimension in the life experience of the person?

This question is based on an existing theory that suggests an understanding of the personal expression of people's personal beliefs and attitudes.

In the interpretive-constructivist approach, the question may appear in the following formulation:

How many times does the word "I" appear in the text, how many times does the word "we" appear, and how do lifestyle changes affect people's discourse?

This question attempts to link between linguistic characteristics of the text and changes in the social system of the community, and to offer an explanation of the nature of the changes in the community.

In a critical approach, the question may appear as follows:

How many times does the word "I" appear in the text, how many times does the word "we" occur, and what is the meaning of each of these expressions on the existence of people in the community?

This question seeks to explore not only the visible data, but also the hidden social forces - what affects people to express something in one way or another. Researchers rely on critical theories that refer their attention to the overt and covert elements of the discourse, with the aim of extracting the text from its overt context and offering the research participants a revised process of understanding-- and as a consequence, to emancipate them from the accepted meaning.

The Study Design and Population Selection and/or Field Research

Critical-focused methodology is characterized, as stated, in respect to three different approaches—the objectivist, interpretive-constructivist, and critical-- which reflect different assumptions about the phenomena under study and about the relationship between researchers and research subjects. The study design of each of the three approaches is derived according to its respective nature: the objectivistic approach is characterized by a linear array of research in which the theoretical perspective is determined at the beginning of the study and continuously directs all other components of the research, whereas studies employing the interpretive-constructivist or critical approach are characterized by greater flexibility in creating a study design.

While the objectivist approach seeks to explore the nature of spoken and/or written texts almost regardless of their context, the other two approaches attempt to understand the texts in light of their context, which is more prominent in studies of critical approach. This has implications for the choice of the "study population." The phrase "study population" in structural-focused methodology is more complex than in the methodologies presented in previous chapters. While all other methodologies focus on a direct encounter with the participants and on research sites with social activities, in structural-focused methodology, the research subjects can be the written texts, without a meeting between the researchers and the speakers or the writers of the texts. However, there is also the option to base the study on observations and interviews with the participants and to obtain the "text" directly from them.

Researchers employing critical-focused methodology look for the most appropriate "field study," which is not only (or necessarily) the site of a social event, but also the accumulation of relevant textual sources (Stake, 2005). In the objectivist approach, we can consider a random sample, in principle. But as in all qualitative research, the research subjects are expected to be limited in scope, what makes insisting on a random sample irrelevant. In the case of evaluation research, the "research subjects" were set earlier by the initiators of the research project. If the research is focused only on a textual analysis, without meeting with live subjects, it is quite possible and even desirable to present a random sample of the texts. In interpretive-constructivist and critical approaches, the purposeful sample is preferred, whether a "sample" of existing texts, research participants, or both. This is due to two main reasons. The first is the concept of construction that characterizes the two approaches, which argues that every "case" is unique (Stake, 1995), thus it is important to consider each case as a separate entity. The second reason for preferring a purposeful sample is the relatively limited scope of the texts or participants (compared to quantitative research), which renders the statistical random sample meaningless.

As in previous qualitative research methodologies, the structural-focused methodology researcher needs the cooperation of the studied population (which constitutes another reason for a purposeful, not random, sample). Sometimes this can mean direct cooperation, such as observation, interviews or focus-groups, and sometimes it denotes indirect cooperation, such as when the participants' agreement to use their texts is needed. It is clear that investigators are required to pay strict attention to maintaining a high ethical standard (Dushnik & Sabar - Ben-Yehoshua, 2001). Actually, it is very simple and even tempting to use existing texts without obtaining permission from the authors. Obviously the researchers should not use texts without full agreement, in accordance to the regulations of the Helsinki Agreements. (King, 2010). It is not only crucial to guarantee participants full confidentiality and ensure the fulfillment of this condition, but also to avoid invading their privacy and maintaining a channel of communication with them, even after the data collection phase and/or obtaining permission to use their oral or written texts.

Data Collection Methods

The research in structural-focused methodology is somewhat unique, according the type of data that is used. The distinction made by Silverman (2006) between "naturally occurring data" and "initiated data" is very relevant. In this methodology, the proportion of "natural data" is significant, with some studies even solely based upon it. Naturally occurring data refers to documents, diaries, articles, literature and theoretical literature, "media data" (from the mass media such as radio, television, newspapers, etc.) and "Internet data" (from various websites, including online forums, etc.). (Gibson & Riley, 2010) The origin of the "initiated data" is mainly from interviews or observations.

As emphasized in previous chapters, the data collected should be preserved and documented for later analysis, since there is a tendency among some researchers to be influenced by impressions rather than keeping to strictly analytical processes. The importance of keeping meticulous documentation is that it preserves the texts just as they were created and in the context they were created, and thus ensures the qualitative kind of validity and reliability of the research process (Greenwood & Levin, 2005; Guba & Lincoln, 2005). It seems that when it comes to structural-focused methodology, the sources' conservation is so obvious that there is little need to emphasize the importance of accurate documentation. It is impossible to imagine how structural analysis can be conducted without using text that is documented verbatim.

Data collection in structural-focused methodology can also include visual data and non-verbal text including movies, art (prints, posters, cartoons, etc.), photos, videos, graphic images, designed products, architectural constructions, and cultural products such as clothing, graffiti, etc. (Grbich, 2007). There may also be relevant analyses of musical segments (Gilboa & Ben-Simon, 2010), which under a broad definition of text can even be considered as text and narrative expression. Qualitative research in general, and research in structural-focused methodology in particular, considers these types of data as significant evidence of research, and therefore specific methods have been developed for analyzing non-verbal materials (Bach, 2007). This book does not devote a special section to the analysis of non-verbal texts, but relevant methodological sources can be found. However, it is understood that the terms delineating principles and approaches to most issues presented in this chapter and in the entire book are also relevant to non- verbal text. Some of the non-verbal sources were originally accompanied by verbal aspects which can stand on their own (movies, videos, etc.), and sometimes researchers are interested only in the literal text. In this case, in regard to the literal text, all that is mentioned in this chapter and others is fully relevant.

Research Categories/Themes

The process of creating categories in structural-focused methodology, and particularly the timing of constructing the array of categories, is largely related to the research approach. In the objectivist research approach, the categories are created in the early stages of research, subject to the theoretical perspective of the study. This category system is used by researchers to locate relevant texts and/or initiate data. There is no doubt that the same categories will be used to set the analysis of the data collected. In interpretive-constructivist and critical approaches, the creation of the category array will be implemented sometime between the data collection stage and the data analysis stage, as explained above.

Figure 17C illustrates an array of categories in research employing the structural-focused methodology.

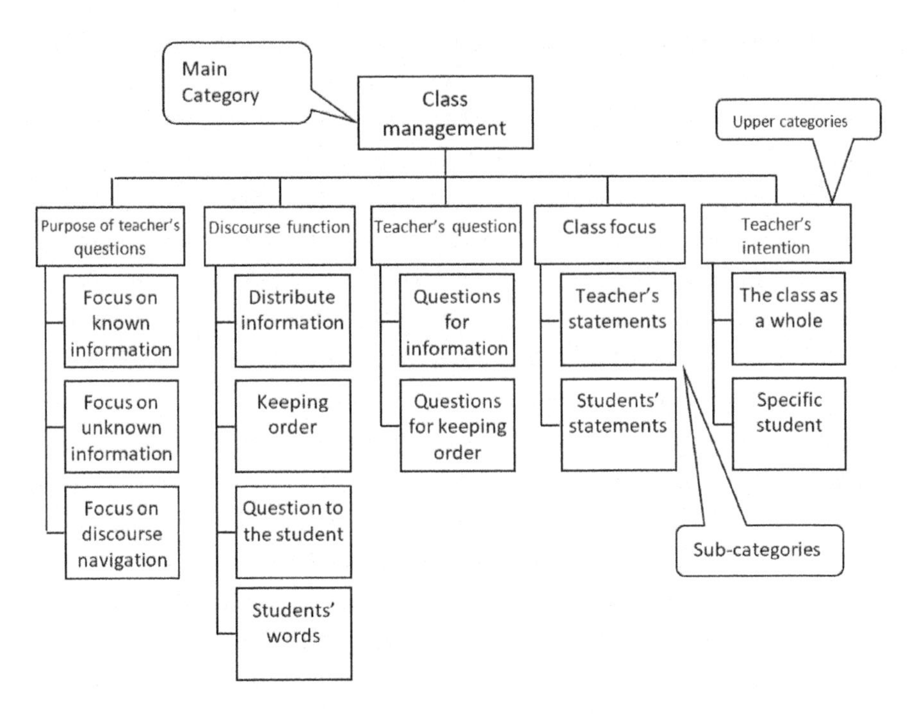

Figure 17C: Category Array of Structural-focused Methodology

The category array presented in Figure 17C focuses on the linguistic-structural aspects of the research and gives answers to questions (or parts of questions) dealing with first-order issues. Basically, qualitative researchers dealing with issues of social sciences are particularly interested in the answers to the second-order issues. The answers to the first-order issues are just a means to obtain insight into the social-cultural issues.

Data Analysis

As mentioned in previous sections, structural-focused methodology allows us to distinguish between two stages of analysis, the first-order and second-order. First-order analysis is the analysis of the linguistic structures of the texts. This analysis is usually grounded in clear, defined categories taken from structural models and theories. At the end of the structural analysis, the second-order analysis begins. At this stage, researchers give their explanation and meaning to the structural analysis. As noted, the second-order analysis is grounded in theories from the social sciences (sociology, psychology, etc.), and suggests a link between linguistic structure and understanding the functioning of humanity and society. It appears that the second-order analysis is characterized by providing a theoretical explanation for the picture resulting from the categorical array of the structural analysis.

Researchers employing the structural-focused methodology sometimes approach the process of structural analysis with a complete category array or one that is nearly complete (primarily in the objectivist approach), and sometimes the category array emerges in the process of data collection and analysis (such as in the interpretive-constructivist and critical approaches). Figure 17D illustrates an example of a category array for analysis which refers to the evaluation component of the story's structure (Labov, 1972). The evaluation component focuses on those elements in the story in which the narrator/author presents his/her point of view. The evaluation component includes several elements, some of which were noted, for example, as upper categories in the category tree. During the analysis process, researchers enter the text segments of the data into the appropriate categories.

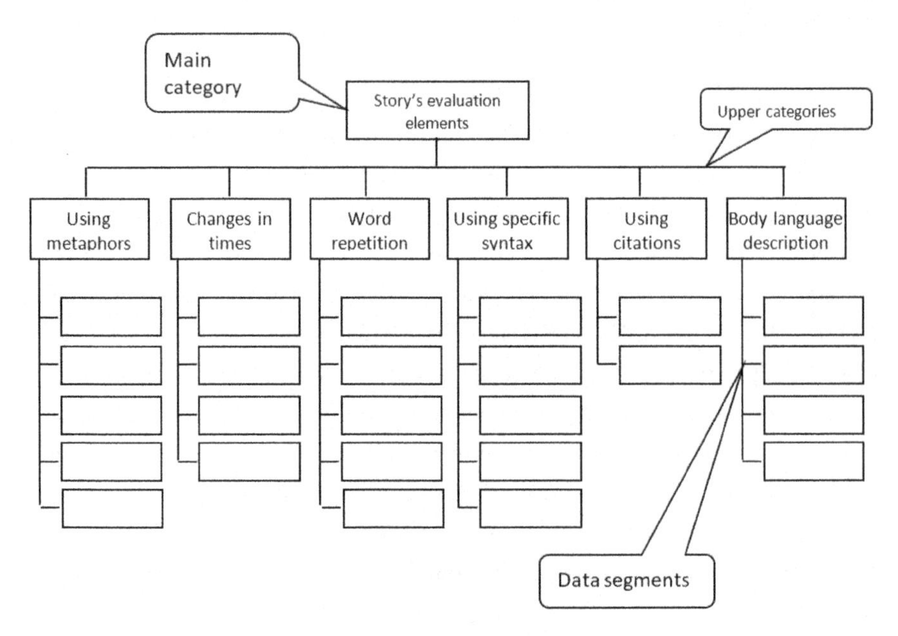

Figure 17D Inserting Data Segments into a Category Array in the Structural-focused Methodology

It is obvious that the "real" tree could be more complex, with additional categories and levels of categories. In the objectivist approach to the analysis, the category array remains relatively stable, and data segments are introduced into the appropriate categories. In the process of analysis within the interpretive-constructivist and critical approaches, the "encounter" between the data and the categories may change the category array by adding, omitting, splitting, or joining categories. In other words, the nature of these approaches causes the process of creating the category array to continue into the analysis process.

Research Report

The research report in structural-focused methodology will include all the components of an acceptable research report, as presented in previous chapters, but will express the unique characteristics of this methodology which substantially differentiate it. We can identify two major elements that distinguish the structural-focused methodology: first, that this methodology encompasses several different research types representing at least three major research paradigms. Secondly, the distinguishing element of structural-focused methodology is the fact that all three approaches rely on two groups of interconnected theories: the linguistic structural and the social science theories. Those who read research reports using a structural-focused methodology are impressed by the enormous variance between them, over and above the other methodology patterns. This book attempts to present common methodological patterns to help researchers and student-researchers find order in the great sea of qualitative research. It must be said that in the case of structural-focused methodology, the various research types may create a confusion and difficulty in finding a common ground between them.

While the final reports will include all the standard elements of the research report, the nature of the various approaches and the process of the creation of each study may also affect the writing of the report. Based on Marshall McLuhan's (1964) well-known observation that "the medium is the message," each separate approach attempts to deliver a different message about the status of linguistic structures and their connection to the phenomena of man and society. A research report employing the objectivistic approach will probably take the form of a "traditional" linear report. In contrast, reports employing the interpretive-constructivist and critical approaches will be characterized by flexibility and creativity, rather than a linear account of the research stages. The reports may flow with the message that they wish to represent. This can result, among other things, in a mixture of the components of the report, such as a combination of presenting descriptions with theories that interpret and explain the phenomenon.

As in each of the research report styles presented above, the most challenging element in writing is the component of the findings. Not only is it the largest component in terms of quantity, it is the core of the report. The findings consist, in fact, of two elements, one that is connected to the linguistic-structural theories, and the other connected to theories from the social sciences. It appears that the appropriate way of writing should be a focused-categorical description (presented in Chapter 4) with a narrative-theoretical description, i.e., a description that combines a narrative-descriptive picture with theoretical aspects (Shkedi, 2003, 2005). A focused-categorical description is a description in which the categories are the backbone of the description (Merriam, 1998), best suited for a linguistic-structural style. The narrative-theoretical description displays the interpretive aspects of the description that link between the linguistic-structural findings and the theoretical aspects originating from the social sciences.

Research Quality Standards

Each of the three research approaches illustrated in this chapter is characterized by different standards for determining research quality. However, all studies of this methodology are based on objectivist linguistic-structural aspects, and criteria derived from valid linguistic-structural theories. It seems that in terms of linguistic-structural aspects, it is possible to use conventional terms of "validity" and "reliability," and to set the objective for a standard of quality of these aspects of the study (Lincoln & Guba, 2000; Schwandt, 2000; Smith & Hodkinson, 2005). All references to the concepts of "validity" and "reliability" presented in the chapter dealing with criteria-focused methodology are also relevant here (see Chapter 6).

With regard to the cultural-social aspects and their link to the linguistic-structural aspect, the sources of the theories in each of the three approaches are different, which affects the standards for quality research in kind. In objectivist-approach studies, where the theoretical perspective is defined at the beginning of the research regarding the area of social sciences in addition to the linguistic-structural area, we can use the terms "validity" and "reliability" in their objective form. The picture is more complex regarding the other two approaches.

In the interpretative-constructivist and critical approaches, the concept of "objective" does not apply, and all knowledge, descriptions and explanations are necessary perspectives, i.e., they depend on the perspective of the researchers (Lincoln & Guba, 2000; Schwandt, 2000; Smith & Hodkinson, 2005). On this basis, we can apply the concepts of "validity" and "reliability" in their constructivist sense. In order to argue for validity and reliability, the researchers should present the research theoretical perspectives alongside the chain of research evidence (interviews, observations, documents, etc.) for each stage of the study, to examine the correlation of the analysis process with the theoretical perspectives (Huberman & Miles, 1994; Mason, 1996; Shaw, 2010a).

As in the case of validity and reliability, the concept of "generalization" (external validity) has a unique meaning in structural-focused methodology. As suggested in previous chapters, we can find relevancy in the concept of "generalization" in qualitative research. As such, it appears that there are three types of qualitative research generalizations: case to case, analytic, and generalized to the population

(Firestone, 1993). Generalization from case to case may be relevant because the accurate analysis used in this methodology is the basis for constructing a wide, in-depth research description that gives researchers and readers a basis to compare and examine the case under study to other cases taken from their daily lives or literature.

Generalization to population will be relevant as the population sample grows, which sometimes becomes possible in this methodological pattern primarily when using the objectivistic approach (Arksey & Knight, 1999). Analytic generalization would also be relevant, because this methodology is based on a clear, transparent theoretical concept which guides the research. In the objectivist approach, it is a predetermined theoretical system, and in the two other approaches, the theoretical perspective is developed throughout the study. Thus the study findings may also be valid in other cases that can be explained according to the same theoretical concepts or be compared to other cases explained by the same theoretical system (Dey, 1993; Firestone, 1993; Marshall & Rossman, 1989).

Triangulation of sources and methods of data collection – i.e., "to use different methods in different combinations" (Fontana & Frey, 2005, p. 722) may enrich the quality of the research and strengthen the argument for the validity, reliability and qualitative generalization.

CHAPTER 18

SELF-FOCUSED METHODOLOGY

In many aspects, self-focused methodology is an expression of a revolution in thinking about research methods. The positivist research thinking, which led the research domain of human and social sciences for decades and remains the dominant paradigm among many scholars, completely rejects the idea that a person can investigate him/herself. If objectivity is the required standard for valid research, it is difficult to consider how one can investigate himself, because in self-focused methodology one cannot claim any kind of objectivity. Choosing self-focused methodology means abandoning objectivity as a central and required element in the research, and perhaps even questioning the possibility of any objectivity whatsoever in the study.

Beginning in the 1970's, changes evolved in the perceptions of research that served to create the groundwork for the growth of self-focused methodology and its legitimacy in the research community (Bullough & Pinnegar, 2001). Lawrence Stenhouse (1988) calls the practitioner-researcher an "artist." (Stenhouse, 1988). The term "participant observer" (Jorgensen, 1989) has become largely a characteristic expression giving legitimacy to the involvement and empathy of the researcher with the research population, as reflected primarily in the participant and critical-focused methodologies presented in previous chapters. The research conception of self-focused methodology is not satisfied with the "observer as a participant," but positions the "participant as an observer" as the focus for the research. Participant observation is seen as not having enough involvement in the phenomenon under study, because the observers are not an authentic part of the phenomenon but are involved only for research purposes (Florio-Ruane, 2001).

As shown in Chapter 12, critical-constructivist focused methodology is based on the assumption that participants are largely unaware of much of the power and knowledge that directs their actions and perceptions. The researchers in critical-focused methodology are considered to have an advantage over the research participants by possessing the opportunity to expose the pervading cultural-political-social facts and forces, and thus act to emancipate the people and their discourse from their bounds to these forces. Self-focused methodology, however, is based on the assumption that the participants themselves have an advantage over the researchers, and they are those who can best display their own conceptions, feelings and knowledge, including revealing their unconscious knowledge and perhaps even exposing the political-social forces that influence their thinking and actions. Self-focused methodology, therefore, seeks to suggest a research methodology that helps the "self" to consider his/her own self-knowledge and what lies behind the actions and ways of thinking, in order to improve them. The researchers in self-focused methodology "turn the analytic lens on themselves and their interactions with other." (Chase, 2005, p. 660)

Although researchers in self-focused methodology focus on their self-knowledge as an object of the study, various researchers differ in how they perceive their self-knowledge. There are those who assume that they are entering into the research with an evident theoretical perspective that reflects their view of their self-knowledge, while others believe that this theoretical perspective grows and becomes clarified throughout the research process. It is unlikely that the researchers in self-focused methodology are bereft of any preconceived theoretical perspective about themselves or their own knowledge. We can reasonably assume that everyone carries a certain theoretical identity as part of his/her personal and professional identity, and this clearly affects the perception of their research. However, sometimes self-focused methodologists act like researchers employing participant-focused methodology and inquire about their theoretical perspective in the course of the research process. Indeed, it is unlikely to believe that there are researchers that enter the study with a clear theoretical perspective, as characterizes the criteria-focused methodology.

As for the tension between the "recruitment" of intuitive inquiry skills and analytical research skills, it appears possible to position self-focused methodology in a certain range between the two poles with a tendency toward intuitive inquiry skills. There are researchers who tend to give greater expression to their analytic capabilities, i.e., researchers who enter into the research with predetermined criteria that reflect their perception, while other researchers may tend to give more expression to their intuitive research skills. Between these two extremes are a great variety of research options.

Self-focused methodology brings the concept of "the human as an instrument" (Lincoln & Guba, 1985) which identifies the human as the ultimate research tool. As explained in Chapter 1, the perception of the human as a research tool assumes that there is no objective external research tool as sensitive as humans themselves (Hogarth, 2001). This view expresses the concept of "connoisseurship" that Eisner associates with qualitative research (Eisner, 1991). Eisner sees the researcher as a human with an expertise for most sensitive natural research characteristics, to a degree which no objective tool can reach. This is the most personal kind of expertise achieved by experience that is very difficult to transfer to others, such as the expertise of wine tasters (Davis, 1996). Professionalism and expertise are acquired through definitive participation and involvement in a certain area. This criterion provides the professional practitioner-as-researcher priority over all external researchers, even those who adopt methods that involve them in the phenomenon being studied, such as "participant observers" (Jorgensen, 1989).

Empowering the Professionals over the Academic Researchers

The legitimacy of studies based on self-focused methodology also derives from the protest against the hegemony of the university and its academic researchers over the sources of knowledge and its creation (Cochran-Smith & Lytle, 1999). This tendency of protesting against the hegemony on the knowledge was the foundation for two different qualitative research traditions, the critical tradition, clearly focused on the effects of hegemonic forces and how to become emancipated from them (see Chapter 7), and the tradition of self-research, which seeks to give expression to the emancipation of practitioners (Carr & Kemmis, 1986; Davis, 1996). This research tradition embodies "a recognition that the generation of knowledge about good practice and good institutions is not the exclusive property of universities and research and development centers, a recognition that practitioners have theories too, that can contribute to the knowledge that informs the work of practitioner communities." (Zeichner, 1993, p. 204).

The tendency to see practitioners as the researchers of their practice is also directed by the idea that it's worthy to enhance the world of practitioners, promote them and give them more power (Kemmis & McTaggart, 2005; Lauriala & Syrjala, 1995). The assumption is that turning practitioners into researchers focused on their practice would improve their self-image and also their public image (Cochran-Smith & Lytle, 1999). Self-focused methodology is thus "a reaction against a view of practitioners as technicians who merely carry out what others, outside the sphere of practice, want them to do, a rejection of top-down forms of reform that involve practitioners merely as passive participants" (Zeichner, 1993, p. 204).

Further justification for the legitimacy of self-study is the desire to give expression to the authentic "voice" of the practitioners. The assumption is that the "voice" which is heard in theoretical and research literature is that of the academic experts who claim to faithfully represent the voice of the practitioners (Guba & Lincoln, 2005). The phrase "voice" is a metaphoric use taken from feminist studies, where it is argued that the existing studies do not represent the authentic voices of women (Lloren, 1994). In self-focused methodology, the professionals make their voices directly heard, without the mediation of researchers (Elbaz-Luwisch, 2002).

Theoretical Paradigms in Self-focused Methodology

The research paradigms that characterize self-focused methodology may be quite varied, similar to that of structural-focused methodology, as described in the previous chapter. There can be an objectivist, interpretive-constructivist, or critical approach, depending on the perspective of the professional researchers. When researchers arrive at the study with a clear theoretical approach and predetermined criteria, their perception will be very close to the positivistic or post-positivist paradigm, namely the objective perspective (Denzin & Lincoln, 2000, 2005a; Heikkien, 2002; Lincoln & Guba, 2000). The choice of a research perspective that is closer to the positivist approaches can result from at least two orientations. One is the belief of the practitioner-researcher that appropriate research needs to be accepted by the academic community. This assumption sees the academic community as the authority in matters of research and theoretical issues, and seeks to gain legitimacy from this community. A second possibility for choosing an objective approach is that the practitioner-researcher adopts the common theoretical framework without reflective examination or even an awareness of the appropriate way to understand the phenomenon under investigation. In these two cases, the practitioner-researcher arrives at the research project, whether consciously or unconsciously, with the positivistic point of view.

In other cases, the practitioner-researcher will indeed come with a stable theoretical framework, but with a greater willingness to examine it and be flexible. In these cases, we can say that their paradigmatic approach is close to the post-positivistic perspective (Guba & Lincoln, 1998; Philips & Burbules, 2000). In all cases discussed, the practitioner-researcher adopts a set of guideline criteria at the beginning of the research process. Those who adhere to a positivistic perception will refer to stable, unchanged criteria that guide the research throughout, while those who adhere to a post-positivistic perception will be more flexible and allow the examination and revision of the predetermined criteria throughout the research project as they receive feedback from the research field.

The interpretive-constructivist approach is characterized by a willingness of the researchers to be attentive to the phenomenon being studied and to use the collected data to deduce the whole picture and the theoretical framework that reflects the phenomenon under investigation. Despite the fact that researchers do not enter the study as a blank slate, but with some kind of theoretical assumptions, they are willing to examine and modify the theoretical framework according to what emerges from the data of the study. As emphasized in Chapter 5 vis-à-vis participant-focused methodology, there is a process of dialogue between the pre-assumed perspective of the researchers and that of the gathered data, with researchers forming their views accordingly. The case of self-focused methodology is much more complex, since it is a dialogue within the practitioner- researcher's self. The practitioner-researcher enters the research project with a particular theoretical approach, sometimes consciously or less consciously, or perhaps not aware at all. It seems that the degree of their openness to examine their pre-assumptions toward the phenomena under inquiry is actually determined by their willingness to adopt either the interpretive-constructivist or the objective approach.

Self-focused methodology, characterized by a critical-theoretical perspective, assumes that the spontaneous observation of the world is characterized by erroneous assumptions that are affected by the pervading political-cultural-social powers (Carr & Kemmis, 1986; Guba & Lincoln, 1998; Kemmis & McTaggart, 2005). Adopting this assumption, they believe that their role as practitioner-researchers is to identify the forces that distort and limit free expression. Awareness of the existence of false ideas and ideologies is a necessary first step towards the emancipation from these inhibitory forces (Phillips & Hardly, 2002; Travers, 2001).It seems that it is easier to be aware of the influence of hegemonic powers when investigating a phenomenon that is outside the internal world of the researchers than to accept the existence and influence of these forces on the researcher himself.

Due to the practitioner-researcher's difficulty to recognize his motives for thought and action, which are hidden in most circumstances, the self-study research becomes a participatory research in many cases. As such, the professionals that want to investigate their activities join together with experts from academia and conduct collaborative research (Kemmis & McTaggart, 2005), a seemingly perfect winning combination. The objects of study, the practitioner-researchers, guarantee absolute proximity and involvement in relation to the phenomenon under study (the "self" of the practitioner-researchers), and the academic experts ensure the distance, examination and reflection necessary for research. The problem that might arise in studies based on such a partnership is the lack of a guarantee to keep the symmetrical collaboration between the two collaborators. In many cases, the academics, because of their professional authority and prestige, take over the research, even if unwittingly, and the idea to raise the authentic voices of the practitioner-researchers is not fulfilled. In this case, we cannot consider this to be self-focused methodology.

Self-study Genres

There are three salient genres of self-study: autobiography, self-case study, and action research. Each genre is unique to self-focused methodology, and each may be expressed in the three research approaches outlined above - objectivist research (positivistic or post - positivistic), interpretive-constructivist research, and critical research. This creates a very large range of methodologies.

Autobiography is a biography that people write by themselves and about themselves. The autobiography can be seen as a genre of narrative research, research that is based on human's verbal reports on their self-experiences (Freeman, 2007; Josselson, 2004). The central axis of the autobiography is descriptive writing, where the authors present their personal experiences as they see them at the time the autobiography is written (Elbaz-Luwisch, 2002). Although the writers rely on documentation they possess or collect specifically for the purpose of writing, the autobiography is not expected to show a historical or scientific truth, but something that could be called "narrative truth" (Bruner, 1990; Gudmundsdottir, 1995). The authors wish to give a story structure to their life experiences by writing chronologically, presenting a plot as a connecting story line or presenting an academic conceptualization which creates a theoretical observation angle as a linking story line (Elbaz-Luwisch, 2002). Autobiography, which is based on the human's self-reflection, is perceived as a genre that may empower the writer (Davis, 1996). Bookcases are full of autobiographies written by politicians, artists and others who summarize their life experiences, in many cases at the close of their professional, public or artistic career. With the rise of constructivist-narrative research approaches, autobiographies receive more and more legitimacy as a research genre that fills an important aspect of enhancing knowledge in human and social sciences. Another self-biographical genre equivalent to an autobiography is a personal diary, which documents the experiences and thoughts of the writers in real time. The personal diary is written chronologically, but may be edited for publication.

Readers of autobiographies are not generally exposed to the methodological explanation of the compilation of an autobiography, as is common in any research product. Similarly, there are generally no unique theoretical chapters clarifying the research perspective of the autobiography, as in any other research report. In this respect, it is difficult to classify the autobiography as a research method. In Part One, we emphasized the difference between a document that can be defined as research and one that does not meet the criteria. To be classified as research, the main criterion is the existence of a transparent control system. In this aspect, it seems that many autobiographies may not meet the conditions for research. By releasing the autobiography to public scrutiny, accompanied by authentic documents, references to authentic resources, and all other necessary methodology requirements, the initial conditions for becoming a research work are filled (Whitehead & Fitzgerald, 2007). It also appears that there are many autobiographies based on memories and impressions and less on authentic sources, which makes it very difficult to identify them as research. In any case, it appears that among the three genres of self-study, the autobiography is the weakest research genre. The autobiography walks a fine line - sometimes more and sometimes obviously less clear - between research and a literary production. A literary work based on autobiographical elements of the writer, as opposed to an autobiography, is committed to the "literary truth" of the author. Autobiography, by contrast, when considered as a research work should illuminate the author's narrative truth, and should be visible, explicit and subject to the criticism of readers.

The Self-Case Study is a research that focuses on and investigates the activities that practitioner--researchers experience by themselves. Although these characteristics have a significant impact on the structure and process of conducting the study, this kind of research is a case-study, with the features of case studies applying here as in any other non-self case study. "The case study is an intensive description and analysis of a phenomenon or social unit such as an individual, group institution or community" (Merriam & Simpson, 1984, p. 89). "Case studies are differentiated from other types of qualitative research in that they are intensive descriptions and analyses of a single unit or bounded system." (Merriam, 1998, p. 19). The case study in general and self-case study in particular focus on a space that is limited in scope, unlike ethnography, for example, that deals with society and culture in the broadest sense of the possible reference. A case study can investigate a program, event, or occurrence and be focused on people, organizations or societies (Stake, 2005; Yin, 1994). Unlike the autobiography which is generally done in retrospect, a self-case study, like most research methodologies, is constructed in pre-planned stages.

The written report of a self-case study is similar in structure to the "conventional" case study. This structure will "set a problem, gather data, analyze the data, and interpret the data for the purposes of reaching some conclusions about the problem set." (LaBoskey, 1994, p. 22). All the common elements of the research report will be included here: a literature review, research or assumptions and/or hypotheses, research questions, a methodological explanation, findings, discussion, summary and conclusions (Arksey & Knight, 1999). But unlike the "conventional" case studies in which the subject of the research and its motivations arise primarily from the interest of the academic community or what the researchers belonging to the academic community perceive as interesting and relevant, the self-case studies grow out of the self-interest of the practitioner-researchers. This is usually a result of their self-coping with personal-practical problems, although in some cases it arises from theoretical interest. In conclusion, the self-case study is characterized by all the criteria that define any research. The difference between a self-case study and a "conventional" case study lies in the methodology, as will be shown later in this chapter.

Action research is defined by Zeichner as a "systematic inquiry by practitioners about their own practices." (Zeichner, 1993, p. 200). A more focused definition which distinguishes between action research and other genres of self-focused methodology is provided by Carr & Kemmis (1986): "a self-reflective enquiry undertaken by participants in social situations in order to improve the rationality and justice of their own practices" (p. 162). Action research was initially associated with social psychologist Kurt Lewin, who designed the circular structure of systematic action research, centering on dealing with practical problems and a link between action and research. The action research type of the Kurt Lewin School is focused on the scientific-positivist approach (Zellermayer, 2001; Kemmis & McTaggart, 2005; McNiff, 1988). Action research is distinguished from other genres of self-focused methodology and other research in general by being motivated and starting with a reference to the practitioners and practitioners' activities, rather than beginning with a theoretical problem and ending in relation to practical activities. The practical activity can be accompanied by theoretical insights, but they are not generally the motivation for the action research process. Unlike "conventional" qualitative or quantitative researchers "who aim to understand the significance of the past to the present, action researchers aim to transform the present to produce a different future." (Carr & Kemmis, 1986, p. 183)

The first phase of action research is to identify the problem existing in the practitioner's action (Nixon, 1987). This is a crucial step in the dynamics of action research: a proper identification of the problem will allow the practitioner-researcher to correctly design the case study, to deal with the problem and to improve it. On the contrary, an erroneous identification of the problem may shift the action in a non-productive way. It is important to note that the process of identifying the problem is challenging and complex. Sometimes what appears to be one problem may be only a visible symptom of a deeper hidden problem (Schwab, 1969). As with any self-focused methodology, the paradigmatic assumptions of the practitioner-researcher become the basis for exposing the problems. If the assumptions of the researchers are positivistic or post-positivistic, they will judge the practical problems according these perspectives, or with other perspectives if the assumptions are interpretive-constructivist or critical (Carr & Kemmis, 1986; Kemmis & McTaggart, 2005). The practitioner-researchers are required to exert a major effort, both intellectual and emotional, to expose the real problems. As practitioners (as distinct from researchers), they are not accustomed to adopting the observation angle of the researcher, which should be characterized by the ability to look at the phenomenon under study simultaneously from both close and distant points of view (Lloren, 1994).

After the definition of the problem, the spiral direction of the research action begins at the planning stage. The planning stems directly from identifying the problem, and seeks to deal practically with the identified problem. The next stage is the action itself, or in other words, the application stage. During the application stage, the data is collected for carrying out the next step of action research, the stage of reflection and evaluation. The resulting circle line is apparently a linear track in which each step is derived from the previous one. But "in reality the process is likely to be more fluid, open, and responsive. The criterion of success is not whether participants have followed the steps faithfully, but rather whether they have a strong and authentic sense of development and evolution in their practices, their understandings of their practices, and the situation in which they practice." (Kemmis & McTaggart, 2005, p. 563). This is actually a spiral two-way track. In each stage, the practitioner-researcher can go back and identify the data and the decisions and change them if necessary. The reflective and evaluation phase ends the first cycle. However, action research may include several cycles, and one cycle's conclusions lead to the beginning of another cycle (see Figure 18A). The additional cycle starts with constructing a new plan, and sometimes with a re-examination of the problem, followed by other steps of action research. The number of the case study cycles is determined by the practitioner-researchers according to their understandings and needs. (Lauriala & Syrjala, 1995; McNiff, 1988).

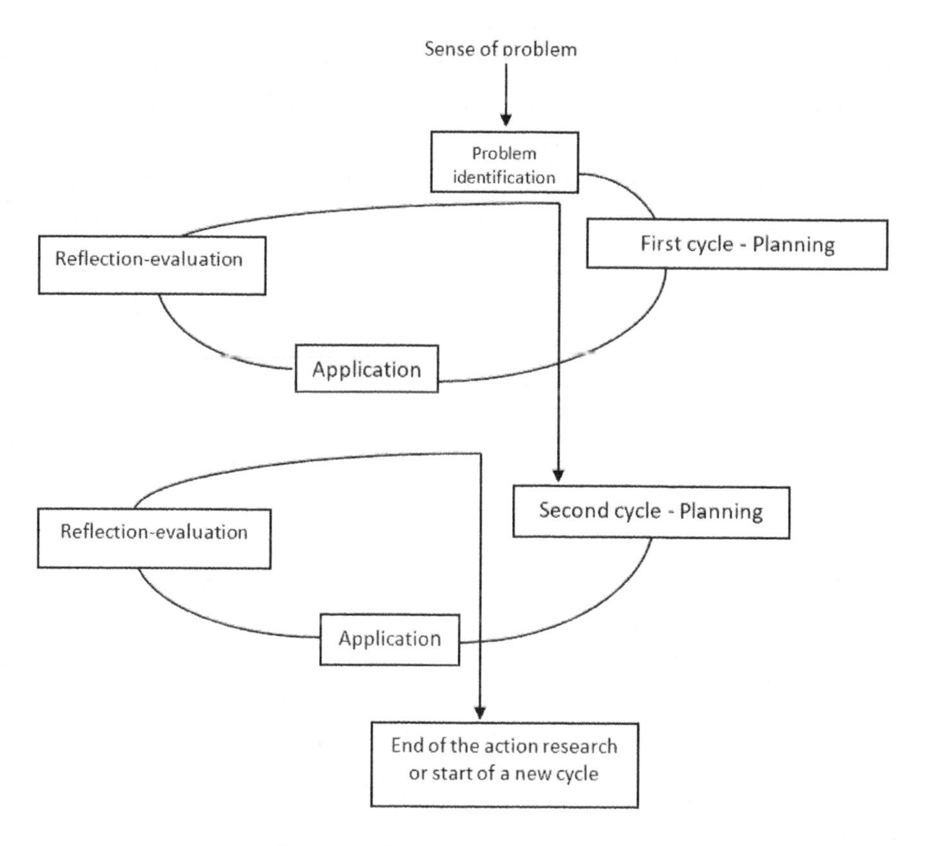

Figure 18A: The Action Research Cycles

In addition to self-action research, performed entirely by the practitioner working in the field, there is a pattern of collaborative action research which combines practitioners with academics (Kemmis & McTaggart, 2005; Lauriala & Syrjala, 1995). As mentioned above, it is doubtful whether we can continue to consider it as self-focused methodology if the academic partners do not explore their own self-actions and thought, but, as in other "conventional" studies, are external elements to the process occurring, even if their involvement in the research process is particularly remarkable.

The Study's Theoretical Perspective

Studies employing a self-focused methodology seemingly appear to lack a theoretical perspective, or at least display no visible theoretical perspective. This is especially true for those who choose the genre of autobiography or action research. Of course, this assumption is incorrect. Qualitative researchers in each methodology do not begin the study with an empty slate, but always approach the research with pre-assumptions. Our perception system is such that we always look at the phenomena around and within us with some theoretical perspective, even if we are not always aware of the fact. This theoretical perspective derives from the researcher's personal and professional life, from previous studies and previous readings, stemming from a personal world view and ideology and even from prejudices (Charmaz, 1995; Strauss & Corbin, 1990).

Customarily, all research begins with a literature review - a stage in which the researchers conceptualize their theoretical perception - even if this literature review may change during the course of the study. Researchers who have conducted a **self-case study** develop their own research, as is common in "conventional" studies, by presenting a theoretical perspective. In the case of **autobiography,** the researcher-writers probably completely skip the process of displaying a theoretical perspective. But even the autobiography's authors, as other researchers and authors, are guided by theoretical conceptions of which they are generally unaware. It seems that proper autobiographers, if they want their work to meet the criteria of research and not only of a life story, should identify the theoretical perspectives that guide them. They should do this both for themselves, so they can examine what they wrote, and to represent it to the readers.

The case of **action research** is slightly more complex. Action research begins, as stated, with a problem originating from practice, not theory, and focuses on finding a practical solution. Identifying the problem is a process that clearly involves a theoretical-conceptual consideration, and different theories may lead to a different identification. It thus seems that awareness and definition of a theoretical perspective, while aided by a rich literature review, will help in defining the problem and will continue to assist the entire action research process (Carr & Kemmis, 1986; Guba & Lincoln, 1998).

Self-focused methodology, contrary to "conventional" research dealing with external subjects, is a process of reflection in which the practitioner-researchers conduct a dialogue with themselves about themselves. It seems that this is an extremely challenging process, and the use of a theoretical perspective, which will be constantly identified, may contribute to the quality of the process.

The practitioner-researcher in self-focused methodology that adopts positivistic or post-positivistic approaches will tend to identify the theory before starting the research process, and use external theoretical criteria. Practitioner-researchers carrying out the research with an interpretive-constructivist approach will tend to conduct a dialogue between the problems and issues arising in the field and a possible theoretical framework. Of course, they will not enter the research with a blank slate, and probably possess theoretical insights which guide them in finding the meaning of the situation, but they are willing to examine their theoretical assumptions and change them in light of insights arising from the field. Practitioner-researchers carrying out the research with a critical approach will adopt a critical theory that enables them to look beyond what's seen at first glance. However, as constructivist researchers, they will be ready to confront the theoretical examination through which they look at reality.

Formulation of Research Questions and Assumptions

The **autobiography** may seem totally devoid of research questions. It appears that autobiographers write their autobiographical life stories spontaneously without any deliberate questions. This does not, of course, distort the picture in any way. Autobiographers actually have research questions that direct their writing, even if these questions are not placed before the reader. Generally these questions are not formulated as research questions, and in many cases are not articulated at all, but they definitely guide the autobiographers in their work, albeit often without their awareness. In many cases, these questions are formulated in the introduction of the autobiography in a type of general statement indicating the goals of the autobiography. It seems that if the autobiographers want their work to be perceived as a kind of research (as opposed to impressions and associative memories), it should be centered around questions or at least around a table of contents.

In a **self-case study,** it is common to formulate research questions as in other research methodologies. Therefore, the research questions of a self-case study will be similar to those of their counterparts in "conventional" studies, as explained in previous chapters. The nature of **action research** suggests a more complex process of postulating research questions, different from all other qualitative methodologies. As mentioned above, action research, as distinct from "conventional" studies, begins with a concrete practical problem rather than theoretical questions. The practitioner-researchers invest full attention in identifying the problem. They may employ various epistemological approaches-- objectivistic (positivistic or post-positivistic), interpretive-constructivist or critical-- and they all assign a different significance to the theoretical perspective in identifying the problem. Adherents of positivist or post-positivistic approaches examine the problems with theories that originate from academic literature, and will rely on these theories to try to understand and define the problem. Researchers who employ critical approaches try to look at the problem with critical theories. The interpretive-constructivist researcher will conduct a dialogue between the practical problem and theoretical concepts, until achieving adequate problem identification.

Problem identification in action research is the appropriate way for question formulation. It is obvious that various practitioner-researchers may define the research problem differently from their colleagues dealing with what seems to be an identical problem. The unique personal experience of each of practitioner-researcher, the context in which the action is studied, and the research approaches employed are the basis for each researcher to identify the problem and formulate the questions. It is reasonable that in the beginning the practitioner-researchers will formulate the research questions broadly, and as the action research advances they will sharpen their research questions and formulate more specific questions. An example of a broad question:

How can we increase the motivation of students to learn mathematics?

Such a question could be formulated after the practitioner-researcher has identified the learning problem to be one of student motivation. In the process of the action research, it may become clear that the sources of the motivation problem are connected to students' cognitive difficulties, and the research question can be honed and changed: **How can we help the students overcome their cognitive difficulties involved in learning mathematics?**

The Study Design and Population Selection and/or Field Research

This issue seems completely irrelevant to **autobiography**, since autobiography is a way of retrospectively observing past life experiences. As for the **self-case study**, the issue of study design and population selection and/or field research will be largely identical to that of the "conventional" methodology presented in previous chapters. The study design and population selection and/or field research appropriate to **action research** are very unique. The first step is to set an action research field whose population is adjacent to the workable area of the practitioner. After defining the research problem, research questions, and the action research area and its population, the practitioner-researcher is ready to determine the array of the action research. As shown earlier in Figure 9B, the array of action research is a spiral, and consists of one or more cycles which each include three components: planning, action (implication) and evaluation. The practitioner-researchers can initially determine if they wish to create one or several cycles, but in many cases the decision on the scope of the action research will be delayed, depending on the picture resulting from the first cycle of the action research. The first phase in each action research cycle is the planning stage, based on the problem and questions.

After completing the planning stage, the practitioner-researcher is ready to implement it in the field, which may require a great deal of flexibility. Practitioner-researchers may change their planned activities in light of the experiential feedback received from the field, and perhaps as a result of reflection during the activities. Thus the cycle of action research not only moves forward in circles, but also reverses in light of the immediate feedback obtained. The action stage is accompanied by documentation which can be carried out in different ways: log- entries, recording or photography of these activities, collecting relevant documents, and more. When the action phase ends, or in many cases when part of the action ends, the practitioner-researcher prepares for the next step, the phase of reflection-evaluation. At this stage, the activities are analyzed with a tight reference to the identified problems and questions, in order to examine whether and how the performed activity actually succeeded in coping with the problem and questions

In fact, feedback received from reflection-evaluation determines the continuation of the action research and shapes the future of the research design. Here there may be several options. Suppose reflective feedback shows that the action is successfully coping with the problem. In this case, the practitioner-researchers may decide to end the action research. But perhaps the action satisfies the practitioner-researcher, but he nevertheless decides to go through another cycle of action research, for example, to examine the relevance of the same activities for another population (e.g., teachers may apply the action research in more classes). In other cases, the practitioner-researcher may find, for example, that the action did not successfully deal with the problem. In such a case they may plan a second cycle with different activities, totally or in part. In other cases, the practitioner-researcher may come to the conclusion that the problem identification was incorrect, in part or in full, and return to the previous step to try to re-identify the problem. The second-circuit components are similar to those of the first circuit. Changes may apply not only in planning and action, but also in the manner of documentation and data collection (to be discussed in the next section). Similarly, the practitioner-researchers can continue to a third cycle of action research and even more, until they feel that the action research has achieved its purpose (i.e., has managed to cope with the problem) and/or exhausted itself.

Data Collection Methods

The concept of "the human as an instrument" (Lincoln & Guba, 1985) reaches maximum expression in self-focused methodology, which places the practitioner-researcher as the main research tool. In the case of self-focused methodology, the boundaries between formal data collection tools and non-formal processes for data collection are indistinct. Data collected by practitioner-researchers often seems to have been gleaned through impressions, insights and feelings, and obtained from the internal dialogue held within the practitioner-researchers themselves.

Researchers involved in writing an **autobiography** base their work on data accumulated in the past, primarily documents or personal notes, memories preserved in their minds, or sometimes with the aid of contextual data acquired from books, the Internet or even conversations or interviews with relevant people. The autobiographers themselves determine which data will be collected, its significance and how to present it, all according to the autobiographers' internal discourse. In order to ensure the control of the autobiographers over their writing process and assure the quality of the autobiography as research, researchers should be aware of the process of selecting data and interpretation, examine the process and perhaps even introduce it to the readers.

The data collection process in a **self-case study** is largely similar, if not identical to the data collection process in "conventional" case study, as explained in the previous methodology chapters. The fact that researchers focus their research on events in which they themselves were involved, poses a significant challenge in the data collection process. It seems that the researchers in a self-case study should commit to a greater degree of dependence upon formal procedures of data collection, enabling control and re-examination by the researchers themselves and by colleagues so that the process will meet the two criteria suggested in the first part of the book: closeness, involvement and empathy, resulting from intuitive inquiry skills, and detachment, reflection and control, resulting from analytical research skills. (Maykut & Morehouse, 1994; Woods, 1996). Self-case study researchers will utilize common data collection tools in qualitative research-- interviews, observations, documents, artifacts and field notes-- and may add a research journal/log which will show the phenomenon under investigation from a personal perspective.

As shown in previous sections, **action research** is a more distinct methodology than other genres of self-focused methodology. Collecting data in action research has a dual purpose: to serve both the action planned and implemented and to support the associated research. Action research begins as a process of defining the problem, and consequently in planning the activities. This is a process based on the past experience of the practitioner-researcher and perhaps even of colleagues. Since this is indeed information coming from the self but focusing on the period before the action research, it appears that the data is based mostly on the memories of the practitioner-researchers, and less on data collected in formal procedures. They can rely, of course, on relevant documents and objects collected in the past and use them as data for making decisions. It seems that the inner dialogue within the practitioner-researcher characterizes this phase of data collection more than any other phase of action research.

The next steps of action research, where the practitioner-researcher is required to reflect upon and evaluate the planning and the action that took place, could be based on initiated processes of data collection (Silverman, 2006) which are more formal by nature. Here we use recording devices and/or video cameras to document the processes, and collect the documents and objects. Moreover, action research can also be accompanied by interviews with participants, in order to get information on how they perceive the process (Flick, 1998; Hugh-Jones, 2010; Seidman, 1991). In any stage of action research, the practitioner-researcher should be encouraged to utilize data collection tools such as personal diaries, notes and impressions, and assign this data a significant place.

The fact that self-case study and action research focus on the practitioner-researcher's "self" requires them to conduct a strict data collection process and be able to make a connection between their uncontrolled self-impressions and the formally-controlled data collection processes. This can be done by making certain to record all data collected, including photography and/or recorded interviews and observations, transcriptions of interviews and observations, and word-processed copies of diaries and all other lists/records used as data sources (Clandinin & Connelly, 2000; Jorgensen, 1989; Shaw, 2010a). This procedure grants legitimacy to the research process.

Research Categories/Themes

In the process of writing an **autobiography,** it is doubtful whether the authors consider the purpose of stated categories/themes as a goal for themselves (and perhaps for the readers). However, as noted above, it can't be assumed that the autobiographers are not using any kind of guide for collecting, sorting, and examining data, and writing the autobiographical story. It is assumed that many of the processes of creating the categories/themes presented in the book reflect the actual work of autobiographers, even if they are unaware. Conductors of **self-case study**, however, act as any qualitative researchers, and set categories throughout the research process. The timing of creating categories (before data collection or following the analysis process) and the extent of the static or dynamic state of the categories is determined according to the research paradigm of the researchers and to the nature of the case study they have investigated (Polkinghorne, 1995). See previous chapters for an explanation of the relevant process for creating categories.

The process of creating categories in **action research** is different from the process in "conventional" qualitative studies. Unlike the cases described above, the categories refer not only to the research process, but are also related to the action. Categories are formulated together with identifying the problems or due to the identification of the problems. They should reflect the problems' components and serve as the foundation for the design of the action and its examination. Thus, in action research the categories are determined at the beginning of the process, at the stage of identifying the problem, and emerge from the discourse between the theoretical perspective and the problems' components. Choosing a theoretical perspective may be different in various research approaches: those adhering to a positivist or post-positivistic approach will come to the data with a pre-formulated theory, and will utilize the theory to create the categories and match them to the data. Those employing the critical approaches will look at the data using critical theories, and those adhering to an interpretive-constructivist approach will conduct a dialogue between the data and possible theories until they reach the necessary balance. In the end, along with identifying the problem, the array of categories will be formulated, even if different research paradigms may lead to a different process of creating categories.

Once the categories are formulated, they will continue to accompany the action research process during all its stages. The categories will form the basis for planning and examining the actions and activities, and through them the practitioner-researcher can conclude if the action research dealt properly with the problem and/or whether it is necessary to continue to another cycle of action research. Even though the array of action research categories is determined at the beginning of the process alongside identifying the problem, the categories may change during the study according to feedback received from the planning, action and evaluation of the action. Figure 18B describe the changes in the category array during the action research process.

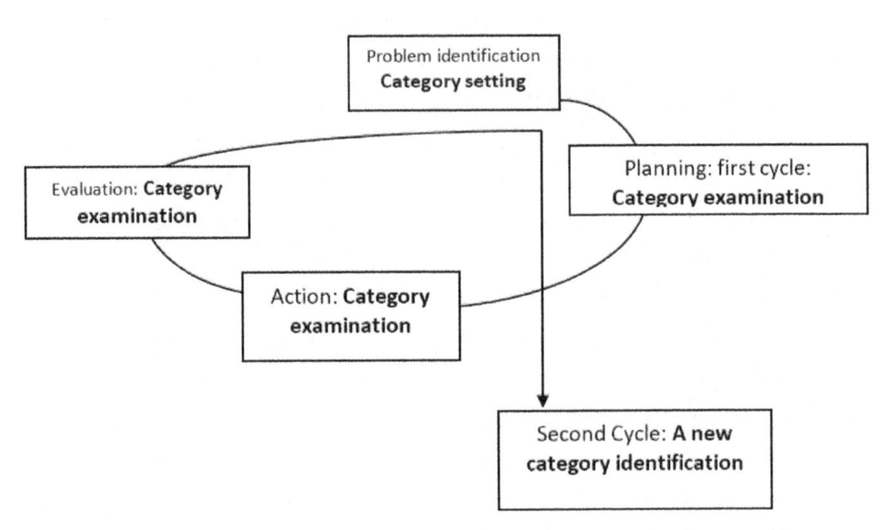

Figure 18B: Conducting Changes in the Categories during the Action Research Process

Data Analysis

In the process of writing an **autobiography,** it is difficult to point to a formal process of analyzing data (although obviously autobiographers analyze the materials, though not in a formal and transparent manner, and in many cases without being aware that they are actually performing an analysis process). In **self-case studies,** the process of data analysis is compatible with the researcher's paradigm, and is described in the previous methodological patterns. The main point that distinguishes the self-case study from "conventional" methodologies is the emphasis on self-interpretation of data of which the researcher is a part. This requires the researcher to maintain maximum transparency within the process, to prevent conducting an analysis process based only on impressions and lacking a sufficient analytical foundation.

Data analysis in **action research** is already conducted in the initial stage of identifying the problem, and continues in other stages of action research. At the problem identification stage, the practitioner-researcher analyzes the data obtained from past experiences along with creating an array of categories. This category array is used to analyze the next stages. Thus, from this point the analysis process will be substantially similar to that explained in Chapters 4 and 6. This is a process in which the data is analyzed according to a given category array, with limited options for changes. The differences between the varied research approaches will be revealed in response to the data analysis in the reflection-evaluation stage. If the resulting picture from the previous (planning and action) stages of the action research does not satisfy the practitioner-researchers, they will react differently towards the research approach that guides them. With a positivist and to some extent a post-positivistic approach, the practitioner-researcher will remain loyal to the guiding theoretical perspective and categories and seek to improve the process of the planning and action in order to be more efficient. With interpretive-constructivist and critical approaches, the researcher will tend to re-examine and modify the theoretical perspective and the categories that guide the action research process, and redesign the action research.

Research Report

As in any qualitative research, the purpose of the final research report is two-fold: to present the findings of the study with the highest quality standards, and to be communicative with the readers. The stage of report writing is the near-final stage of all qualitative research genres, except the writing of an **autobiography**. It seems that the distinction between the process of conducting the research and writing the final report, characteristic to all genres of research including in the self-focused methodology, is not characteristic of writing an autobiography. Even if the first step of writing an autobiography focuses on gathering data, it is much more integrated within the activity of writing than in any other research genre. The final research report in the **self-case study** will be identical to the reports of "conventional" case studies presented in previous chapters. The final report of **action research** will be completely different from all "conventional" research reports, including the reports of the two other genres of self-focused methodology.

The final report of action research will reflect its unique character, which combines action with research. The report will describe both the action and the accompanying research. This report may include all sections common to research reports – a literature review, research questions, methodology, presentation of the population and the study field, a description of the action research, discussion and conclusions - albeit with some flexibility as opposed to the usual "conventional" studies. At the center of the action research is, of course, a description of action and research combined. It seems that the best way to display the description of action research will be by viewing the chronological order of the process. The description will start by presenting the problem that prompted the action research and ways to identify the problem, including criteria (categories/themes) for its identification. Many of the writing principles presented in the previous sections also apply here.

However, distinct from the "conventional" studies, where we focus on the description of the findings of the analysis, an action research description must include the analysis as well as the activities, in tandem. It seems that a good way to present an integrated description and analysis will be by placing them side by side on the same page: the description on the left, with the analysis in a "margin" on the right. Figure 18C is as an example of a short excerpt from a description of a teachers' workshop. The practitioner-researcher who led the workshop and conducted the action research described and analyzed the workshop.

Description of the workshop:	**Analysis:**
The workshop session was devoted to the issue of clarifying values in the Biblical text "Do not rejoice when your enemy falls" At the beginning of the workshop, I asked the participants to read the text in full and to suggest how to teach it.	I start the workshop by presenting the text with **general guidance**
Tom: (one of the participants) "I think this is a most appropriate text. Many times I see the student so delighted to see the failures of others, even if the "others" are not exactly their enemies ..."	Tom has actually determined the direction of the discussion on the issue of **to what extent is the text suitable to the students.**
Sharon (another participant) interrupts: "So this is exactly why... I think this text is really not suitable for our students. They once again see us as the advocate of morality. On the contrary, I was looking for moral texts that they can identify with."	Sharon continues the discussion in the same route, but with the opposite conclusion: **the text is not suitable for students.**
Ronnie: "Why do you think that students cannot identify with this text? I'm sure they like it and it will enhance their appre-	Ronnie continues the same issue but believes that **it is very appropriate to the student.**

ciation of the Bible."

Sharon: "Try to ask them [the students] what is meant by "enemies." They all think about people from the area close to them. Now tell them that the intention is to enemies from other nations - see how they react. We will bring the class to blows and get the opposite effect."	Sharon maintains her opinion, and this time raises an additional reason: **the student may not be ready to apply the values to every "enemy"**

Figure 18C: Description and Analysis of Action Research

From the above analysis, it is clear that the criterion by which the practitioner-researcher examined the activities is "the degree of suitability to the students." This is just an example designed to illustrate the need to accompany the activity description with transparent, descriptive analysis. One might present the activities and their analyses in other ways as well, provided that the links between the activity description and the corresponding analysis are visible and clear.

Research Quality Standards

Applying the concepts of "validity" and "reliability" in self-focused methodology is irrelevant at first glance, even in terms of qualitative research. The "conventional" researchers, as presented in previous chapters, investigate external objects - that is, dealing not within themselves, but through other perspectives. There is, then, a possibility for both the researchers and their colleagues and readers of the research to examine the work from an external point of view, positioning the theoretical perspective against the data and examining their compatibility. However, in self-focused methodology, the researchers focus on their "self." Apparently, anything that the researchers tell about themselves reflects them without the possibility for judgment by an external perspective. Nevertheless, it seems that despite the difficulties, or perhaps because of the uniqueness of the self-focused methodology, applying quality standards for research is essential, and a condition for defining the products of self-focused methodology as research, as emphasized in Part One. The border between research and non-research lies in the transparency of the process (not just of its products), applying the entities of distance, examination and reflection, and the use of analytical skills of the researchers. Researchers employing the self-focused methodology may find themselves, even unintentionally, exuded from the "space" in which any research is conducted, and focusing on presenting stories that may not meet the standards of research. We have no better example than literary works, in which authors openly and explicitly tell the story of their lives. These are not works of research, even if the writers did some preliminary research. Such works do not seek to examine reality by research standards, but by literary standards. But those conducting self-focused methodology present their work as research, and it is essential that they themselves and others will review those works according to high research standards for quality.

The term "validity" deals with the question of to what extent the research findings are consistent with the theoretical perspective and their correct interpretation, or put in another way, to what extent the researchers actually studied what they wanted to investigate (Huberman & Miles, 1994; Mason, 1996). It seems that ensuring transparency of the research process allows the application of this concept in self-focused methodology. The term "reliability" is about the possibility of repeating the same process of research and getting the same results, even if the research is carried out by other researchers (Yin, 1994). In light of this definition, it is clear that the term "reliability" in its "classic" definition is hard to apply to self-focused methodology, even more than to any other qualitative methodology (Merrick, 1999; Schofield, 1989). But if we do understand the term "reliability" as a possibility to re-examine the research process and findings from the original perspective of the practitioner-researcher, then the concept of reliability becomes fully relevant. In order to ensure the application of the concepts of "validity" and "reliability" in their qualitative meaning, the practitioner-researcher is required to show the chain of evidence (interviews, observations, documents, actions, protocols, etc.) for each stage of the study, for their self-examination and the examination of colleagues and readers (Huberman & Miles, 1994; Mason, 1996).

It would appear that the term "generalization" is even more difficult to apply in self-focused methodology. This is a research method that by definition declares to focus on the research itself. Why, then, is there a need to examine its relevance to other situations and cases? However, it seems that just as we found the relevance of the concept of "generalization" to other qualitative methodologies, so we can find its relevance here. Of the three types of qualitative research generalizations of qualitative research (Firestone, 1993), generalization to population will be irrelevant in this type of research. Self-focused methodology research restricts itself to dealing within the context of a specific practitioner-researcher, and therefore, by definition, does not deal with a large population. Self-case studies may be different in this context. If engaged in a significant number of research cases, the study may be relevant for generalization on a broader population. Generalization from case to case would be very relevant to this methodology (Schofield, 1989). Because self-focused methodology is characterized by rich descriptions of the phenomenon under investigation, new cases can be examined in light of the case being studied. Analytic generalization (generalization through theory) will be possible if the research is not satisfied by descriptive pictures, but will present a theoretical picture (Arksey & Knight, 1999; Dey, 1993; Firestone, 1993; Marshall & Rossman, 1989).

Like other qualitative methodologies, as the researchers expand the scope of the triangulation of data sources and methods of data collection, this will strengthen the arguments for the validity, reliability and qualitative generalization (Fontana & Frey, 2005).

CHAPTER 19

MIXED METHODOLOGIES

Literature dealing with mixed methodologies points to the problematic nature of such combinations between qualitative and quantitative methodologies, due to the different characteristics of each methodology. Previous chapter of this book show that significant differences exist not only between qualitative and quantitative research, but also among the various qualitative approaches and methodologies. These differences pose a challenge to researchers seeking to integrate methodologies. The methodological challenge of combinations that will be discussed in the following sections is therefore valid for combining various qualitative methodologies as well as for combining qualitative with quantitative methodologies.

The Challenge of Integrating Qualitative Research Types

In order to examine the possible combination of various qualitative methodologies (and with quantitative methodologies as well), it is necessary to distinguish between four key concepts characteristic of the various qualitative research types:

A. The research paradigm or research approach (e.g., constructivist, positivistic, post-positivistic, critical, and those marking other paradigms)

B. The research genre (for example, case studies, grounded theory, ethnography, life story, etc.)

C. The research methodology (e.g., the qualitative methodological patterns presented in previous chapters, as well as quantitative methodology)

D. The research methods (research tools such as the structured interview, in-depth interview, participant observation, focus groups, etc.)

Considering the possibility of mixed methodologies in qualitative research, we have to address each of the four concept groups and see if we can combine different research types within the same concept group (of paradigm, or methodologies etc.), which contexts can be combined, in which contexts the combining is more problematic, and what options are open to the researchers. Can we execute a research combination based on two different paradigms--for example, a positivistic paradigm with a constructivist paradigm—and if so, how can this combination be applied? Is it possible to propose a research combination based on two different methodologies, such as criteria-focused methodology with critical-focused methodology, and if so, how to do so? Is it possible to combine different research genres, such as a case study with a life story? Is it possible to offer a combination of different research methods, such as an in-depth interview and structured questionnaire?

As emphasized, over the last two decades many books and articles have appeared that suggest the possibility of combinations in qualitative research. However, most deal with a combination of quantitative and qualitative research, and less on the issue of integrating different types of qualitative studies. In reality, qualitative researchers conduct a combination of various types of qualitative research, sometimes without paying enough attention to this issue. Much of the literature on the combinations of qualitative research with quantitative research is relevant to the issue of integrating different types of qualitative studies. This chapter deals mainly with a combination of qualitative types, but as mentioned, this is also relevant to combinations of qualitative and quantitative research.

Method-Combinations (Research Tool Combinations)

Any research project, qualitative or quantitative, consists of a large number of methods. The term "methods," as distinguished from "methodologies," means research tools. The long list of methods includes various types of interviews, observations, data analysis and much more. There are methods which are more associated with a specific research paradigm than others (for example, the semi-structured interview is largely affiliated with the post-positivistic paradigm, and the in-depth interview with the interpretive-constructivist paradigm). There are methods more associated with one particular genre than with others (e.g., participant-observation is strongly associated with the ethnographic research genre, while the open interview is identified with that of life stories). Of course, there are several methods that are appropriate to a number of methodologies, to several research paradigms or several research genres.

At first glance, it seems that a combination of different methods is the simplest option. There are those who argue that the difference between the methods is basically a matter of emphasis and prominence of a particular property over another property, and thus we should identify the methods as located on a sequence (Hammersley, 1996; Stake, 1995). According to this view, the difference between the structured interview and open interview is only a matter of emphasis, and both are in the same sequence. Thus it seems that there is no real difficulty to integrate them. Miles & Huberman (1984) recommended a combination of high structure methods in qualitative research as part of the desire to ensure a higher level of rigor in qualitative research. Assuming that the research methods are indeed on the same sequence and the difference between them is only a difference in rank, there appears to be no practical difficulty to integrate them. From this, quantitative and qualitative methods may be combined easily in qualitative research.

However, we cannot avoid pointing out some principle difficulties. Even if the methods are in the same sequence and it is technically possible to incorporate different methods, we cannot ignore the fact that certain methods are identified with a specific research paradigm and/or methodology and/or genre, and sometimes a combination of methods associated with different types of qualitative studies would be problematic, if not unlikely. For example, an in-depth interview is associated with the constructivist research approach and participant-focused methodology or critical-focused methodology. It is not suitable to integrate an in-depth interview in criteria-focused methodology, where, of course, a structured interview is preferable. Technically, it is possible to combine such an interview, but the significant question is whether it fits. The opposite is also problematic: a combination of structured interviews, typical to criteria-focused methodology, when used in participant-focused methodology may limit the research picture.

"Triangulation" is a key concept in understanding the importance of combining different research methods in the same research project. Many researchers emphasize that the triangulation of methods and sources of information, expressed in a multiplicity and variety of methods, may strengthen the validity of the study, and therefore they encourage the use of various methods (Alpert, 2010; Denzin & Lincoln, 2005a; Hammersley, 1996; Merriam, 1998; Stake, 2005). But with all the importance of triangulation, even here there are limits and restrictions. It does not seem likely to use research methods which do not fit the methodological pattern of the study. Researchers are thus faced with a dilemma as to where to draw the line between the desire to enrich the triangulation by multiple methods and the need to adapt the appropriate methods to the research type.

The issue of method integration is focused largely upon the issue of mixed qualitative and quantitative methods. The sharp distinction between qualitative research and quantitative research is evident on the method level. Several methods are based on numbers, the mathematical language, while others are based on the language of words. To what extent can one combine quantitative data and methods in a qualitative research work? Until the 1970's, the tendency was to emphasize the differences between qualitative and quantitative research, and see one type as an alternative to the other (Woods, 1996). In recent years, however, the tendency is rather to look for a combination of the two. This is reflected, for example, in the desire of many qualitative researchers to use better, neater and more systematic methodologies, which often motivates them to combine quantitative methods in qualitative research (Miles & Huberman, 1984).

There are also researchers who suggest that the differences between quantitative and qualitative methods and data are not so extensive. They argue that many qualitative researchers combine certain quantitative elements to their reports, even without being aware of doing so. For example, qualitative researchers use many common expressions such as "often," "rarely," "sometimes," "in general," "always," and the like. The fact that the researchers use words, not numbers, does not change the fact that those are actually quantitative characterizations. In the same way, quantitative researchers express verbal references, because their quantitative description is always accompanied by a verbal description. From this standpoint, it seems that the difference between quantitative and qualitative methods is a matter of degree more than a difference of principle, and thus there is no reason to avoid combining methods characteristic of quantitative research in qualitative research (Hammersley, 1996; Miles & Huberman, 1994; Stake, 1995, 2005).

A Combination of Methodologies

A second possibility of combining types of qualitative research refers to research methodology. Methodology is the research operational design. It reflects the research structure and its components, and consists of a combination of different research methods into one research entity. In this book, six methodological patterns have been presented, each reflecting a wide range of qualitative research methodologies.

As described in previous chapters, the methodological patterns can be positioned on a continuum between two poles which extend from an emphasis on expressing intuitive inquiry skills to an emphasis on analytical research skills. The fact that we can put all the methodologies on one continuous line has caused some researchers to entertain the possibility of combining the methodological patterns into one research system according to how closely the methodology patterns are located on the sequence line.

The location of the methodologies on a continuum relating to specific methodological components (for example, an external theoretical perspective vs. an internal theoretical perspective) allows the combination of methodologies that are "close" to each other in the sequence line. Thus, for example, it is possible to combine the critical-focused methodology with participant-focused methodology. There may be a combination of structural-focused methodology, the specific type based on an interpretive-constructivist research paradigm, with participant-focused methodology, because the two methodologies are adjacent to the pole of intuitive research skills and the internal theoretical perspective. However, according to the location, the sequence on the continuum line indicates that methodologies that are "distant" from each other are quite difficult to combine, such as mixing criteria-focused methodology with participant-focused methodology (Charmaz, 2000; Kuckartz, 1995; Tashakkory & Teddlie, 1998).

By the same logic of the "close" mixed methodologies, it seems that we can smoothly integrate quantitative methodologies with qualitative methodologies, in such cases where the analytical research skills of the qualitative methodology dominate the research process. This is mainly appropriate for qualitative methodologies which adopt approaches close to positivistic or post–positivistic research. Such is the case of a combination of quantitative methodologies with criteria-focused, partial criteria-focused, and sometimes also self or structural-focused methodologies where they adopt a close approach to positivist or post-positivistic research paradigms. A combination of quantitative methodologies with participant and/or critical-focused methodologies seems much less appropriate and more complex and problematic, although this combination is possible by "across methodologies," (Tashakkory & Teddlie, 1998) as will be explained below.

A Combination of Approaches (Paradigms)

A combination of different research approaches (paradigms) is perhaps the most problematic among all the possible combinations. When we speak of research approaches, the intention is, as stated, to assumptions about the nature of the phenomena under study, the beliefs about the nature of the reality, and the relationship between reality and the people observing it (Denzin & Lincoln, 2000, 2005a; Guba & Lincoln, 1998, 2005; Maykut & Morehouse, 1994; Sciarra, 1999). Many researchers emphasize that such a combination is completely impossible. Bruner (1985, 1996) mentions the fundamental differences between the positivist approach (or in his words, "the logico-scientific") and the constructivist approach (what he calls "narrative"). These differences are seen as essential, and Bruner claims that an argument based on one research approach cannot substantiate or refute arguments derived from another research approach. Guba & Lincoln (1998, 2005) also emphasize the differences between the varied research paradigms, arising from different assumptions. According to these researchers, it is generally impossible to integrate the approaches. However, it is possible to integrate a positivist paradigm with a post-positivist paradigm, because they are based on close assumptions and we can see the connections between them across a continuum.

Tashakkory & Teddlie (1998) suggest another way to relate to a combination of approaches (paradigms) in qualitative research. They are aware of the differences between the different research approaches, and recognize that according to the nature of these approaches based on different epistemological assumptions, it is almost impossible to combine them. But to overcome this difficulty, they suggest the "pragmatic approach" to bridge the existing approaches. Tashakkory & Teddlie argue that instead of relying on metaphysical truths as a basis for determining differences between the approaches, pragmatists consolidate the approaches according to the question of "what works." The pragmatic approach practically ignores the epistemological differences between various research approaches, and is actually based on a combination of methods and methodologies. Given this, the principles of integration according to the pragmatic approach are the same as those explained in the previous two sections: a combination of methods and methodologies.

Combinations of Genres

Another option for integration in qualitative research is a combination on the level of research genres. The term "research genre" means the general framework of the study, reflected in the nature of the final report. Research genres include case studies, ethnography, life stories, action research and so on. Integration at the level of genres refers to at least two options. One is a combination of different research units characterized by the same research genre. Thus, the same research project can be combined into one research study in two different sections (or more), each of which is, for example, a unique and different life story--for instance, a research project containing parallel life stories of several social workers. The second option is a combination of different research genres into one research project. For example, we can combine action research with a case study or incorporate a life story with ethnography. By such combinations, we obtain a rich research picture, emphasizing certain aspects of the same phenomenon under investigation in different ways.

***

To summarize the options for integrating different types of qualitative research or a combination of qualitative with quantitative research, it appears that the simplest option is a combination of methods within the same study. The integration of methodologies will be simplified if there is proximity between the nature of the methodologies and they are close along the continuum. It will be more problematic if the methodologies are distant from one another. A combination of genres might be simpler if the research is based on the same paradigm and/or the same research methodology, and it will be more complex if based upon approaches and methodologies that are far from each other. The combination of different research approaches seems most problematic, because each research approach is based on different epistemological assumptions. It is important to note that a distinction is made here between simple combinations and more complex combinations, while avoiding the presentation of any combination as impossible. This leads to further consideration of the issue of integration, based on the order of the integration, as shown in the following section.

Integrated and Across Combinations

If it so fraught with problems, how can we explain the popularity of mixed methodologies in qualitative and quantitative research and its growing presence in many studies? It seems that the answer lies in the way one overcomes the problems associated with the combination. This is reflected in the significance assigned to the order of the combined research components. From this point of view, we can distinguish between two types of combinations: integrated and across-combinations.

Integrated Combinations

Integrated combinations are cases in which the same research unit will integrate various research components together. The most typical example is a combination of research methods and different research data within the same research unit. For example, in a case study which is mainly based on in-depth interviews, we can combine survey data, qualitative or quantitative, conducted with the same interviewed population. Here there is a method combination that does not substantially change the character of the case study, and data from various sources and methods only strengthens the emerging picture (Creswell, 1998). There is an integrated combination of research methodologies located next to each other across the continuum line (illustrated in Figure 10A). The resulting methodology is integrated and does not suffer from substantial methodological contrasts. Examples of such an integrated combination include the integration of structural-focused methodology with criteria-focused methodology and a combination of structural-focused methodology with critical-focused methodology, and so on.

By nature, integrating a combination of methodologies that are far from each other will be more complex if not problematic, as explained above (Charmaz, 2000; Kuckartz, 1995; Tashakkory & Teddlie, 1998). A similar problem exists in integrating research units based on different approaches (Bruner, 1985, 1996, Guba & Lincoln, 1998). As mentioned, the possibility exists, to some extent, to create an integrated combination of research approaches which are close to each other. Based on the principles outlined above, and with regard to the difficulties presented, the option of an integrated combination of various qualitative research units is quite limited.

Across-Combination

An across-combination (Tashakkory & Teddlie, 1998) occurs when researchers conduct different research units side by side, so each unit exists separately and retains its character and uniqueness, but together the units are connected to the same research project and deal with the same research issue. The most appropriate way to combine different types of research is by the across-combination. It appears that this type of combination meets most of the integration challenges presented in the previous sections. In practice, we actually have several different research units in parallel that are involved in the same research topic, even sharing several identical, common data sources.

There are different types of across-combinations, differing in the significance assigned to each research unit, its position along the continuum of time, and its purpose. It is possible to conduct an across combination of not only different qualitative research units, but also of qualitative research units with quantitative research units, almost without limit. The following are several possibilities of across-combining, characterized in regard to the dimensions of time and function of the various research units.

The Time Dimension

A reference to the nature of the time that any research unit is conducted allows us to distinguish between several combination options:

- A continuous stage-based research: In this case, the researchers are not engaged simultaneously in several different research units, but finish one research unit and then move to another research unit (Creswell, 1998; Silverman, 2006). For example, researchers can open the research with a qualitative (or even a structured quantitative) questionnaire survey over a wide population studied (criteria-focused methodology), then conduct a qualitative case study with a small, limited population (participant or critical-focused methodology).
- Simultaneous equal research units: In this case, the researchers conduct two (or more) research units separately, in parallel and with a similar weight and focus (Creswell, 1998). For example, researchers who want to explore the processes of

teaching a particular subject can investigate the issue from two perspectives at the same time. One research unit will be based on participant-focused methodology and examine the processes of teaching as perceived by the teachers. This research unit is based on in-depth interviews and observations with those teachers. The parallel research unit examines the process of teaching on the basis of a given educational theory, and may be based on those interviews and observations, but analyzed according to the principles of criteria-focused methodology. As such, we receive a parallel research picture from two points of view, enabling us to compare between them.

- Simultaneous research units where one dominates the other: For example, a research study which examines the work of family doctors. The central research unit will explore the work of many physicians with relatively structured questionnaires based on quantitative research methodology. At the same time, the researchers will also conduct in-depth interviews and non-structured observations with a few family doctors, based on the principles of participant-focused methodology. The entire research project will therefore be based on the broad quantitative picture, while creating a more limited-depth qualitative picture.
- Research based on a single research unit, by combining a limited component of another research unit (Creswell & Plano Clark, 2007). This is a research based on one central methodology that combines several research methods (research tools) compatible with different methodologies. For example, a study based on structural-focused methodology which examines a text written by the studied population. The researchers can combine a structured questionnaire which gathers background information about the participants. Although this questionnaire is compatible with the criteria-focused methodology and not with the structural-focused methodology, we can add it without any contradiction as a complementary research element to the central research unit.

The Function Dimension

Each of the four options for combinations presented above focuses on particular research functions. We can indicate three such functions:

- Combination as support: This is a research project in which one research unit serves as a source or basis of assumptions or hypotheses for another research unit. In other words, one research unit supports another research unit. For example, researchers can use one research unit for exploratory research (pilot research), and use its findings as a basis for the main research project, to be carried out later in another research unit (Creswell & Plano Clark, 2007; Hammersley, 1996, Silverman, 2006). Or, conducting participant-focused methodology for an exploratory research unit and continuing with partial criteria-focused methodology as the main research unit.

- Combination as enrichment: This is a situation where the researchers use several research units in parallel to achieve rich findings that reinforce each other (Hammersley, 1996). In this case, the same phenomenon is investigated in parallel by different research units, each of which illuminates the phenomenon from a different angle. For example, one research unit may display the life story of participants as it emerges from interviews conducted with them, based on participant-focused methodology. A parallel research unit will focus on a structural analysis of the participants interviewed, based on a structural-focused methodology. In this way, we receive a rich picture of the phenomenon under investigation from both content and structural angles (Tashakkory & Teddlie, 1998).
- Combination as an enlargement: In this case, the researchers conduct various research units, which ultimately complement each other (Hammersley, 1996). For example, the researchers start with a qualitative case study in participant or critical-focused methodology, and then broaden the research picture by conducting a qualitative survey with an enlarged population using criteria-focused methodology or even quantitative methodology. The limited number of case studies acquires depth and detail over a small group of participants, while the survey enlarges the picture into a wide population (Silverman, 2006).

One of the fertile mixed methodologies is a combination of research units employing the content-based analysis tradition along with research units based on the structural analytic tradition. This combines research units of the structural-focused methodology with one of the five content-based methodology patterns. These two types of research are very different from one another, with each looking at the phenomenon from a different perspective (Lieblich, Tuval - Mashiach and Zilber, 2010; Ryan & Bernard, 2000). Putting them together in an "across combination," if according to "support" "enrichment" or "enlargement," may constitute quite a fruitful combination. These are mixed methodologies based on the same research data for each of the research units (Fielding & Lee, 1998). Sometimes the combination will be reflected in the study of the phenomenon by using two different analysis methods in parallel, and sometimes with one analysis method based on another analysis method. For example, a participant-focused analysis will be conducted prior to a structural-focused analysis, in order to locate the content that will be analyzed in the structural manner (Rosenthal, 1993).

It is important to understand that a combination of different research units for the purpose of support, enrichment or enlargement does not eliminate the differences between the research units in terms of approach, methodology, genre or methods. The resulting picture of each of the research units is not necessarily a harmonious picture, i.e., a picture whose total parts are necessarily coherent as a whole. Sometimes the results from one research unit which reflects a certain perspective will be inconsistent with the picture that emerges from another research unit, observed from another perspective. In such cases, researchers will be required to present explanations that allow readers to see the two seemingly contradictory points of view together, without reaching the conclusion that one nullifies the other, or with the quality of the research being diminished due to this combination.

APPENDIXES

APPENDIX 1

ORGANIZE A DATABASE BY THE NARRALIZER

When the data analysis process and organization is carried out through the **Narralizer**, the researcher can utilize word processed documents, web documents, e-mail documents or any other texts (but not pictures of text). For the effectiveness of the analysis process, there is an option to bind together the raw data documents (which are named as "sub-cases") into an array of software directory databases.

The following is an explanation of how to create a **Narralizer** database:

The **Narralizer** "produces" two types of documents: Research Documents, which execute the analysis process, and Source Documents, where raw data documents are stored.

Upon opening the **Narralizer**, the software will "produce" a **Research document** by default. To create **Source documents**, you must give a special command, as follows:

1. Choose the command **"New"** in the "**File**" menu. As a result, a new box is opened with two options: "Research Document" and "Source Document." The default of the software is "Research Document." In order to "produce" a source document, you must **mark** the second row, "**Source Document**" as shown in Figure AP1.

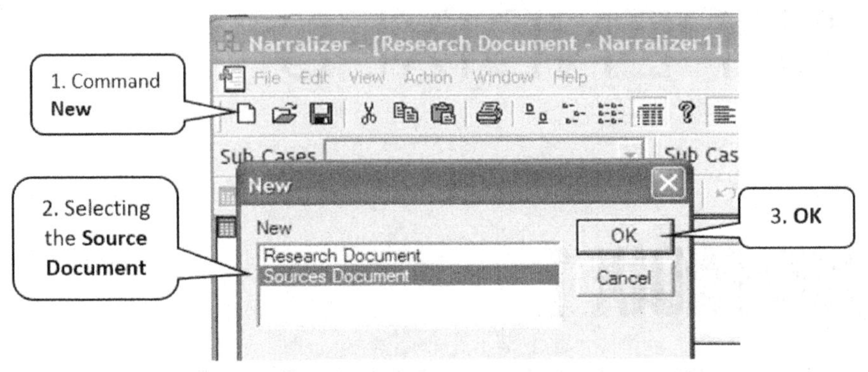

Figure AP1: Selecting a Source Document

2. After clicking the **OK** button, a screen of Document Sources will open (Figure AP2). Be certain that the icon is changed from "Research Document" to "Source Document," as shown in Figure AP2.

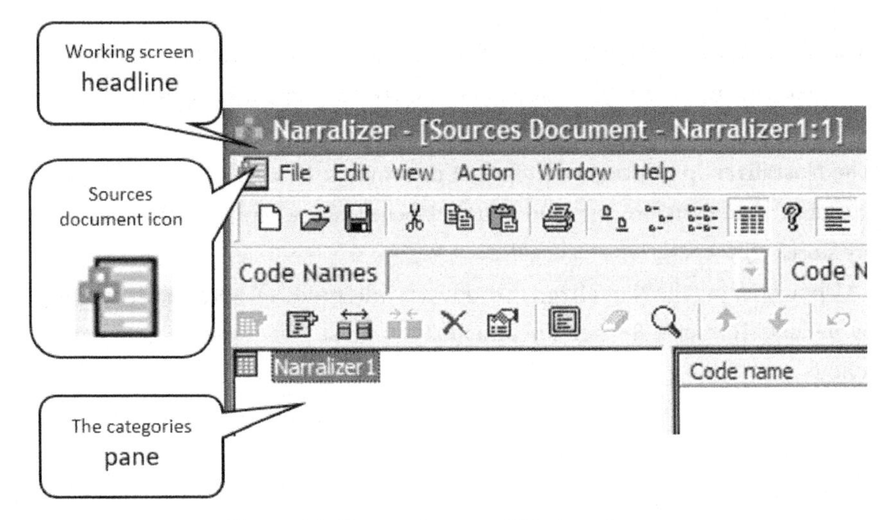

Figure AP2: "Source Document" Working Screen

3. With the opening of the **Narralizer** working screen, the temporary name "**Narralizer 1**" will appear in the category pane. (Additional documents will open with temporary names "Narralizer 2," "Narralizer 3" and so on) To change the temporary name to the name of the source document (probably the name of the "case" under investigation), see instructions in Chapter 8.

Organization of the Data Document Directory

Following the definition of the data document directory is the stage of placing all the data documents in the same directory. Each data document is identified by a name that will allow its characteristics to be recognized and to be distinguished from other documents, emphasizing the common and the varying traits. Each such document is identified as a "**sub-case.**" Thus, there may be many sub-cases in one study project. Each sub-case will be identified by a short name to facilitate its quick identity and retrieval if necessary. As mentioned in Chapter 12, the naming system will be uniform. For example, if you are dealing with a group of participants, the first sign that identifies them is likely to be their name, the second identifying sign may be the type of data (interview, observation and so on), and the third would be the date of receipt of the information. For example "Jacob", interview, 12.8," "Rachel, observation No. 1," "Sharon, diary, 1st year," etc.

Insert Data Documents to Source Directory in the **Narralizer**

This is a process to import data documents into a **Narralizer** Source Document.

1. Open the external document in the word processor (or other electronic text program), mark the entire text or parts which you wish to insert into the **Narralizer**, and perform the **Copy** command. Then, place the cursor in the **categories pane** of the **Narralizer** on the **name of the research project** (in Figure AP3: "Teachers' Approaches Sources") and **click the left mouse button**.
2. Choose the command "**Add Text Source...**"

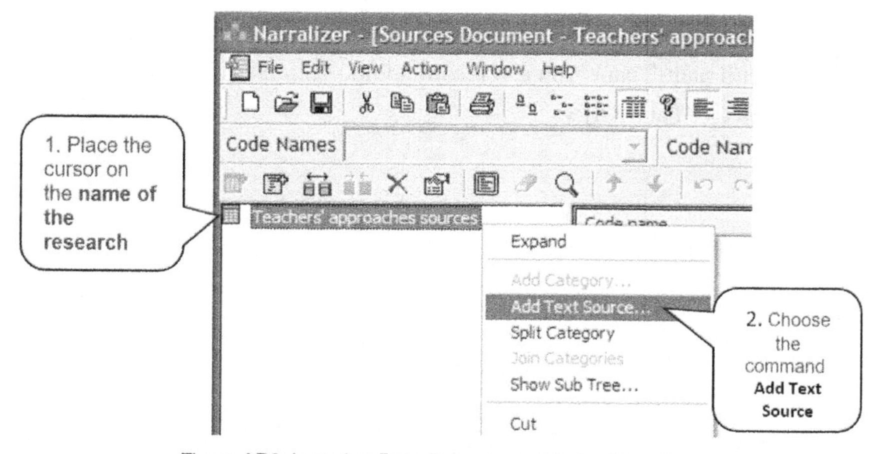

Figure AP3: Inserting Data Documents into the Narralizer

3. Clicking on the command "**Add Text Source**" opens the box **"New text box source for ..."** Place the cursor in the large pane with the command **Paste** (see Figure AP4) and the copied text is inserted into the box.

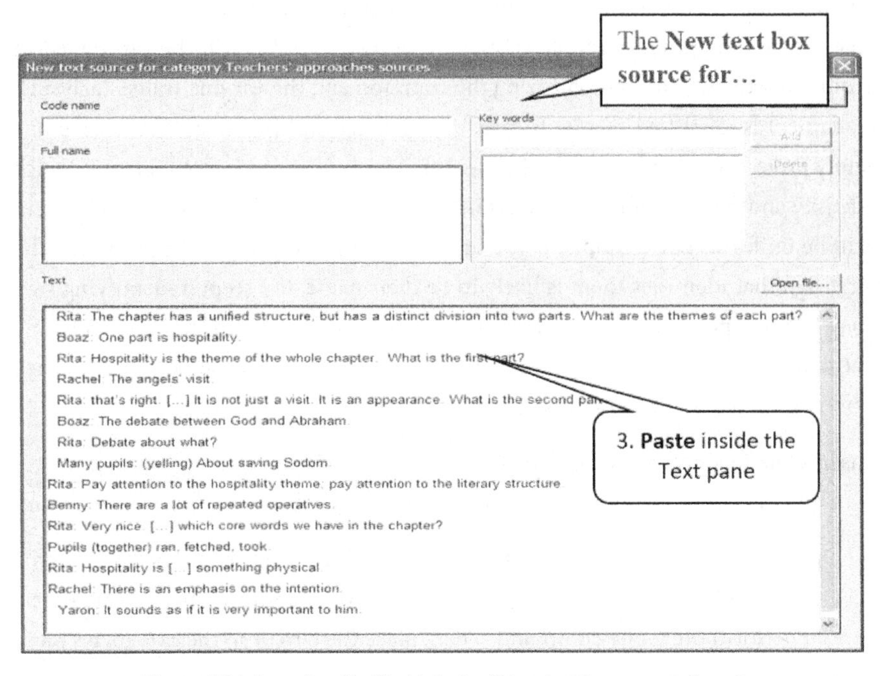

Figure AP4: Inserting the Text into the "New text box source for ..."

4. In the top part of the **"New text box source for ..."** under the title Code name, enter the **name of the sub-case**. (In Figure AP5, the name is "Rita interview")

5. In the "Full name" pane, you can record additional information about the data document (see Figure AP5)

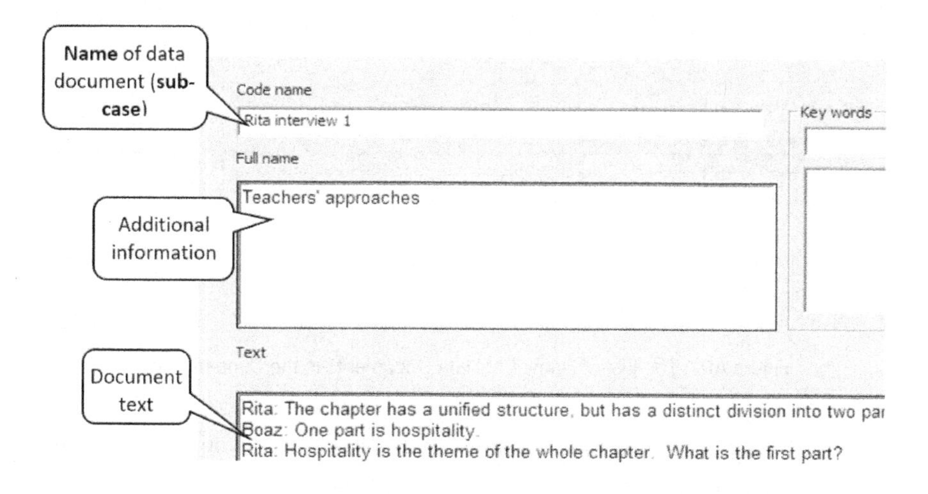

Figure AP5: The Definitions of the Sub-Case (Data Document)

6. An OK command in the box inserts the data into the source directory. After inserting the data document, the name of the sub-case will appear in the content pane under the "**Code name,**" and the first words of the text will appear under the Text column. (Figure AP6)

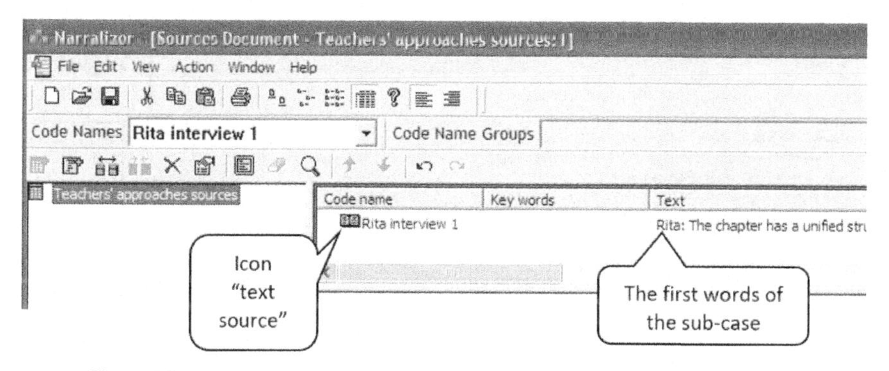

Figure AP6: Indication of the Sub-Case (Data Document) in the Content Pane

7. Similarly, you may insert more data documents, as shown in Figure AP7.

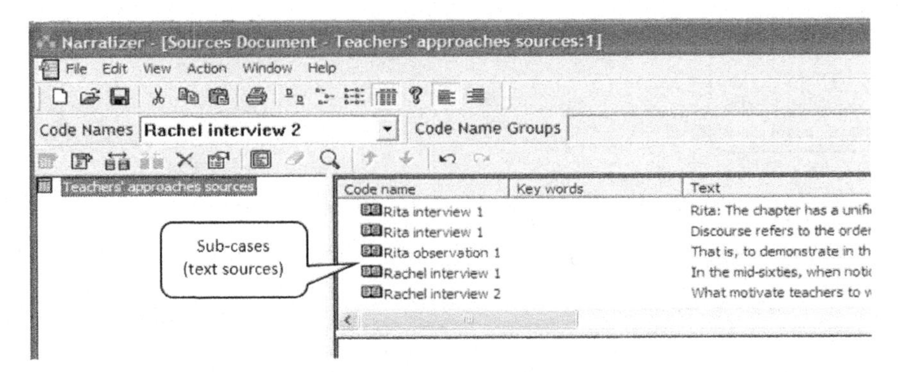

Figure AP7: The List of Names of Data Documents in the Content Pane

8. In the top right corner of the box in the pane "Key words," you may record the main themes in the document (sub-case) as will be explained below. The list of the main themes is a non-obligatory list which allows researchers to keep the overall picture of data.

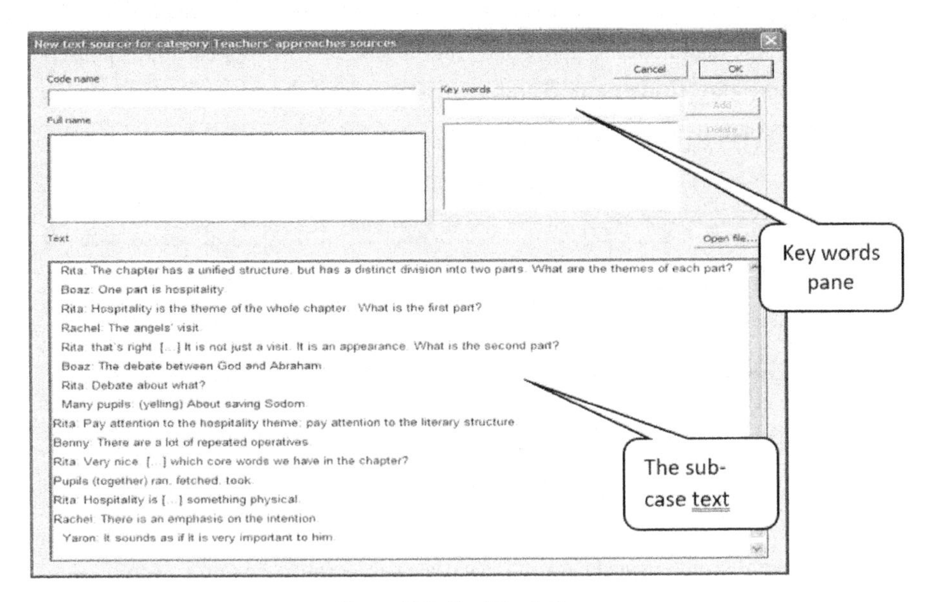

Figure AP8: Key Words Pane

Listing the main themes of the data is performed as follows:

A. The concept (one of the main themes) is entered in the **Key words pane**. In Figure AP7, the main theme is "classroom discussion."

B. **Using the Add** command, insert the main theme from the key words pane into the main themes pane.

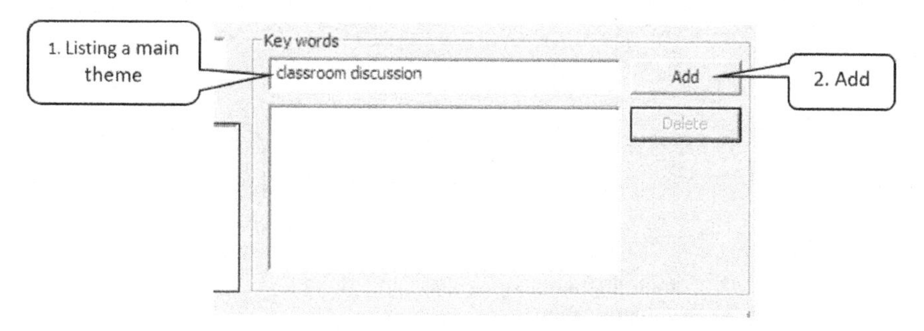

Figure AP9: Inserting One of the Main Themes

C. Similarly, you may insert the other main themes to the Key Words list.

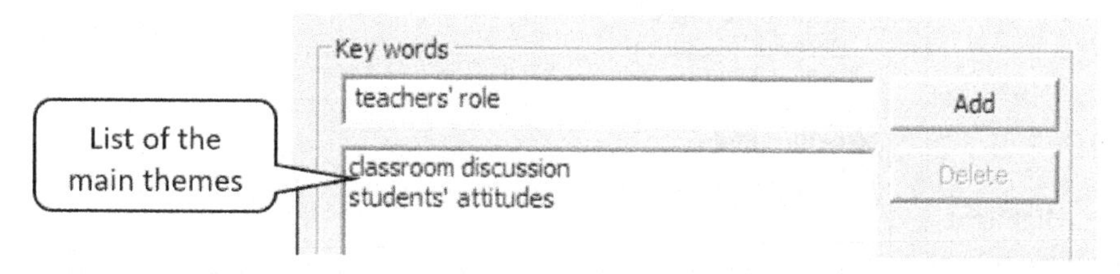

Figure AP10: List the Main Themes in the Key Words Pane

D. After completing inserting the main themes, click **OK** and the box is closed. Figure AP11 illustrates the record on the Content Pane.

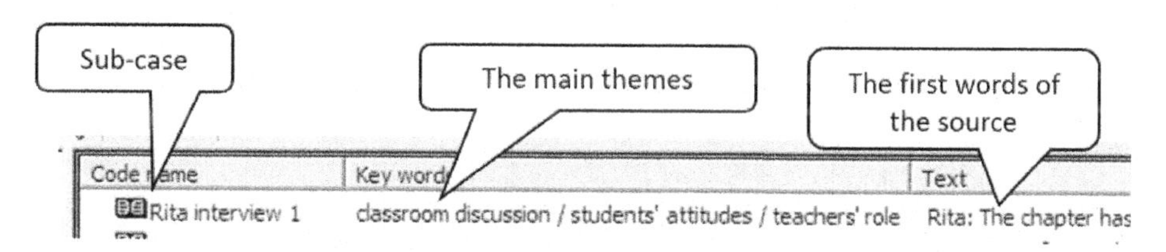

Figure AP11: Listing in the Content Pane

It should be noted that all documents belonging to the research project should be inserted to the same sources document. The main reason is, of course, organizational:

the goal is to enable researchers to easily access all the data documents. It also allows creating a list of themes, cumulative for the entire research project. All the themes are in the list of the document sources, and the list can be examined at any time. Click on the triangle arrow on the pane to open the entire Key Words list. This allows the use of the same themes (key words) in a new data document, as shown in Figure AP12.

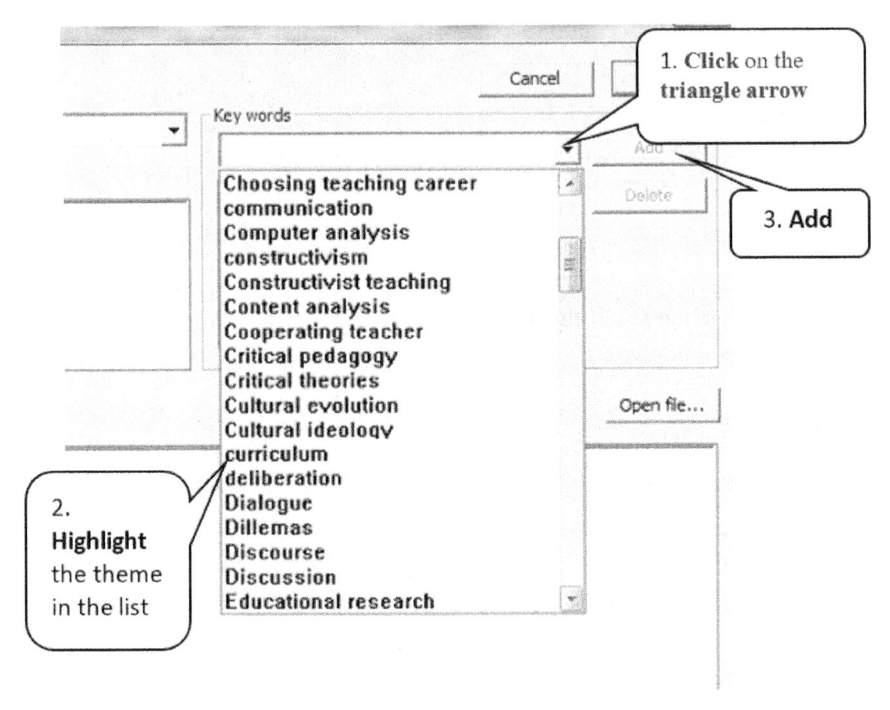

Figure AP12: List of Themes in the Key Words Pane

APPENDIX 2

STORING BIBLIOGRAPHIC DATA

In order to create a bibliographic database, use the Text sources documents created in Appendix 1.

The current chapter contains two sections, each concerned with the research requirements:

A. Creating a bibliographic database

B. Using the bibliographic database to produce a literature review

A. Creating a Bibliographic Database

1. Open a **Text sources** document, as explained in Appendix 1, and give the document a name that indicates that the document is being used as a bibliographic database (in the example below, the name Bibliography was chosen).

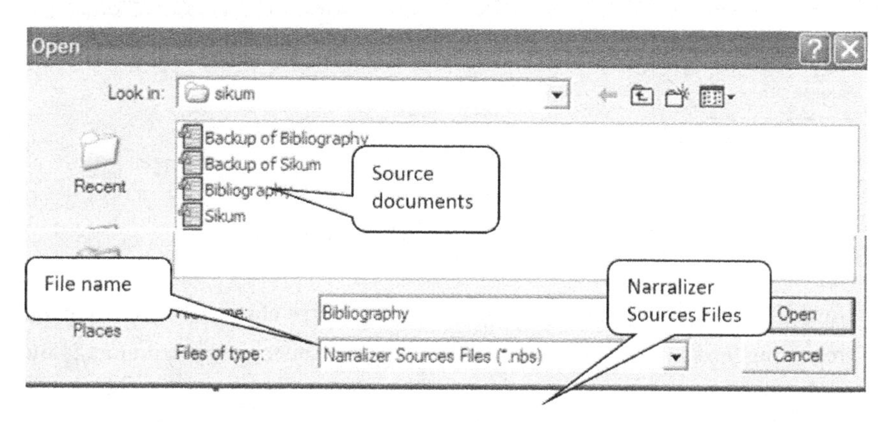

Figure BP1: Giving a Name to a Bibliographic Source Document

2. Click **Add text source** to open the **New text source** box as described in Appendix 1.

Type the full name of the bibliographic source in the **Full name** pane (researchers may use any bibliographic format they choose).

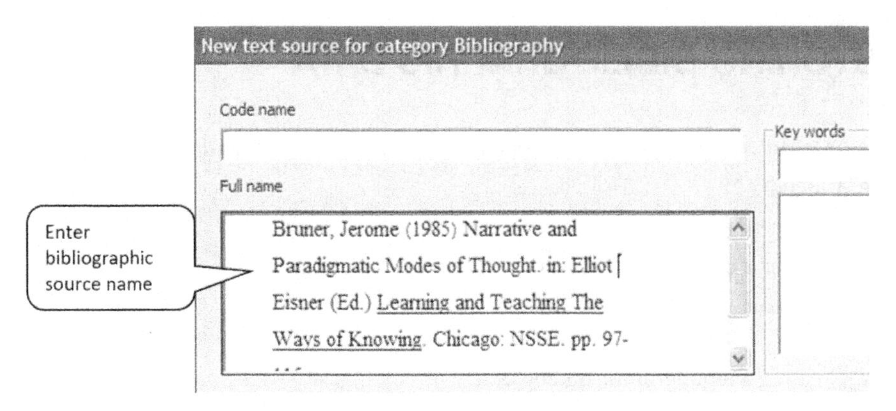

Figure BP2: The Full Name Pane

3. In the **Code name** pane, type a shortened name for the bibliographic source, for easy location. We suggest using the surname(s) of the author(s) and the publication year. In Figure BP3, we used Jerome Bruner's article from 1985, and gave it the **Code name** Bruner, 1985.

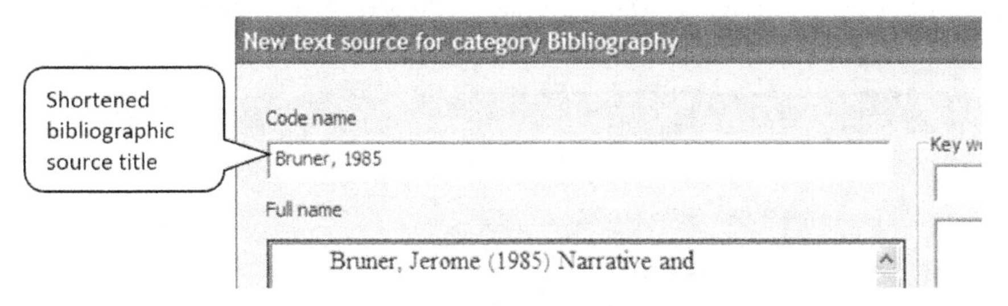

Figure BP3: Giving the Code Name

4. In the **Text pane**, type in (or Copy/Paste) the source abstract or your remarks regarding the source content, whichever is appropriate to your requirements and work style.

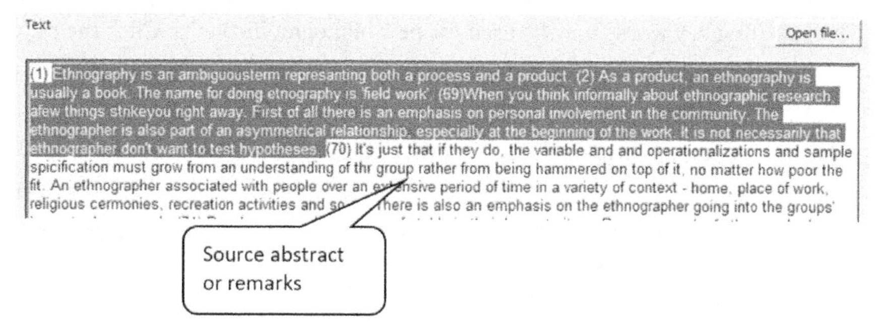

Figure BP4: Inserting Text to the Text Pane

5. In the **Key words** pane, type the key words for the article. The purpose of key words is to remind you which concepts appear in the bibliographic source. Researchers can create a list of key words as they work or use existing ones. The procedure for entering key words for a source document is explained in Appendix 1. See Figure BP5.

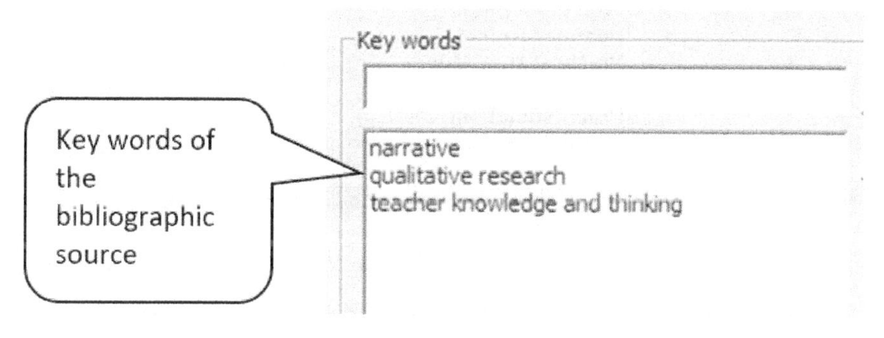

Figure BP5: Typing the Key Words

<u>**Bear in Mind:**</u> **Narralizer** enables the researcher to create a list of key words, which can be applied to all bibliographic sources. This function allows you to add new key words to a constantly growing list and creates a uniform set of criteria for identifying bibliographic sources (naturally, it allows you to "retrieve" bibliographic items by topic).

6. Type the name of the key word in the top pane and click **Add** to automatically

add the key word to the specific bibliographic source and the total keywords list. Existing key words may be used for new bibliographic items. Click the right arrow on the Key words pane to see whether a term is already listed. If it is, highlight the term to place it in the key words pane, then click **Add** to assign it to the new bibliographic item. To find a term or nearby terms, type the first letter of the key word in the top pane. (In Figure BP6 below, "narrative" was chosen from the list of key words.) When the key word is highlighted, it appears automatically in the Key words pane.

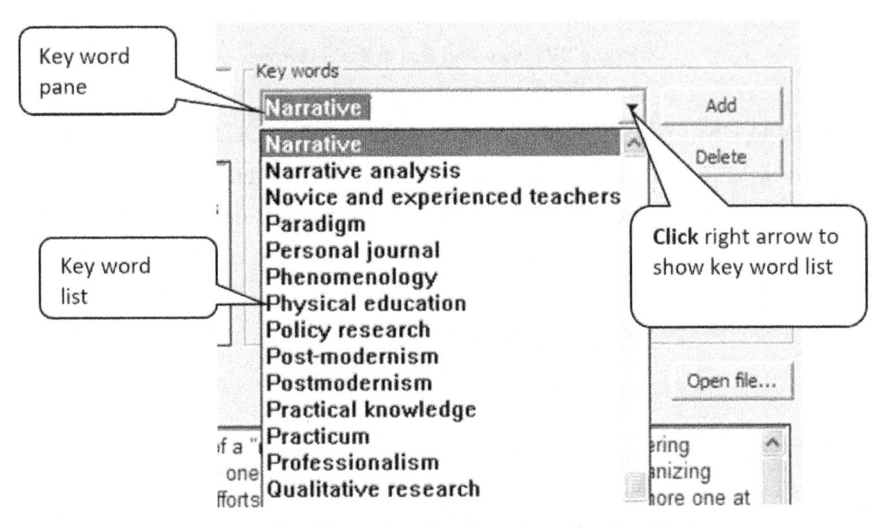

Figure BP6: Choosing a Key Word from Key Word List

7. When all items for the bibliographic source are entered in the **New text source** box, click OK, and the source details will appear in the Content pane.

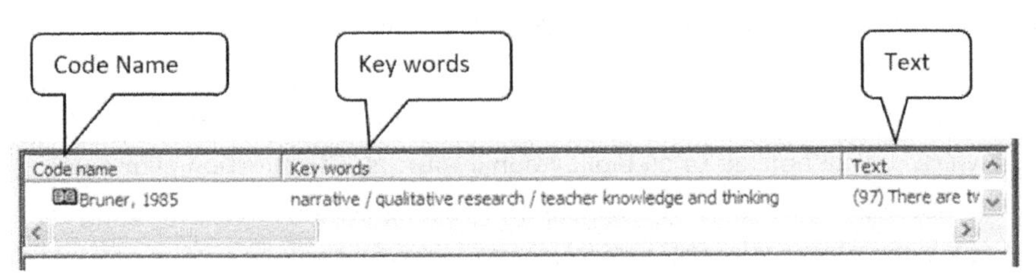

Figure BP7: The Source Detail in the Content Pane

Similarly, the researcher can introduce an unlimited number of sources. Sources are listed in alphabetical order in the Code Names pane in the upper section of the Work Screen, as shown below.

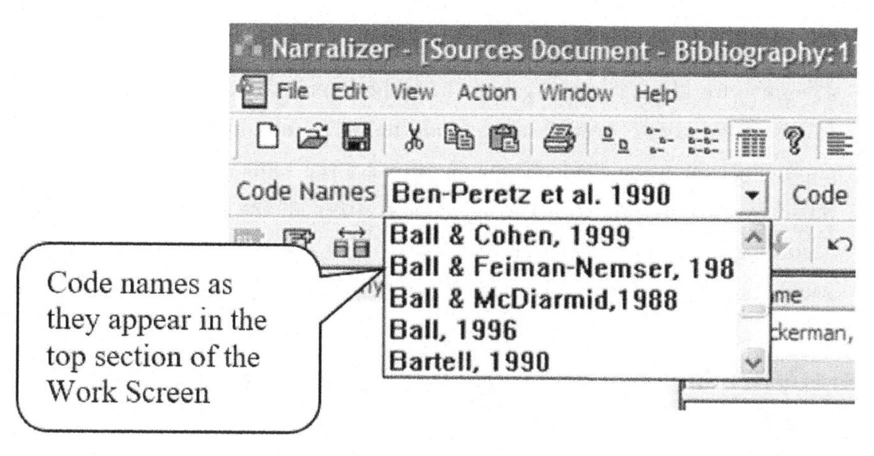

Figure BP8: The Code Name List

8. Data identification and retrieval from the bibliographic database:

   When the Main Category in the **Category Pane** is highlighted, **Narralizer** displays the list of bibliographic sources for that particular specific bibliographic database in the **Content pane**. The sources are listed alphabetically and can be found by scrolling up or down. Point the cursor at the source's **Code name** and select **Properties**, or double click on **Code name** to open the **Properties box** containing all the source information for that specific bibliographic database.

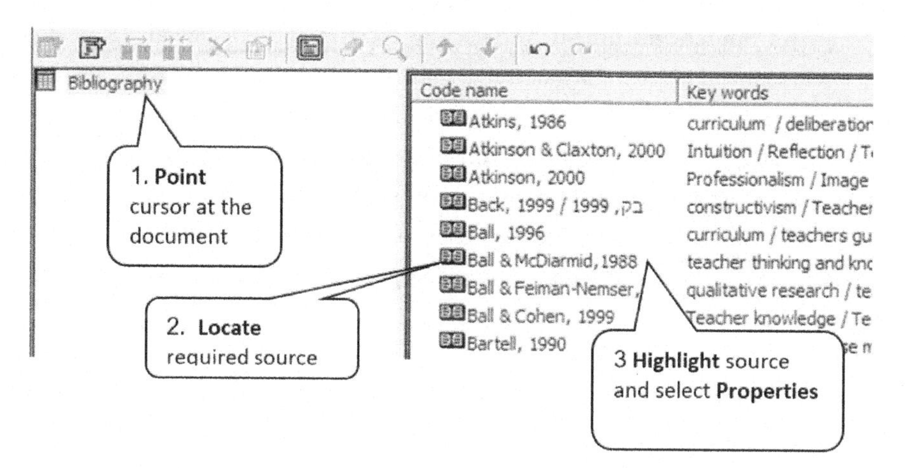

Figure BP9: Data identification and retrieval from the bibliographic database

9. Arranging bibliographic sources in a group:

   The software allows you to organize the bibliographic sources in groups. Sources can be divided into groups using **Code name groups** in the **File** menu. The commands for using **Code Name Groups** are the same as the commands for **Sub-Case Groups**. As with the latter, **Narralizer** allows you to group sources according to different criteria, such as content themes. Remember that one source can appear in several different groups.

B. Using Bibliographic Data to Create a Literature Review

Using **Narralizer,** we can not only store bibliographic data, but use it to produce literature reviews for articles and research papers. The greater the number of abstracts and quotations the researcher places in the software, the more efficient it will be in producing literature reviews. We will now suggest an approach for creating a literature review based on the analysis and assembly functions presented in the earlier chapters:

1. Opening a document and establishing the bibliographic boundary for the literature review:

   Open a **Sources document** by selecting **New** in the **File** menu. Name the document and its main category. If desired, the name should reflect the content of the planned literature review. The document name will be displayed as a main category in the **Categories pane** of the **Work** screen (the main category in Figure BP10 below is "**qualitative analysis**"). Next, create the subsections for the literature review in the **Categories pane** (split the main category, as described in previous chapters). Each subsection is an upper-category (in the example below there are five upper-categories.).

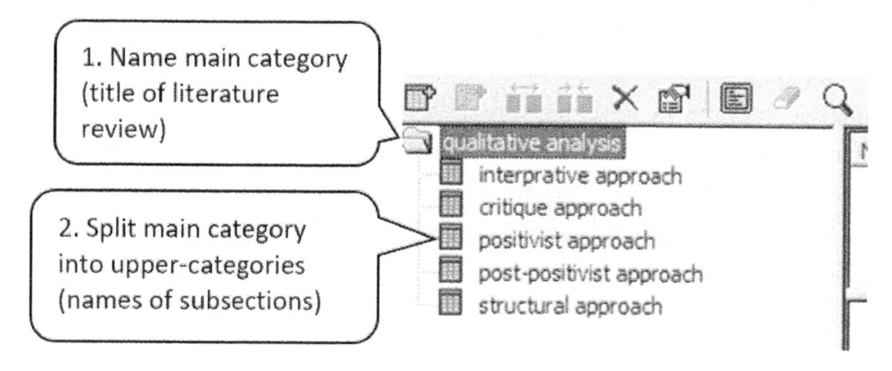

Figure BP10: Creating the Literature Review Document

2. Place the **Narralizer** document for the bibliographic database alongside the **Narralizer** document for the literature review. Now you may move the data from the database document to the newly-created literature review document. In order to move data from one document to another, both documents must be positioned in the Work screen. Use the **Cascade or Tile function** in the **Window** menu, as illustrated in Figure BP11 below:

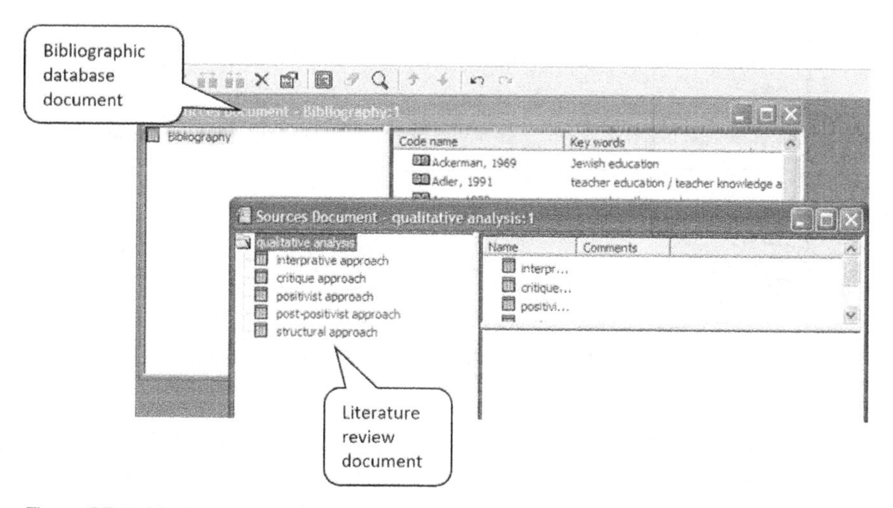

Figure BP11: Moving the data from the database document to the literature review document

3. Identifying relevant bibliographic sources:

Now that since the two documents are stacked behind each other, we can move the bibliographic data from the database document to the literature review document. The researcher's bibliographic database may contain a large number of items, not all of which are relevant to the literature review. The researcher must therefore identify those that are relevant.

To decide which material is relevant, look through the **Content pane** of the **bibliographic database document** containing the alphabetical list of bibliographic items. The important subjects in each item appear in the **Key words** column. The researcher can identify relevant items by their **Key words**. In the following example, the researcher wishes to find items concerning **qualitative research**. [To explain the steps involved, the bibliographic items dealing with "qualitative research" and the relevant key words have been marked with arrows. (Note: these arrows are not part of the software).]

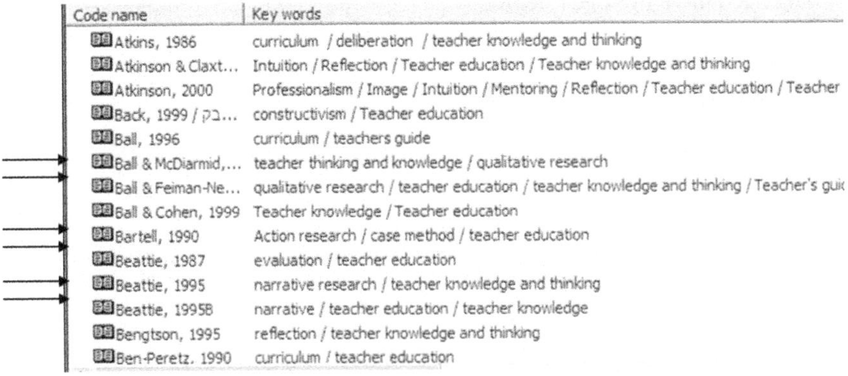

Code name	Key words
Atkins, 1986	curriculum / deliberation / teacher knowledge and thinking
Atkinson & Claxt...	Intuition / Reflection / Teacher education / Teacher knowledge and thinking
Atkinson, 2000	Professionalism / Image / Intuition / Mentoring / Reflection / Teacher education / Teacher
Back, 1999 / בק...	constructivism / Teacher education
Ball, 1996	curriculum / teachers guide
Ball & McDiarmid,...	teacher thinking and knowledge / qualitative research
Ball & Feiman-Ne...	qualitative research / teacher education / teacher knowledge and thinking / Teacher's gui
Ball & Cohen, 1999	Teacher knowledge / Teacher education
Bartell, 1990	Action research / case method / teacher education
Beattie, 1987	evaluation / teacher education
Beattie, 1995	narrative research / teacher knowledge and thinking
Beattie, 1995B	narrative / teacher education / teacher knowledge
Bengtson, 1995	reflection / teacher knowledge and thinking
Ben-Peretz, 1990	curriculum / teacher education

Figure BP12: Identifying the relevant bibliographic items

4. When both documents (the bibliographic source document and the literature review document) are open – either one on top of the other or one next to the other in **Cascade** or **Tile** mode, the researcher can examine the literature review and find and mark bibliographic items that are relevant. In Figure B14 below, **Agar, 1980** is relevant to the literature review.

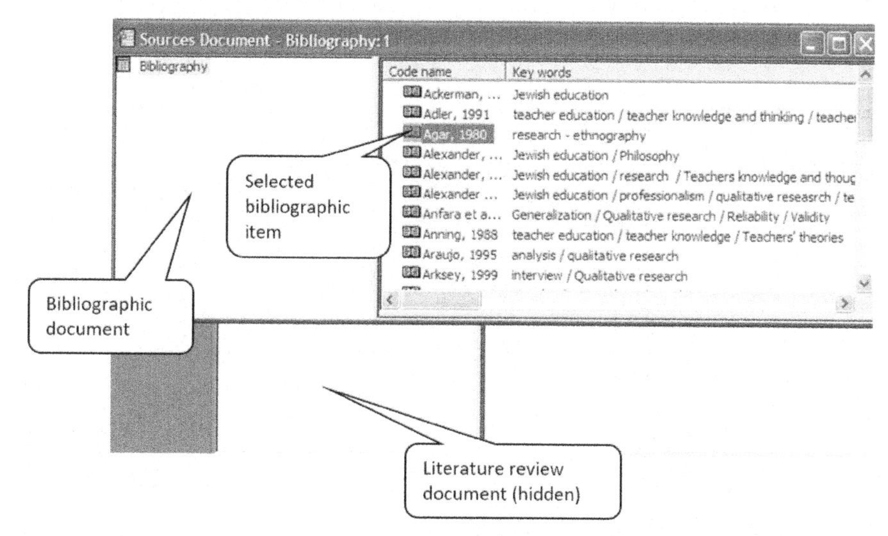

Figure BP13: Marking bibliographic items that are relevant

5. Now, double click the icon next to the bibliographic items of interest (or choose **Properties**) to open the **Properties box**. The main pane in the box is the **Text** pane. As explained earlier, the **Text** pane contains summaries or selected segments of the article. The next step is to highlight and **Copy** segments relevant to the literature review.

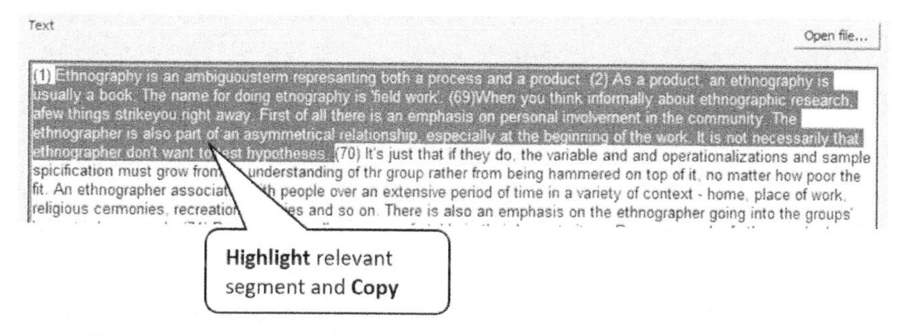

Figure BP14: Highlight and Copy Segments Relevant to the Literature Review

6. After **Copying**, switch to the literature review document and point to the destination category for the segment (in Figure BP15: "interpretive approach"). Then click **Add Text Source** as illustrated in Figure BP16 below.

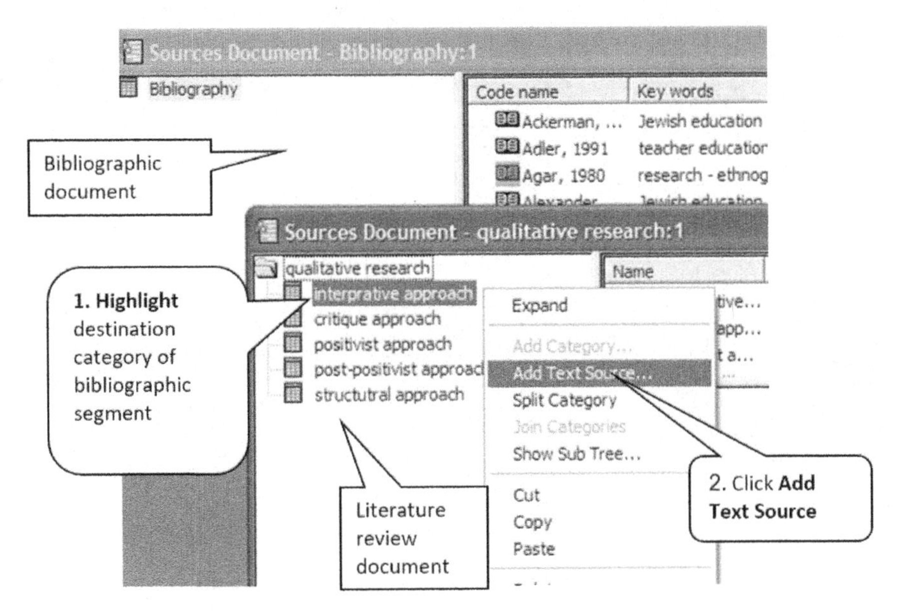

Figure BP15: Add Text Source to the Destination Category

7. After selecting **Add Text Source**, a **New Text Source** box will open and the highlighted and copied text will appear in the **Text** pane of the box. Type an appropriate name in the **Code name** pane. There is no need to write anything in the **Full name** or **Key words** panes.

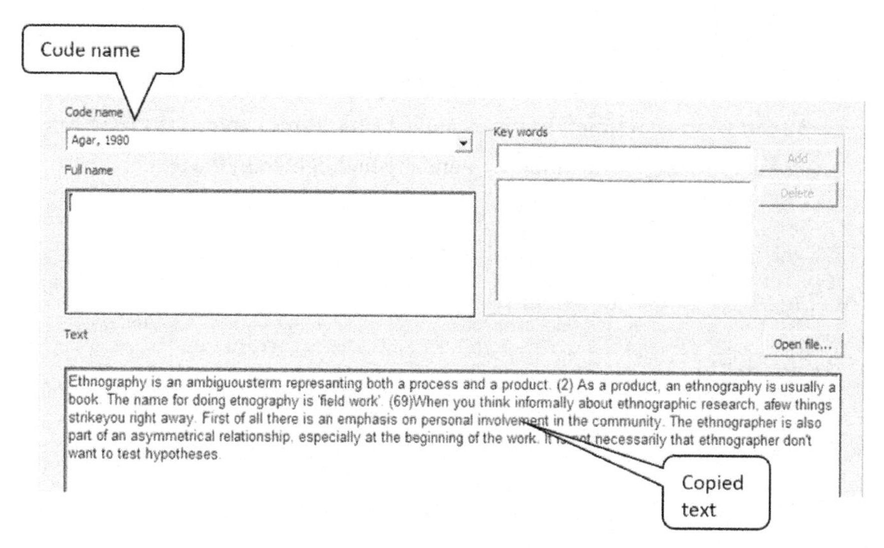

Figure BP16: New Text Source Box

8. Click **OK** to close the box and the segment icon will appear in the **Content pane**. In this way, we can add many different bibliographic segments to each category.

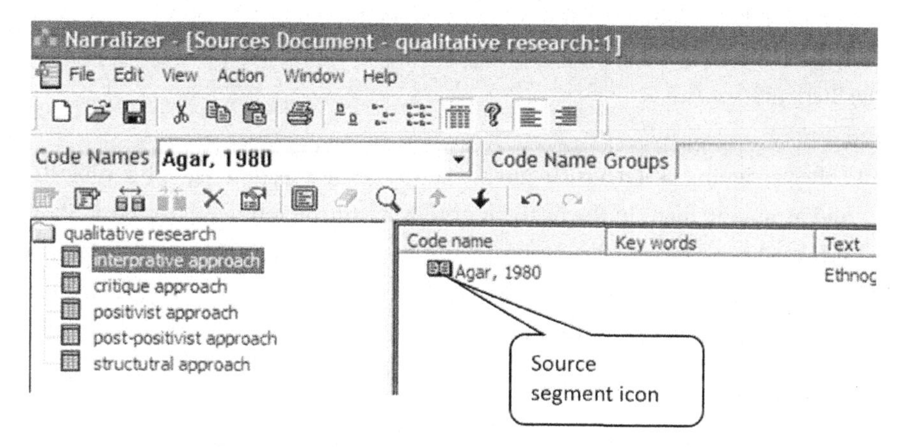

Figure BP17: The Segment Icon in the Content Pane

Point to the source segment icon and select **Properties** (or double click the icon) to open the box containing the bibliographic source segment, which can now be altered.

9. The **Find** function can be used in the bibliographic sources document. Click **Show sub-tree** to display the bibliographic sources document on the **Display Screen.** When the text appears on the **Display Screen,** use the **Find** function (1-3 expressions at a time). In the example below, three expressions were sought – "school", "teacher" and "student" – and are highlighted in the text.

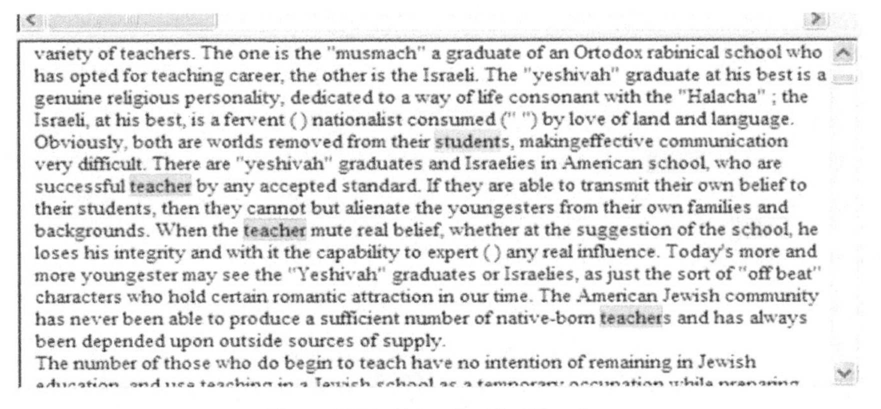

variety of teachers. The one is the "musmach" a graduate of an Ortodox rabinical school who has opted for teaching career, the other is the Israeli. The "yeshivah" graduate at his best is a genuine religious personality, dedicated to a way of life consonant with the "Halacha" ; the Israeli, at his best, is a fervent () nationalist consumed (" ") by love of land and language. Obviously, both are worlds removed from their students, makingeffective communication very difficult. There are "yeshivah" graduates and Israelies in American school, who are successful teacher by any accepted standard. If they are able to transmit their own belief to their students, then they cannot but alienate the youngesters from their own families and backgrounds. When the teacher mute real belief, whether at the suggestion of the school, he loses his integrity and with it the capability to expert () any real influence. Today's more and more youngester may see the "Yeshivah" graduates or Israelies, as just the sort of "off beat" characters who hold certain romantic attraction in our time. The American Jewish community has never been able to produce a sufficient number of native-born teachers and has always been depended upon outside sources of supply.
The number of those who do begin to teach have no intention of remaining in Jewish

Figure BP18: Using the Find Function

Segments can now be examined to determine their relevance to the literature review. If a segment is relevant, copy it from the **Display Screen** using **Copy,** then click on **Add Text Source** to assign it to the appropriate category in the **Category pane** of the literature review.

10. After assigning all relevant segments to the categories (bibliographic subjects) and arranging them in the necessary order for the literature review, click **Show sub-tree** to view the bibliographic segments on the **Display Screen**. The **Export** function will now export these segments to a Word document and the material is ready for the final processing of the literature review.

BIBLIOGRAPHY

Agar, M.H. (1980). *The professional stranger: An informal introduction to ethnography*. New York: Academic Press.

Anfara, V.A., Jr., Brown, K.M., & Mangione, T.L. (2002). Qualitative analysis on stage: Making the research process more public. *Educational Researcher, 31*, 28–38.

Angrosino, M.V. (2005). Recontextualizing observation: Ethnography, pedagogy, and the prospects for a progressive political agenda. In N.K. Denzin & Y.S. Lincoln (Eds.), *The Sage handbook of qualitative research* (3rd ed., pp. 729–745). London: Sage Publications.

Araujo, L. (1995). Designing and refining hierarchical coding frames. In U. Kelle (Ed.), *Computer-aided qualitative data analysis* (pp. 96–104). London: Sage Publications.

Arksey, H., & Knight, P. (1999). *Interviewing for social scientists*. London: Sage Publications.

Bach, H. (2007). Composing a visual narrative inquiry. In J. Clandinin (Ed.), *Handbook of narrative inquiry* (pp. 280–307). London: Sage Publications.

Bar-On, D. & Sheinberg-Taz, M. (2010) Dilemmas in Gathering and Analyzing Qualitative Data. In: L. Kasen & M. Krumer-Nevo (Eds.) *Qualitative Research Analysis* (pp. 465-486). Beer Sheva: Ben-Gurion University of the Negev Press. (in Hebrew).

Baron–Cohen, S. (2003). *The essential difference: Men, women and the extreme male brain*. New York: Basic Books, Inc.

Barthes, R. (1977). Death of the author: Structuralist analysis of narrative. In *Image, music, text* (pp. 142–148). New York: Hill & Wang.

Beattie, M. (1995). *Constructing professional knowledge in teaching: A narrative of change and development*. New York: Teachers College Press.

Bettencourt, A. (1993). The construction of knowledge: A radical constructivist view. In K. Tobin (Ed.) *The practice of constructivism in science education* (pp. 39-50). American Association for the Advancement of Science.

Baiet-Marom & Shahar (2007) *The psychology of intuitive judgment*. Raanana: The Open University of Israel Press. (in Hebrew)

Bishop, R. (2005). Freeing ourselves from neocolonial domination in research. In N.K. Denzin & Y.S. Lincoln (Eds.), *The Sage handbook of qualitative research* (3rd ed., pp. 109–138). London: Sage Publications.

Blackmore, S. (1999). *The meme machine*. New York: Oxford University Press.

Bruner, J. (1985). Narrative and paradigmatic modes of thought. In E. Eisner (Ed.), *Learning and teaching the ways of knowing* (pp. 97–115). Chicago: NSSE.

Bruner, J. (1990). *Acts of meaning*. Cambridge, MA: Harvard University Press.

Bruner, J. (1996). *The culture of education*. Cambridge, MA: Harvard University Press.

Buber, M. (1958) *I and Thou*. New York: Scriber.

Bullough, R.V., & Pinnegar, S. (2001). Guidelines for quality in autobiographical forms of self-study research. *Educational Researcher, 30*(3), 13–21.

Carr, W., & Kemmis, S. (1986). *Becoming critical: Education, knowledge and action research*. London: The Falmer Press.

Carter, K. (1993). The place of story in the study of teaching and teacher education. *Educational Researcher, 22*(1), 5–12.

Charmaz, K. (1983). The grounded theory method: An explication and interpretation. In R.M. Emerson (Ed.), *Contemporary field research* (pp. 109-126). Illinois: Waveland Press.

Charmaz, K. (1990). 'Discovering' chronic illness: Using grounded theory. *Social Science and Medicine, 30*(11), 1161–1172.

Charmaz, K. (1995). Grounded theory. In J.H. Smith, R. Harre & V. Langenhove (Eds.), *Rethinking psychology* (pp. 27–42). London: Sage Publications.

Charmaz, K. (2000). Grounded theory: Objectivist and constructivist methods.

In N.K. Denzin & Y.S. Lincoln (Eds.), *Handbook of qualitative research* (2nd ed., pp. 509–535). London: Sage Publications.

Charmaz, K. (2005). Grounded theory in the 21st century: Application for advancing social justice studies. In N.K. Denzin & Y.S. Lincoln (Eds.), *The Sage handbook of qualitative research* (3rd ed., pp. 507–535). London: Sage Publications.

Charmaz, K. (2006). *Constructing grounded theory: A practical guide through qualitative analysis*. London: Sage Publications.

Chase, S.E. (2005). Narrative inquiry. In N.K. Denzin & Y.S. Lincoln (Eds.), *The Sage handbook of qualitative research* (3rd ed., pp. 651–679). London: Sage Publications.

Clandinin, D.J. (2007). Preface. In D.J. Clandinin (Ed.), *Handbook of narrative inquiry: Mapping a methodology* (pp. IX–XVII). London: Sage Publications.

Clandinin, D.J., & Connelly, F.M. (2000). *Narrative inquiry: Experience and story in qualitative research*. San Francisco: Jossey-Bass Publishers.

Clandinin, D.J., & Rosiek, J. (2007). Mapping a landscape of narrative inquiry. In D.J. Clandinin (Ed.), *Handbook of narrative inquiry: Mapping a methodology* (pp. 35–75). London: Sage Publications.

Cochran-Smith, M., & Lytle, S.L. (1999). The teacher research movement: A decade later. *Educational Researcher, 28*(7), 15–25.

Cohen, J.A. (1999). Hermeneutic options for the teaching of canonical texts: Freud, Fromm, Strauss, and Buber read the Bible. *Courtyard: A Journal of Research and Thought in Jewish Education, 1*, 35–65.

Connelly, F.M., & Clandinin, D.J. (1988). *Teachers as curriculum planners: Narratives of experience*. New York: Teacher College Press.

Connelly, F.M., & Clandinin, D.J. (1990). Stories of experience and narrative inquiry. *Educational Researcher, 19*(5), 2–14.

Cortazzi, M. (1993). *Narrative analysis*. London: The Falmer Press.

Craig, C.J. (2001). The relationship between and among teachers' narrative knowledge, communities of knowing, and school reform: The case of "The Monkey's Paw." *Curriculum Inquiry, 31*(3), 303–331.

Creswell, J.W. (1998). *Qualitative inquiry and research design: Choosing among five traditions*. London: Sage Publications.

Creswell, J.W., & Plano Clark, V.L. (2007). *Designing and conducting mixed methods research* (2nd ed.). Thousand Oaks, CA: Sage Publications.

Crowley, C. (2010). Writing up the qualitative methods research report. In M.A. Forrester (Ed.), *Doing qualitative research in psychology* (pp. 229–257). London: Sage Publications.

Damasio, A. (2003). *Looking for Spinoza*. Orlando, FL: Harcourt, Inc.

Dargish, R. & Sabar-Ben Yehoshua, N. (2001) Hermeneutics and hermeneutic research. In: N. Sabar (Ed.) *Genres and tradition in qualitative research* (pp. 77-99). Lod: Dvir, Publishing. (in Hebrew)

Davis, N.T. (1996). Looking in the mirror: Teachers' use of autobiography and action research to improve practice. *Research in Science Education, 26*(1), 23–32.

Dawkins, R. (1986). *The blind watchmaker*. New York: W.W. Norton & Co.

Dawkins, R. (1989). *The selfish gene* (2nd ed.). London: Oxford University Press.

Dawkins, R. (2006). *The God delusion*. Boston: Houghton Mifflin.

Dawkins, R. (2009). *The greatest show on earth*. New York: Free Press.

Deacon, T. (1997). *The symbolic species: The co-evolution of language and the human brain*. London: Penguin.

Denzin, N.K. (1995). Symbolic interactionism. In J.H. Smith, R. Harre & V. Langenhove (Eds.), *Rethinking psychology* (pp. 43–58). London: Sage Publications.

Denzin, N.K. (2000). The practices and politics of interpretation. In N.K.

Denzin & Y.S. Lincoln (Eds.), *Handbook of qualitative research* (2nd ed., pp. 897–922). London: Sage Publications.

Denzin, N.K., & Lincoln, Y.S. (1994). Introduction: Entering the field of qualitative research. In N.K. Denzin & Y.S. Lincoln (Eds.), *Handbook of qualitative research*. (pp. 1–17). London: Sage Publications.

Denzin, N.K., &. Lincoln, Y.S. (Eds.) (1998). *The landscape of qualitative research*.

London: Sage Publications.

Denzin, N.K., & Lincoln, Y.S. (2000). Introduction: The discipline and practice of qualitative research In N.K. Denzin & Y.S. Lincoln (Eds.), *Handbook of qualitative research* (2nd ed., pp. 1–28). London: Sage Publications.

Denzin, N.K., & Lincoln, Y.S. (2005a). Introduction: The discipline and practice of qualitative research. In N.K. Denzin & Y.S. Lincoln (Eds.), *The Sage handbook of qualitative research* (3rd ed., pp. 1–32). London: Sage Publications.

Denzin, N.K., & Lincoln, Y.S. (2005b). Preface. In N.K. Denzin & Y.S. Lincoln (Eds.), *The Sage handbook of qualitative research* (3rd ed., pp. IX–XIX). London: Sage Publications.

Derrida, J. (1976). *Of grammatology*. Baltimore, MD: Johns Hopkins University Press.

Derrida, J. (1978). *Writing and difference*. Baltimore, MD: The John Hopkins University Press.

De Waal, F. (2005). *Our inner ape*. New York: Riverhead Books.

Dey, I. (1993). *Qualitative data analysis*. London: Routledge.

Dey, I. (1999). *Grounding grounded theory: Guidelines for qualitative inquiry*. San Diego: Academic Press.

Diamond, J. (1992). *The third chimpanzee: The evolution and future of the human animal*. New York: Harper Perennial.

Diamond, J. (1999). *Guns, germs and steel: The fates of human societies*. New York: W.W. Norton & Co.

Dunbar, R. (1997). *Grooming, gossip and the evolution of language*. Cambridge, MA: Harvard University Press.

Dushnik, L. & Sabar-Ben Yehushua, N. (2001) Ethics of qualitative research. In: N. Sabar (Ed.) *Genres and tradition in qualitative research* (pp. 368-343). Lod: Dvir, Publishing. (in Hebrew)

Eagleton, T. (1991). *Ideology: An introduction*. London: Verso.

Eisner, E.W. (1979). Recent developments in educational research affecting art education. *Art Education, 32*(4), 12–15.

Eisner, E.W. (1985). *The art of educational evaluation*. London: The Falmer Press.

Eisner, E.W. (1991). *The enlightened rye: Qualitative inquiry and the enhancement of educational practice*. New York: Macmillan Publishing Co.

Ekman, P. (2003). *Emotions revealed: Recognized faces and feeling to improve communication and emotional life*. New York: Owl Books.

Elbaz, F. (1991). Research on teacher's knowledge: The evolution of a discourse. *Journal of Curriculum Studies, 23*(1), 1–9.

Elbaz-Luwisch, F. (2002). Writing as inquiry: Storying the teaching self in writing workshops. *Curriculum Inquiry, 32*(4), 403–428.

Ely, M. (2007). In-forming re-presentation. In J. Clandinin (Ed.), *Handbook of narrative inquiry* (pp. 567–598). London: Sage Publications.

Etcoff, N. (1999). *Survival of the Prettiest*. New York: Anchor Books.

Evans, J. St. B.T. (2003). In two minds: Dual process accounts of reasoning. *Trends in Cognitive Sciences, 7*, 454–459.

Fetterman, D.M. (1989). *Ethnography: Step by step*. London: Sage Publications.

Fielding, N.G., & Lee, R.M. (1998). *Computer analysis and qualitative research*. London: Sage Publications.

Firestone, W.A. (1993). Alternative arguments for generalizing from data as applied to qualitative research. *Educational Researcher, 22*(4), 16–23.

Flanders, N.A. (1970). *Analyzing teacher behavior*. Reading, MA: Addison-Wesley Publishing.

Flick, U. (1998). *An introduction to qualitative research*. London: Sage Publications.

Florio-Ruane, S. (2001). *Teacher education and the culture imagination: Autobiography, conversation and narrative*. Mahwah, NJ: Lawrence Erlbaum Associates, Publishers.

Fontana, A., & Frey J.H. (2000). The interview: From structured questions to negotiated text. In N.K. Denzin & Y.S. Lincoln (Eds.), *Handbook of qualitative research* (2nd ed., pp. 645–670). London: Sage Publications.

Fontana, A., & Frey, J.H. (2005). The interview: From neutral stance to political involvement. In N.K. Denzin & Y.S. Lincoln (Eds.), *The Sage handbook of qualitative research* (3rd ed., pp. 695–727). London: Sage Publications.

Forrester, M.A. (2010). QM4: Conversation analysis. In M.A. Forrester (Ed.), *Doing qualitative research in psychology* (pp. 202–226). London: Sage Publications.

Foucault, M. (1988) *Madness and Civilization: A History of Insanity in the Age of Reason.* New York: Vintage House

Foucault, M. (1980). *Power/knowledge: Selected interviews and other writings, 1972–1977*. Brighton: Harvester.

Freeman, M. (2007). Autobiographical understanding and narrative inquiry. In J. Clandinin (Ed.), *Handbook of narrative inquiry: Mapping a methodology* (pp. 120–145). London: Sage Publications.

Gadamer, H.G. (1975). *Truth and method*. London: Sheed & Ward.

Gadamer, H.G. (1986). *The relevance of the beautiful and other essays*. Cambridge: Cambridge University Press.

Gall, M.D., Borg, W.R., & Gall, J.P. (1996). *Educational research*. White Plains, NY: Longman.

Gallagher, S. (1992). *Hermeneutics and education*. Albany, NY: State University of New York Press.

Geertz, C. (1973). *The interpretation of cultures*. New York: Basic Books, Inc.

Gibson, S., & Riley, S. (2010). Approaches to data collection in qualitative research. In M.A. Forrester (Ed.), *Doing qualitative research in psychology* (pp. 59–76). London: Sage Publications.

Gill, R. (1996). Discourse analysis: Practical implementation. In J.T.E. Richardson (Ed.), *Handbook of qualitative research methods for psychology and the social sciences* (pp. 141–156). Leicester: The British Psychological Society Books.

Glaser, B.G. (1978). *Theoretical sensitivity: Advances in the methodology of grounded theory*. Mill Valley, CA: Sociology Press.

Glaser, B.G., & Strauss, A.L. (1967). *The discovery of grounded theory: Strategy for qualitative research*. New York: Aldine Publishing Co.

Goldberg, E. (2007). *The wisdom paradox: How your mind can grow stronger as your brain grows older*. London: Pocket Books.

Goleman, D. (1995). *Emotional intelligence*. New York: Bantam Books.

Goleman, D. (2006). *Social intelligence: The new science of human relationships*. New York: Bantam Books.

Gordon-Finlayson, A. (2010). QM2: Grounded theory. In M.A. Forrester (Ed.), *Doing qualitative research in psychology* (pp. 154–176). London: Sage Publications.

Goulding, C. (2002). *Grounded theory*. London: Sage Publications.

Grbich, C. (2007). *Qualitative data analysis: An introduction*. London: Sage Publications.

Greenwood, D.J., & Levin, M. (2005). Reform of the social sciences and of universities through action research. In N.K. Denzin & Y.S. Lincoln (Eds.), *The Sage handbook of qualitative research* (3rd ed., pp. 43–64). London: Sage Publications.

Grimmett, P.P., & Mackinnon, A.M. (1992). Craft knowledge and the education of teachers. In G. Grant (Ed.), *Review of research in education* (pp. 385–456). Washington, D.C.: American Educational Research Association.

Guba, E.G., & Lincoln, Y.S. (1989). *Fourth generation evaluation*. Newbury Park, CA: Sage Publications.

Guba, E.G., & Lincoln, Y.S. (1998). Competing paradigms in qualitative research. In N.K. Denzin, & Y.S. Lincoln (Eds.), *The landscape of qualitative research* (pp. 195–220). London: Sage Publications.

Guba, E.G., & Lincoln, Y.S. (2005). Paradigmatic, controversies, contradictions, and emerging confluences. In N.K. Denzin & Y.S. Lincoln (Eds.), *The Sage handbook of qualitative research* (3rd ed., pp. 191–215). London: Sage Publications.

Gudmundsdottir, S. (1995). The narrative nature of pedagogical content knowledge. In H. McEwan & K. Egan (Eds.), *Narrative in teaching, learning and research* (pp. 24–38). New York: Teachers College Press.

Gudmundsdottir, S. (1996). The teller, the tale and the one being told: The narrative nature of the research interview. *Curriculum Inquiry*, *26*(3), 293–306.

Habermas, J. (1971). *Knowledge and human interests.* Boston: Beacon Press.

Hammersley, M. (1996). The relationship between qualitative and quantitative research: Paradigm loyalty versus methodological eclecticism. In J.T.E. Richardson (Ed.), *Handbook of qualitative research methods for psychology and the social sciences* (pp. 159–174). Leicester: The British Psychological Society Books.

Harris, Judith R. (1998). *The nurture assumption: Why children turn out the way they do.* New York: A Touchstone Book.

Harris, Judith R. (2006). *No two alike: Human nature and human individuality.* New York: W.W. Norton & Co.

Hawkins, J., & Blakeslee, S. (2004). *On intelligence*. New York: Times Books, Henry Holt & Co.

Heikkien, H.L.T. (2002). Whatever is narrative research. In R. Huttunen, H.L.T. Heikkien & L. Syrjala (Eds.), *Narrative research: Voices of teachers and philosophers* (pp. 13–28). Jyvaskyla: SoPhi.

Henwood, K.L. (1996). Qualitative inquiry: Perspectives, methods and psychology. In J.T.E. Richardson (Ed.), *Handbook of qualitative research methods for psychology and the social sciences* (pp. 25–40). Leicester: The British Psychological Society Books.

Hogarth, R.M. (2001). *Educating intuition*. Chicago: University of Chicago Press.

Hollingswort, S., & Dybdahl, M. (2007). Talking to learn: The critical role of conversation in narrative inquiry. In J. Clandinin (Ed.), *Handbook of narrative inquiry* (pp. 146–176). London: Sage Publications.

Holstein, J.A., & Gubrium, J.F. (2005). Interpretive practice and social action. In N.K. Denzin & Y.S. Lincoln (Eds.), *The Sage handbook of qualitative research* (3rd ed., pp. 483–505). London: Sage Publications.

Huberman, A.M., & Miles, M.B. (1994). Data management and analysis methods.

In N.K. Denzin & Y.S. Lincoln (Eds.), *Handbook of qualitative research* (pp. 428–444). Thousand Oaks, CA: Sage Publications.

Hugh-Jones, S. (2010). The interview in qualitative research. In M.A. Forrester (Ed.), *Doing qualitative research in psychology* (pp. 77–97). London: Sage Publications.

Idan, A. (1990) *The concept of ideology*. Tel Aviv: Dvir Publishing House. (In Hebrew)

Jacob, F. (2001) *Of flies, mice, and men* (translated by G Weiss) Cambridge, Massachusetts: Harvard University Press.

Johnson, S. (2004). *Mind wide open*. New York: Scribner.

Jorgensen, D.L. (1989). *Participant observation: A methodology for human studies.*

London: Sage Publications.

Josselson, R. (2004). The hermeneutics of faith and the hermeneutic of suspicion. *Narrative Inquiry, 14*(1), 1–28.

Josselson, R. (2006). Narrative research and the challenge of accumulating knowledge. *Narrative Inquiry, 16*(1), 3–10.

Jovchelovitch, S., & Bauer, M.W. (2000). Narrative interviewing. In M.W. Bauer & G. Gaskell (Eds.), *Qualitative researching with text, image and sound* (pp. 57–74). London: Sage Publications.

Kahneman, D (2011) *Thinking, fast and slow*. New York: Farrar, Straus and Giroux.

Kahneman, D., & Frederick, S. (2002). Representativeness revisited: Attribute substitution in intuitive judgment. In T. Gilovich, D. Griffin & D. Kahneman (Eds.), *Heuristics and biases: The psychology of intuitive judgment* (pp. 49–81). New York: Cambridge University Press.

Kamberelis, G., & Dimitriadis, G. (2005). Focus groups. In N.K. Denzin & Y.S. Lincoln (Eds.), *The Sage handbook of qualitative research* (3rd ed., pp. 887–907). London: Sage Publications.

Kemmis, S., & McTaggart, R. (2005). Participatory action research. In N.K. Denzin & Y.S. Lincoln (Eds.), *The Sage handbook of qualitative research* (3rd ed., pp.559–603). London: Sage Publications.

Kincheloe, J.L., & McLaren, P. (2005). Rethinking critical theory and qualitative research. In N.K. Denzin & Y.S. Lincoln (Eds.), *The Sage handbook of qualitative research* (3rd ed., pp. 303–342). London: Sage Publications.

King, N. (2010). Research ethics in qualitative research. In M.A. Forrester (Ed.), *Doing qualitative research in psychology* (pp. 98–118). London: Sage Publications.

Kirk, J., & Miller, M.L. (1986). *Reliability and validity in qualitative research.* Beverly Hills, CA: Sage Publications.

Kuckartz, U. (1995). Case-oriented quantification. In U. Kelle (Ed.), *Computer-aided qualitative data analysis* (pp. 158–166). London: Sage Publications.

Kuhn, T.S. (1962). *The structure of scientific revolutions.* Chicago: University of Chicago Press.

LaBoskey, V.K. (1994). *Development of reflective practice: A study of preservice teachers.* New York: Teachers College Press.

Labov, W. (1972). *Language in the inner city: Studies in the black English vernacular.* Philadelphia: University of Pennsylvania Press.

Ladson-Billings, G., & Donnor, J. (2005). The moral activist role of criticalrace theory scholarship. In N.K. Denzin & Y.S. Lincoln (Eds.), *The Sage handbook of qualitative research* (3rd ed., pp. 279–301). London: Sage Publications.

Lam, Z. (2000) Ideology and educational philosophy. In Y. Harpaz (Ed.): *Pressure and resistance in education* (pp. 217-254). Tel Aviv: Sifriat Poalim Publishing House Ltd. (in Hebrew)

Lampert, A. (2007) *Naked soul.* Tel Aviv: Yedioth Ahronot Books. (In Hebrew)

Lauriala, A., & Syrjala, L. (1995). The influence of research into alternative pedagogies on the professional development of prospective teachers. *Teachers and Teaching: Theory and Practice, 1*(1), 101–118.

Leakey, R. (1994). *The origin of humankind.* New York: Orion Publishing Group Ltd.

LeDoux, J. (1998). *The emotional brain: The mysterious underpinnings of emotional life*. New York: Touchstone.

Lieblich, A, Tuval-Mashiach, R. & Zilber, T. (2001). Between the whole and its parts and between the content and the structure. In: L. Kasen & M. Krumer-Nevo (Eds.) *Qualitative Research Analysis* (pp. 21-42). Beer Sheva: Ben-Gurion University of the Negev Press. (in Hebrew)

Lieblich, A., Tuval–Mashiach, R., & Zilber, T. (1998). *Narrative research*. London: Sage Publications.

Lincoln, Y.S., & Guba, E.G. (1985). *Naturalistic inquiry*. Beverly Hills, CA: Sage Publications.

Lincoln, Y.S., & Guba, E.G. (2000). Paradigmatic controversies, contradictions, and emerging confluences. In N.K. Denzin & Y.S. Lincoln (Eds.), *Handbook of qualitative research* (2nd ed., pp. 163–188). London: Sage Publications.

Lincoln, Y.S., & Guba, E.G. (2002). Judging the quality of case study reports. In A.M. Huberman & M.B. Miles (Eds.), *The qualitative researcher's companion* (pp.205–215). Thousand Oaks, CA: Sage Publications.

Lloren, M.B. (1994). Action research: Are teachers finding their voice? *The Elementary School Journal*, *95*(1), 3–10.

Loewenstein, G., & Lerner, J.S. (2002). The role of affect in decision making. In R.J. Davidson, K.R. Scherer & H.H. Goldsmith (Eds.), *Handbook of affective sciences* (pp. 619–642). New York: Oxford University Press.

Lortie, C.D. (1975). *Schoolteacher*. Chicago: University of Chicago Press.

Lyons, N. (2007). Narrative inquiry: What possible future influence on policy or practice. In J. Clandinin (Ed.), *Handbook of narrative inquiry: Mapping a methodology* (pp. 600–631). London: Sage Publications.

Marble, S. (1997). Narrative visions of schooling. *Teaching and Teacher Education*, *13*(1), 55–64.

Marshall, C., & Rossman, G.B. (1989). *Designing qualitative research*. London: Sage Publications.

Mason, J. (1996). *Qualitative researching*. London: Sage publications.

Maykut, P., & Morehouse, R. (1994). *Beginning qualitative research: A philosophic and practical guide*. London: The Falmer Press.

McLuhan, M. (1964). *Understanding media*. Cambridge, MA: MIT Press.

McNiff, J. (1988). *Action research: Principles and practice*. London: Macmillan Education.

Merriam, S.B. (1985). The case study in educational research: A review of selective literature. *Journal of Educational Thought*, *19*(3), 204–217.

Merriam, S.B. (1998). *Qualitative research and case study applications in education*. San Francisco: Jossey-Bass Publishers.

Merriam, S.B., & Simpson, E.L. (1984). *A guide to research for educators and trainers of adults*. Malabar, FL: Robert E. Krieger Publishing Co.

Merrick, E. (1999). An exploration of quality in qualitative research. In M. Kopala & L.A. Suzuki (Eds.), *Using qualitative methods in psychology* (pp. 25–36). London: Sage Publications.

Miles, B.M., & Huberman, A.M. (1984). Drawing valid meaning from qualitative data: Toward a shared craft. *Educational Researcher*, *13*(5), 20–30.

Miles, M.B., & Huberman, A.M. (1994). *Qualitative data analysis: An expanded sourcebook* (2nd ed.). Thousand Oaks: Sage Publications.

Mills, S (2003) *Michel Foucault*. London: Routledge.

Milo, D (2009) *The invention of tomorrow*. Tel Aviv: Hakibbutz Hameuchad Publishing House Ltd. (in Hebrew)

Mishler, E.G. (1986). *Research interviewing: Context and narrative*. Cambridge, MA: Harvard University Press.

Moilanen, P. (2002). Narrative, truth and correspondence: A defence. In R. Huttunen, H.L.T. Heikkien & L. Syrjala (Eds.), *Narrative research: Voices of teachers and philosophers* (pp. 91–104). Jyvaskyla: SoPhi.

Moss, P.A. (1996). Enlarging the dialogue in educational measurement: Voices from interpretive research traditions. *Educational Researcher*, *25*(1), 20–28.

Nixon, J. (1987). The teacher as researcher: Contradiction and continuities.

Peabody Journal of Education, 64(2), 20–32.

Noblit, G.W., & Hare, R.D. (1988). *Meta-ethnography: Synthesizing qualitative studies*. London: Sage Publications.

Ofir, A. (1996) Postmodernism: A philosophical position. In I Gur-Ze'ev (Ed.) *Education in the era of the postmodern discourse*. Jerusalem: The Magness Press. (in Hebrew)

Perakyla, A. (2005). Analyzing talk and text. In N.K. Denzin & Y.S. Lincoln (Eds.), *The Sage handbook of qualitative research* (3rd ed., pp. 869–886). London: Sage Publications.

Peshkin, A. (1993). The goodness of qualitative research. *Educational Researcher, 22*(2), 23–29.

Peshkin, A. (2000). The nature of interpretation in qualitative research. *Educational Researcher, 29*(9), 5–9.

Peters, M.A., & Burbules, N.C. (2004). *Poststructuralism and educational research*. London: Rowman & Littlefield, Inc.

Philips, D.C., & Burbules, N.C. (2000). *Postpositivism and educational research*. New York: Rowman & Littlefield Publishers, Inc.

Phillips, N., & Hardly, C. (2002). *Discourse analysis: Investigating processes of social construction*. Thousand Oaks: Sage Publications.

Pidgeon, N. (1996). Grounded theory: Theoretical background. In J.T.R. Richardson (Ed.), *Handbook of qualitative research methods for psychology and the social sciences* (pp. 75–85). Leicester: The British Psychological Society Books.

Pidgeon, N., & Henwood, K. (1996). Grounded theory: Practical implementation. In J.T.R. Richardson (Ed.), *Handbook of qualitative research methods for psychology and the social sciences* (pp. 113–124). Leicester: The British Psychological Society Books.

Pink, D.H. (2005). *A whole new mind*. New York: Penguin Books.

Pinker, S. (1997). *How the mind works*. New York: W.W. Norton & Co., Inc.

Pinker, S. (2002). *The blank slate*. New York: Penguin Books.

Pinnegar, S., & Daynes, G. (2007). Locating narrative inquiry historically: Thematics in the turn to narrative. In J. Clandinin (Ed.), *Handbook of narrative inquiry: Mapping a methodology* (pp. 3–34). London: Sage Publications.

Plummer, K. (2005). Critical humanism and queer theory. In N.K. Denzin & Y.S. Lincoln (Eds.), *The Sage handbook of qualitative research* (3rd ed., pp. 357–373). London: Sage Publications.

Polanyi, M. (1967). *The tacit dimension*. Chicago: University of Chicago Press.

Polkinghorne, D. (1995). Narrative configuration in qualitative analysis. *Qualitative Studies in Education*, *8*(1), 5–23.

Popper, K.R. (1959). *The logic of scientific discovery*. London: Routledge.

Posner, G.J. (1998). Modes of curriculum planning. In L.E. Beyer & R.W. Apple (Eds.), *The curriculum* (pp. 79–97). Albany, NY: State University of New York Press.

Potter, J. (1996). Discourse analysis and constructionist approaches: Theoretical background. In J.T.E. Richardson (Ed.), *Handbook of qualitative research methods for psychology and the social sciences* (pp. 125–140). Leicester: The British Psychological Society Books.

Rachel, J. (1996). Ethnography: Practical implementation. In J.T.R. Richardson (Ed.), *Handbook of qualitative research methods for psychology and the social sciences* (pp. 113–124). Leicester: The British Psychological Society Books.

Ratey, J.J. (2002). *A user's guide to the brain*. New York: Vintage Books.

Restak, R. (2006). *The naked brain*. New York: Harmony Books.

Richards, T., & Richards, L. (1995). Using hierarchical categories in qualitative data analysis. In U. Kelle (Ed.), *Computer-aided qualitative data analysis* (pp.80–95). London: Sage Publications.

Richardson, L., & St. Pierre, E.A. (2005). Writing a method of inquiry In N.K. Denzin & Y.S. Lincoln (Eds.), *The Sage handbook of qualitative research* (3rd ed., p.959–978). London: Sage Publications.

Ricoeur, P. (1970). *Freud and philosophy: An essay on interpretation*. New Haven, CT: Yale University Press.

Ridley, M. (2003). *Nature via nurture: Genes, experience and what makes us humans.* New York: HarperCollins Publishers.

Riessman, C.K. (1993). *Narrative analysis.* London: Sage Publications.

Riessman, C.K., & Speedy, J. (2007). Narrative inquiry in the psychotherapy professions: A critical review. In J. Clandinin (Ed.), *Handbook of narrative inquiry: Mapping a methodology* (pp. 426–456). London: Sage Publications.

Rizzolatti, G., & Arbib, M.A. (1998). Language within our grasp. *Trends in Neuroscience, 21*, 188–194.

Roberts, B. (2002). *Biographical research.* Buckingham: Open University Press.

Rock, A. (2004). *The mind at night: The new science of how and why we dream.* New York: Basic Books, Inc.

Rosenthal, G. (1993). Reconstruction of life-stories: Principles of selection in generating stories for narrative interview. In R. Josselson & A. Lieblich (Eds.), *The narrative study of lives* (Vol. 1, pp. 59–91). Newbury Park, CA: Sage Publications.

Rosenthal, R., & Jacobson, L. (1968). Pygmalion in the classroom: Teacher expectations and pupil's intellectual development. *The Urban Review, 3*(1), 16–20.

Ryan, G.W., & Bernard, H.R. (2000). Data management and analysis methods. In N.K. Denzin & Y.S. Lincoln (Eds.), *Handbook of qualitative research* (2nd ed., pp.769–802). London: Sage Publications.

Ryan, W., & Pitman, W. (2000). *Noah's flood: The new scientific discoveries about the event that changed history.* New York: Touchstone.

Said, E.W. (1978). *Orientalism.* New York: Vintage Books.

Saukko, P. (2005). Methodologies for cultural studies: An integrative approach. In N.K. Denzin & Y.S. Lincoln (Eds.), *The Sage handbook of qualitative research* (3rd ed., pp. 343–356). London: Sage Publications.

Schofield, J.W. (1989). Increasing the generalizability of qualitative research. In E.W. Eisner & A. Peshkin (Eds.), *Qualitative inquiry in education* (pp. 201–232). New York: Teachers College Press.

Schon, D.A. (1987). *Educating the reflective practitioner.* San Francisco: Jossey-Bass Publishers.

Schon, D.A. (1983). *The reflective practitioner: How professionals think in action.* New York: Basic Books, Inc.

Schwab, J.J. (1969). The practical: A language for curriculum development. *School Review, 78*(1), 1–23.

Schwab, J.J. (1973). The practical 3: Translation into curriculum. *School Review, 81*(4), 501–522.

Schwandt, T.A. (1998). Constructivist, interpretivist approaches to human inquiry. In N.K. Denzin & Y.S. Lincoln (Eds.), *The landscape of qualitative research* (pp. 221–259). London: Sage Publications.

Schwandt, T.A. (2000). Three epistemological stances for qualitative inquiry. In N.K. Denzin & Y.S. Lincoln (Eds.), *Handbook of qualitative research* (2nd ed., pp.189–214). London: Sage Publications.

Sciarra, D. (1999). The role of the qualitative researcher. In M.S. Kopele & L.A. Sucuki (Eds.), *Using qualitative methods in psychology* (pp. 37–48). London: Sage Publications.

Seidel, J., & Kelle, U. (1995). Different functions of coding in the analysis of textual data. In U. Kelle (Ed.), *Computer–aided qualitative data analysis* (pp. 52–61). London: Sage Publications.

Seidman, I.E. (1991). *Interviewing as qualitative research.* New York: Teachers College Press.

Sfard, A. (2008) The mismeasure of man. *Hed Hachinuch* 82 (4): 60-63. (in Hebrew)

Shaw, R. (2010a). Conducting literature reviews. In M.A. Forrester (Ed.), *Doing qualitative research in psychology* (pp. 39–55). London: Sage Publications.

Shaw, R. (2010b). QM3: Interpretive phenomenological analysis. In M.A. Forrester (Ed.), *Doing qualitative research in psychology* (pp. 177–201). London: Sage Publications.

Sheinberg-Taz, M. (2010) Narrative Thinking. In Tuval-Mashiach, R. & Spector-

Mersel G. (Eds.) *Narrative Research: Theory, Creation and Interpretation.* (pp. 81-105). Jerusalem: Magness Press. (In Hebrew)

Shor, P& Sabar Sabar-Ben Yehushua, N. (2010) From "report" to "story". In Tuval-Mashiach, R. & Spector-Mersel G. (Eds.) *Narrative research: Theory, creation and interpretation*. (pp. 197-220). Jerusalem: Magness Press. (In Hebrew)

Shkedi, A. (1995). Teachers' attitudes toward a teachers' guide: Implications for the roles of planners and teachers. *Journal of Curriculum and Supervision, 10*(2), 155–170.

Shkedi, A. (2001). Studying culturally valued texts: Teachers' conception vs. students' conception. *Teaching and Teacher Education, 17*, 333–347.

Shkedi, A. (2003) *Words of meaning: Qualitative research - Theory and practice.* Tel Aviv: Ramot Publishing House, Tel Aviv University. (In Hebrew)

Shkedi, A. (2004). Second–order theoretical analysis: A method for constructing theoretical explanation. *International Journal of Qualitative Studies in Education, 17*(5), 627–646.

Shkedi, A. (2005). *Multiple case narrative: A qualitative approach to studying multiple populations*. Amsterdam: John Benjamins Publishing Co.

Shkedi, A. (2010) Narrative-based theory. In: L. Kasen & M. Krumer-Nevo (Eds.) *Qualitative Research Analysis*. (pp. 436-461) Beer Sheva: Ben-Gurion University of the Negev Press. (in Hebrew)

Shkedi, A & Nisan, M. (2006) Teachers' cultural ideology: Patterns of curriculum and teaching culturally valued texts. *Teachers College Record*, 108 (4): 687-725.

Shulman, L.S. (1986). Those who understand: Knowledge growth in teaching. *Educational Researcher, 15*(2), 4–14.

Silverman, D. (2006). *Interpreting qualitative data — Methods for analyzing talk, text and interaction* (3rd ed.). London: Sage Publications.

Simons, H. (1996). The paradox of case study. *Cambridge Journal of Education, 26*(2), 225–240.

Sloman, S.A. (2002). Two systems of reasoning. In T. Gilovich, D. Griffin & D. Kahneman (Eds.), *Heuristics and biases: The psychology of intuitive judgment* (pp.

379–396). New York: Cambridge University Press.

Smith, J.K., & Hodkinson, P. (2005). Relativism, criteria, and politics. In N.K. Denzin & Y.S. Lincoln (Eds.), *The Sage handbook of qualitative research* (3rd ed., pp.915–932). London: Sage Publications.

Stake, R.E. (1978). The case study method in social inquiry. *Educational Researcher, 7*(2), 5–8.

Stake, R.E. (1995). *The art of case study research*. London: Sage Publications.

Stake, R.E. (2000). Case studies. In N.K. Denzin & Y.S. Lincoln (Eds.), *Handbook of qualitative research* (2nd ed., pp. 435–454). London: Sage Publications.

Stake, R.E. (2005). Qualitative case studies. In N.K. Denzin & Y.S. Lincoln (Eds.), *The Sage handbook of qualitative research* (3rd ed., pp. 443–466). London: Sage Publications.

Stanovich, D.E. (2004). *The robot's rebellion: Finding meaning in the age of Darwin*. Chicago: University of Chicago Press.

Stanovich, K.E., & West, R.F. (2000). Individual differences in reasoning: Implications for the rationality debate. *Behavioral and Brain Sciences, 23*, 645–665.

Stenhouse, L. (1988). Artistry and teaching: The teacher as focus of research and development. *Journal of Curriculum and Supervision, 4*(1), 43–51.

Stokoe, W.C. (2001). *Language in hand: Why sign came before speech*. Washington, D.C.: Gallaudet University Press.

Strauss, A.L. (1987). *Qualitative analysis for social scientists*. Boston: Cambridge University Press.

Strauss, A., & Corbin, J. (1990). *Basics of qualitative research: Grounded theory procedures and techniques*. London: Sage Publications.

Strauss, A., & Corbin, J. (1994). Grounded theory methodology: An overview. In N.K. Denzin & Y.S. Lincoln (Eds.), *Handbook of qualitative research* (pp. 273–285). Thousand Oaks, CA: Sage.

Sullivan, C. (2010). Theory and method in qualitative research. In M.A.

Forrester (Ed.), *Doing qualitative research in psychology* (pp. 15–38). London: Sage Publications.

Tashakkory, A., & Teddlie, C. (1998). *Mixed methodology: Combining qualitative and quantitative approaches*. Thousand Oaks: Sage Publications.

Tattersall, I. (1998). *Becoming human: Evolution and human uniqueness*. New York: Harcourt Brace & Co.

Tedlock, B. (2005). The observation of participation and the emergence of public ethnography. In N.K. Denzin & Y.S. Lincoln (Eds.), *The Sage handbook of qualitative research* (3rd ed., pp. 467–481). London: Sage Publications.

Travers, M. (2001). *Qualitative research through case studies*. London: Sage Publications.

Turner, M. (1996). *The literary mind: The origins of thought and language*. Oxford University Press.

Tversky, A., & Kahneman, D. (1974). Judgment under uncertainty: Heuristics and biases. *Science*, *185*, 1124–1131.

Van Dijk, T.A. (1998). *Ideology: A multidisciplinary approach*. London: Sage Publications.

Van Manen, M. (1990). *Researching lived experience: Human science for an action sensitive pedagogy*. Albany, NY: State University of New York Press.

Weiner, J. (1994). *The beak of the finch: A story of evolution in our time*. New York: Vintage Book.

Weitzman, E.A. (2000). Software and qualitative research. In N.K. Denzin & Y.S. Lincoln (Eds.), *Handbook of qualitative research* (2nd ed., pp. 803–820). London: Sage Publications.

Whitehead, J., & Fitzgerald, B. (2007). Experiencing and evidencing learning through self-study: New ways of working with mentors and trainees in a training school partnership. *Teaching and Teacher Education*, *23*(1), 1–12.

Widdershoven, G.A.M. (1993). The story of life: Hermeneutic perspectives on the relationship between narrative and life history. In R. Josselson & A.Lieblich (Eds.), *The narrative study of lives* (pp. 1–20). London: Sage Publications.

Wiggins, S., & Riley, S. (2010). QM1: Discourse analysis. In M.A. Forrester (Ed.), *Doing qualitative research in psychology* (pp. 135–153). London: Sage Publications.

Winston, R. (2002). *Human instinct*. London: Bantam Press.

Woods, P. (1996). *Researching the art of teaching: Ethnography for educational use*. London: Routledge.

Yin, R.K. (1994). *Case study research: Design and methods*. London: Sage Publications.

Zeichner, K.M. (1993). Action research: Personal renewal and social reconstruction. *Educational Action Research*, *1*(2), 199–219.

INDEX

E

F

G

H

I

K

L

M

N

O

P

Q

R

S

T

V